Lecture Notes in Physics

Volume 823

For further volumes:
http://www.springer.com/series/5304

The Lecture Notes in Physics

The series Lecture Notes in Physics (LNP), founded in 1969, reports new developments in physics research and teaching—quickly and informally, but with a high quality and the explicit aim to summarize and communicate current knowledge in an accessible way. Books published in this series are conceived as bridging material between advanced graduate textbooks and the forefront of research and to serve three purposes:

- to be a compact and modern up-to-date source of reference on a well-defined topic
- to serve as an accessible introduction to the field to postgraduate students and nonspecialist researchers from related areas
- to be a source of advanced teaching material for specialized seminars, courses and schools

Both monographs and multi-author volumes will be considered for publication. Edited volumes should, however, consist of a very limited number of contributions only. Proceedings will not be considered for LNP.

Volumes published in LNP are disseminated both in print and in electronic formats, the electronic archive being available at springerlink.com. The series content is indexed, abstracted and referenced by many abstracting and information services, bibliographic networks, subscription agencies, library networks, and consortia.

Proposals should be sent to a member of the Editorial Board, or directly to the managing editor at Springer:

Christian Caron
Springer Heidelberg
Physics Editorial Department I
Tiergartenstrasse 17
69121 Heidelberg/Germany
christian.caron@springer.com

Giovanni Costa · Gianluigi Fogli

Symmetries and Group Theory in Particle Physics

An Introduction to Space-time and Internal Symmetries

 Springer

Prof. Giovanni Costa
Dipartmento di Fisica
Università di Padova
Via Marzolo 8
35131 Padova
Italy

Prof. Gianluigi Fogli
Dipartimento die Fisica
Università die Bari
Via Amendola 173
70126 Bari
Italy

ISSN 0075-8450
ISBN 978-3-642-15481-2
DOI 10.1007/978-3-642-15482-9
Springer Heidelberg New York Dordrecht London

e-ISSN 1616-6361
e-ISBN 978-3-642-15482-9

Library of Congress Control Number: 2011945259

Printed on acid-free paper

Springer is part of Springer Science+Business Media (www.springer.com)

To our children
Alessandra, Fabrizio and Maria Teresa

Preface

The aim in writing this book has been to give a survey of the main applications of group and representation theory to particle physics. It provides the essential notions of relativistic invariance, space-time symmetries and internal symmetries employed in the standard University courses of Relativistic Quantum Field Theory and Particle Physics. However, we point out that this is neither a book on these subjects, nor it is a book on group theory.

Specifically, its main topics are, on one side, the analysis of the Lorentz and Poincaré groups and, on the other side, the internal symmetries based mainly on unitary groups, which are the essential tools for the understanding of the interactions among elementary particles and for the construction of the present theories. At the same time, these topics give important and enlightening examples of the essential role of group theory in particle physics. We have attempted to present a pedagogical survey of the matter, which should be useful to graduate students and researchers in particle physics; the only prerequisite is some knowledge of classical field theory and relativistic quantum mechanics. In the Bibliography, we give a list of relevant texts and monographs, in which the reader can find supplements and detailed discussions on the questions only partially treated in this book.

One of the most powerful tools in dealing with invariance properties and symmetries is *group theory*. Chapter 1 consists in a brief introduction to group and representation theory; after giving the basic definitions and discussing the main general concepts, we concentrate on the properties of Lie groups and Lie algebras. It should be clear that we do not claim that it gives a self-contained account of the subject, but rather it represents a sort of glossary, to which the reader can refer to recall specific statements. Therefore, in general, we limit ourselves to define the main concepts and to state the relevant theorems without presenting their proofs, but illustrating their applications with specific examples. In particular, we describe the root and weight diagrams, which provide a useful insight in the analysis of the classical Lie groups and their representations; moreover, making use of the Dynkin diagrams, we present a classification of the classical semi-simple Lie algebras and Lie groups.

The book is divided into two parts that, to large extent, are independent from one another. In the first part, we examine the invariance principles related to the symmetries of the physical space-time manifold. Disregarding gravitation, we consider that the geometry of space-time is described by the Minkowski metric and that the inertial frames of reference of special relativity are completely equivalent in the description of the physical phenomena. The co-ordinate transformations from one frame of reference to another form the so-called *inhomogeneous Lorentz group* or *Poincaré group*, which contains the space-time translations, besides the pure Lorentz transformations and the space rotations. The introductory and didactic nature of the book influenced the level of the treatment of the subject, for which we renounced to rigorousness and completeness, avoiding, whenever possible, unnecessary technicalities.

In Chapter 2 we give a short account of the three-dimensional rotation group , not only for its important role in different areas of physics, but also as a specific illustration of group theoretical methods. In Chapter 3 we consider the main properties of the homogeneous Lorenz group and its Lie algebra. First, we examine the restricted Lorentz group, which is nothing else that the non-compact version $SO(3,1)$ of the rotation group in four dimensions. In particular we consider its finite dimensional irreducible representations: they are non unitary, since the group is non compact, but they are very useful, in particle physics, for the derivation of the relativistic equations. Chapter 4 is devoted to the Poincaré group, which is most suitable for a quantum mechanical description of particle states. Specifically, the transformation properties of one-particle and two-particle states are examined in detail in Chapter 5. In this connection, a covariant treatment of spin is presented and its physical meaning is discussed in both cases of massive and massless particles. In Chapter 6 we consider the transformation properties of the particle states under the discrete operations of parity and time reversal, which are contained in the homogeneous Lorentz group and which have important roles in particle physics. In Chapter 7, the relativistic wave functions are introduced in connection with one-particle states and the relative equations are examined for the lower spin cases, both for integer and half-integer values. In particular, we give a group-theoretical derivation of the Dirac equation and of the Maxwell equations.

The second part of the book is devoted to the various kinds of internal symmetries, which were introduced during the extraordinary development of particle physics in the second half of last century and which had a fundamental role in the construction of the present theories. A key ingredient was the use of the unitary groups, which is the subject of Chapter 8. In order to illustrate clearly this point, we give a historical overview of the different steps of the process that lead to the discovery of elementary particles and of the properties of fundamental interactions. The main part of this chapter is devoted to the analysis of *hadrons*, i.e. of particle states participating in strong interactions. First we consider the isospin invariance, based on the group $SU(2)$ and on the assumption that the members of each family of hadrons, almost degenerate in mass but with different electric charge, are assigned to the same irreducible representation. Further analysis of the different kinds of hadrons lead to the introduction of a larger symmetry, now called flavor $SU(3)$ invariance, which allowed the inclusion of different isospin multiplets in the same irreducible representation of the $SU(3)$ group and gave rise to a more complete classification of hadrons. Moreover, it provided a hint to the introduction of *quarks* as the fundamental constituents of matter. Finally, the analysis of the hadronic states in terms of quarks lead to the discovery of a new degree of freedom, called *color*, that gave a deeper understanding of the nature of strong interactions. It was clear from the very beginning that the flavor $SU(3)$ symmetry was only approximate, but it represented an important step toward the more fundamental symmetry of color $SU(3)$.

Chapter 9 is a necessary complement of the previous chapter, since it describes a further successful step in the development of particle physics, which is the introduction of gauge symmetry. After reminding the well-known case of quantum electrodynamics, we briefly examine the field theory based on the gauge color $SU(3)$ group, i.e. quantum chromodynamics, which provides a good description of the peculiar properties of the strong interactions of quarks. Then we consider the electroweak Standard Model, the field theory based on the gauge $SU(2) \otimes U(1)$ group, which reproduces with great accuracy the properties of weak interactions of leptons and quarks, combined with the electromagnetic ones. An essential ingredient of the theory is the so-called spontaneous symmetry breaking, which we illustrate in the frame of a couple of simple models. Finally, we mention the higher gauge symmetries of Grand Unification Theories, which combine strong and electroweak interactions.

The book contains also three Appendices, which complete the subject of unitary groups. In Appendix A, we collect some useful formulas on the rotation matrices and the Clebsh-Gordan coefficients. In Appendix B, the symmetric group is briefly considered in connection with the problem of identical particles. In Appendix C, we describe the use of the Young tableaux for the study of the irreducible representations of the unitary groups, as a powerful alternative to the use of weight diagrams.

Each chapter, except the first, is supplied with a list of problems, which we consider useful to strengthen the understanding of the different topics discussed in the text. The solutions of all the problems are collected at the end of the book.

The book developed from a series of lectures that both of us have given in University courses and at international summer schools. We have benefited from discussions with students and colleagues and we are greatly indebted to all of them.

Padova, 2011   *Giovanni Costa*
Bari, 2011   *Gianluigi Fogli*

Notation

The natural system of units, where $\hbar = c = 1$, is used throughout the book. In this system: $[length] = [time] = [energy]^{-1} = [mass]^{-1}$.

Our conventions for special relativity are the following. The metric tensor is given by

$$g^{\mu\nu} = g_{\mu\nu} = \begin{pmatrix} 1 & 0 & 0 & 0 \\ 0 & -1 & 0 & 0 \\ 0 & 0 & -1 & 0 \\ 0 & 0 & 0 & -1 \end{pmatrix}, \tag{0.1}$$

and the *controvariant* and *covariant* four-vectors are denoted, respectively, by

$$x^{\mu} = (x^0, \mathbf{x}), \qquad\qquad x_{\mu} = g_{\mu\nu}x^{\nu} = (x^0, -\mathbf{x}). \tag{0.2}$$

Greek indices run over $0, 1, 2, 3$ and Latin indices denote the spacial components. Repeated indices are summed, unless otherwise specified.

The derivative operator is given by

$$\partial_{\mu} = \frac{\partial}{\partial x^{\mu}} = \left(\frac{\partial}{\partial x^0}, \nabla \right). \tag{0.3}$$

The Levi-Civita tensor ϵ^{0123} is totally antisymmetric; we choose, as usual, $\epsilon^{0123} = +1$ and consequently one gets $\epsilon_{0123} = -1$.

The complex conjugate, transpose and Hermitian adjoint of a matrix M are denoted by M^*, $\tilde{M}$ and $M^{\dagger} = \tilde{M}^*$, respectively.

Contents

Introduction to Lie groups and their representations

This Chapter consists of a brief survey of the most important concepts of group theory, having in mind the applications to physical problems. After a collection of general notions which apply both to finite and infinite groups, we shall consider the properties of the Lie groups and their representations. We shall avoid mathematical rigour and completeness and, in order to clarify the main aspects, we shall make use of specific examples.

1.1 Basic definitions

The aim of this section is to collect the basic and general definitions of group theory.

Group - A set $\mathcal{G}$ of *elements* $a, b, c, \ldots$ is a group if the following four axioms are satisfied:

1. there is a composition law, called *multiplication*, which associates with every pairs of elements a and b of $\mathcal{G}$ another element c of $\mathcal{G}$; this operation is indicated by $c = a \circ b$;
2. the multiplication is *associative*, i.e. for any three elements a, b, c of $\mathcal{G}$: $(a \circ b) \circ c = a \circ (b \circ c)$;
3. the set contains an element e called *identity*, such that, for each element a of $\mathcal{G}$, $e \circ a = a \circ e = a$;
4. For each element a of $\mathcal{G}$, there is an element a' contained in $\mathcal{G}$ such that $a \circ a' = a' \circ a = e$. The element a' is called *inverse* of a and is denoted by a^{-1}.

Two elements a, b of a group are said to *commute* with each other if $a \circ b = b \circ a$. In general, the multiplication is not commutative, i.e. $a \circ b \neq b \circ a$.

Abelian group - A group is said to be *Abelian* if all the elements commute with one another.

G. Costa and G. Fogli, *Symmetries and Group Theory in Particle Physics*,
Lecture Notes in Physics 823, DOI: 10.1007/978-3-642-15482-9_1,
© Springer-Verlag Berlin Heidelberg 2012

Order of the group - The number of elements of a group is called the order of the group; it can be *finite* or *infinite, countable* or *non-countable* infinite.

Examples
1. *Additive group of real numbers.* The elements are the real numbers; the composition law is the addition and the identity element is zero. The group is Abelian and of infinite non-countable order.
2. *Symmetric group.* The elements are the permutations of degree n
$$\begin{pmatrix} 1 & 2 & ... & n \\ p_1 & p_2 & ... & p_n \end{pmatrix}.$$
The set is a (non-commutative) group of order $n!$ and it is usually denoted by $\mathcal{S}_n$.
3. *Rotation group.* The elements are the rotations in the three-dimensional space. Each rotation can be characterized by three independent parameters, e.g. the three Euler angles (α, β, γ).

Subgroup - A subset $\mathcal{H}$ of the group $\mathcal{G}$, of elements $a', b', ...$, that is itself a group with the same multiplication law of $\mathcal{G}$, is said to be a *subgroup* of $\mathcal{G}$. A necessary and sufficient condition for $\mathcal{H}$ to be a subgroup of $\mathcal{G}$ is that, for any two elements a', b' of $\mathcal{H}$, also $a' \circ b'^{-1}$ belongs to $\mathcal{H}$. Every group has two trivial subgroups: the group consisting of the identity element alone, and the group itself. A non-trivial subgroup is called a *proper* subgroup.

Examples
1. The additive group of rational numbers is a subgroup of the additive group of real numbers.
2. Let us consider the group $\mathcal{S}_n$ of permutations of degree n. Each permutation can be decomposed into a product of *transpositions* (a transposition is a permutation in which only two elements are interchanged). A permutation is said to be *even* or *odd* if it corresponds to an even or odd number of transpositions. The set of even permutations of degree n is a subgroup of $\mathcal{S}_n$ (if a and b are two even permutations, also $a \circ b^{-1}$ is even). It is denoted by $\mathcal{A}_n$ and called *alternating group.*

Invariant subgroup - Let $\mathcal{H}$ be a subgroup of the group $\mathcal{G}$. If for each element h of $\mathcal{H}$, and g of $\mathcal{G}$, the element $g \circ h \circ g^{-1}$ belongs to $\mathcal{H}$, the subgroup $\mathcal{H}$ is said to be *invariant.*

In connection with the notion of invariant subgroup, a group $\mathcal{G}$ is said to be:

Simple - If it does not contain any invariant subgroups;

Semi-simple - If it does not contain any invariant Abelian subgroups.

In the case of continuous groups, finite or discrete invariant subgroups are not to be taken into account in the above definitions.

Factor group - Let us consider a group $\mathcal{G}$ and a subgroup $\mathcal{H}$. Given an element g, different from the identity e, in $\mathcal{G}$ but not in $\mathcal{H}$, we can form the set $G = g \circ h_1, g \circ h_2, ...$ (where $h_1, h_2, ...$ are elements of $\mathcal{H}$), which is not a subgroup since it does not contain the unit element. We call the set $g \circ \mathcal{H}$ *left coset* of $\mathcal{H}$ in $\mathcal{G}$ with respect to g. By varying g in $\mathcal{G}$, one gets different cosets. It can be shown that either two cosets coincide or they have no element in common. The elements $g_1, g_2, \ldots$ of the group $\mathcal{G}$ can be distributed among the subgroup $\mathcal{H}$ and all its distinct cosets $g_1 \circ \mathcal{H}$, $g_2 \circ \mathcal{H}$, ... The group $\mathcal{G}$ is a disjoint union of these sets. In the case in which $\mathcal{H}$ is an invariant subgroup, the sets $\mathcal{H}$, $g_1 \circ \mathcal{H}$, $g_2 \circ \mathcal{H}$ ... themselves can be considered as elements of a group ($\mathcal{H}$ plays the role of unit element for this group) with the following multiplication rule:

$$(g_1 \circ \mathcal{H}) \circ (g_2 \circ \mathcal{H}) = g_1 \circ g_2 \circ \mathcal{H} \ . \tag{1.1}$$

The group is called *factor group* and is denoted by $\mathcal{G}/\mathcal{H}$. The same considerations hold for the *right cosets* $\mathcal{H} \circ g$. Left and right cosets ($g \circ \mathcal{H}$ and $\mathcal{H} \circ g$) with respect to the same element g are not necessarily identical; they are identical if and only if $\mathcal{H}$ is an invariant subgroup of $\mathcal{G}$.

Homomorphism - A mapping of a group $\mathcal{G}$ onto the group $\mathcal{G}'$ is said to be *homomorphic* if it *preserves the products*. Each element g of $\mathcal{G}$ is mapped onto an element g' of $\mathcal{G}'$, which is the image of g, and the product $g_1 \circ g_2$ of two elements of $\mathcal{G}$ is mapped onto the product $g_1' \circ g_2'$ in $\mathcal{G}'$. In general, the mapping is not one-to-one: *several* elements of $\mathcal{G}$ are mapped onto the same element of $\mathcal{G}'$, but an equal number of elements of $\mathcal{G}$ are mapped onto each element of $\mathcal{G}'$. In particular, the unit element e' of $\mathcal{G}'$ corresponds to the set of elements $e_1, e_2, \ldots$ of $\mathcal{G}$ (only one of these elements coincides with the unit element of $\mathcal{G}$), which we denote by $\mathcal{E}$. The subgroup $\mathcal{E}$ is an invariant subgroup of $\mathcal{G}$ and it is called *kernel* of the homomorphism.

Isomorphism - The mapping of a group $\mathcal{G}$ onto the group $\mathcal{G}'$ is said to be *isomorphic* if the elements of the two groups can be put into a one-to-one correspondence ($g \leftrightarrow g'$), which is *preserved under the composition law*. If $\mathcal{G}$ is homomorphic to $\mathcal{G}'$, one can show that the factor group $\mathcal{G}/\mathcal{E}$ is *isomorphic* to $\mathcal{G}'$.

Example
It can be shown that the alternating group $\mathcal{A}_n$ of even permutations is an invariant subgroup of the symmetric group $\mathcal{S}_n$. One can check that there are only two distinct left (or right) cosets and that the factor group $\mathcal{A}_n/\mathcal{S}_n$ is isomorphic to the group of elements $1, -1$.

Direct product - A group $\mathcal{G}$ which possesses two subgroups $\mathcal{H}_1$ and $\mathcal{H}_2$ is said to be *direct product* of $\mathcal{H}_1$ and $\mathcal{H}_2$ if:

1. the two subgroups $\mathcal{H}_1$ and $\mathcal{H}_2$ have only the unit element in common;
2. the elements of $\mathcal{H}_1$ commute with those of $\mathcal{H}_2$;
3. each element g of $\mathcal{G}$ is expressible in one and only one way as $g = h_1 \circ h_2$, in terms of the elements h_1 of $\mathcal{H}_1$ and h_2 of $\mathcal{H}_2$.

The direct product is denoted by $\mathcal{G} = \mathcal{H}_1 \otimes \mathcal{H}_2$.

Semi-direct product - A group $\mathcal{G}$ which possesses two subgroups $\mathcal{H}_1$ and $\mathcal{H}_2$ is said to be *semi-direct product* of $\mathcal{H}_1$ and $\mathcal{H}_2$ if:

1. $\mathcal{H}_1$ is an invariant subgroup of $\mathcal{G}$;
2. the two subgroups $\mathcal{H}_1$ and $\mathcal{H}_2$ have only the unit element in common;
3. each element g of $\mathcal{G}$ is expressible in one and only one way as $g = h_1 \circ h_2$, in terms of the elements h_1 of $\mathcal{H}_1$ and h_2 of $\mathcal{H}_2$.

The semi-direct product is denoted by $\mathcal{G} = \mathcal{H}_1 \circledS \mathcal{H}_2$.

Representation of a group - Let us consider a finite n-dimensional complex *vector space* L_n, and a mapping T which associates with a vector $\mathbf{x}$ a new vector $\mathbf{x}'$ in L_n:

$$\mathbf{x}' = T\mathbf{x} . \tag{1.2}$$

T is a *linear operator*, i.e., for $\mathbf{x}$ and $\mathbf{y}$ in L_n, and α and β two real numbers, it satisfies the relation:

$$T(\alpha\mathbf{x} + \beta\mathbf{y}) = \alpha T\mathbf{x} + \beta T\mathbf{y} . \tag{1.3}$$

If the mapping is one-to-one, the inverse operator T^{-1} exists. For each vector $\mathbf{x}$ in L_n:

$$T^{-1}T\mathbf{x} = TT^{-1}\mathbf{x} = I\mathbf{x} , \tag{1.4}$$

where the *identity operator* I leaves all the vectors unchanged.

Let us now consider a group $\mathcal{G}$. If for each element g of $\mathcal{G}$ there is a correponding linear operator $T(g)$ in L_n, such that

$$T(g_1 \circ g_2) = T(g_1)T(g_2) , \tag{1.5}$$

we say that the set of operators $T(g)$ forms a *linear (n-dimensional) representation* of the group $\mathcal{G}$. It is clear that the set of operators $T(g)$ is a group $\mathcal{G}'$ and in general $\mathcal{G}$ is homomorphic to $\mathcal{G}'$. If the mapping of is one-to-one, then $\mathcal{G}$ is isomorphic to $\mathcal{G}'$.

Matrix representation - If one fixes a *basis* in L_n, then the linear transformation performed by the operator T is represented by a $n \times n$ matrix, which we denote by $D(g)$. The set of matrices $D(g)$ for all $g \in \mathcal{G}$ is called n-dimensional *matrix representation* of the group $\mathcal{G}$. Defining an orthonormal basis $\mathbf{e}_1, \mathbf{e}_2, ..., \mathbf{e}_n$ in L_n, the elements of $D(g)$ are given by

$$T(g)\mathbf{e}_k = \sum_i D_{ik}(g)\mathbf{e}_i \tag{1.6}$$

and the transformation (1.2) of a vector $\mathbf{x}$ becomes:

$$x_i' = \sum_i D_{ik}(g)x_k . \tag{1.7}$$

The set of vectors $\mathbf{e}_1, \mathbf{e}_2, ..., \mathbf{e}_n$ is called the *basis of the representation* $D(g)$.

Faithful representation - If the mapping of the group $\mathcal{G}$ onto the group of matrices $D(g)$ is one-to-one, the representation $D(g)$ is said to be *faithful*.

In other words, different elements of $\mathcal{G}$ correspond to different matrices $D(g)$ and the mapping is isomorphic.

Equivalent representations - If we change the basis of the vector space L_n, the matrices $D(g)$ of a representation are transformed by a non-singular matrix S

$$D'(g) = SD(g)S^{-1} . \tag{1.8}$$

The representations $D(g)$ and $D'(g)$ are said to be *equivalent* and Eq. (1.8) is called *similarity* transformation; the two representations are regarded as essentially the same.

Reducible and irreducible representations - The representation $T(g)$ of $\mathcal{G}$ in L_n is said to be *reducible* if there exists a non trivial subspace L_m of L_n which is left *invariant* by all the operators $T(g)$. If no non-trivial invariant subspace exists, the representation $T(g)$ is said to be *irreducible*. In the case of a reducible representation, it is possible to choose a basis in L_n such that all the matrices corresponding to $T(g)$ can be written in the form

$$D(g) = \left(\begin{array}{c|c} D_1(g) & D_{12}(g) \\ \hline 0 & D_2(g) \end{array} \right) . \tag{1.9}$$

If also L_{n-m} is invariant, by a similarity transformation all the matrices $D(g)$ can be put in block form

$$D(g) = \left(\begin{array}{c|c} D_1(g) & 0 \\ \hline 0 & D_2(g) \end{array} \right) \tag{1.10}$$

and the representation is *completely reducible*. In this case one writes

$$D(g) = D_1(g) \oplus D_2(g) \tag{1.11}$$

and the representation is said to be decomposed into the *direct sum* of the two representations D_1, D_2.

In general, if a representation $D(g)$ can be put in a block-diagonal form in terms of ℓ submatrices $D_1(g), D_2(g), ... D_\ell(g)$, each of which is an irreducible representation of the group $\mathcal{G}$, $D(g)$ is said to be *completely reducible*. If the group $\mathcal{G}$ is Abelian its irreducible representations are one-dimensional.

A test of irreducibility (for non-Abelian groups) is provided by the following lemma due to Schur.

Schur's lemma - If $D(g)$ is an irreducible representation of the group $\mathcal{G}$, and if

$$AD(g) = D(g)A \tag{1.12}$$

for all the elements g of $\mathcal{G}$, then A is multiple of the unit matrix.

Unitary representation - A representation of the group $\mathcal{G}$ is said to be unitary if the matrices $D(g)$, for all the elements g of $\mathcal{G}$, are unitary, i.e.

$$D(g)D(g)^\dagger = D(g)^\dagger D(g) = I \,, \tag{1.13}$$

where $D(g)^\dagger$ is the *adjoint* (i.e. *conjugate transposed*) or *Hermitian conjugate* of $D(g)$. Such representations are very important for physical applications.

Unitary representation of finite groups - In the case of finite groups one can prove that every representation is equivalent to a unitary representation. Moreover, every unitary representation is irreducible or completely reducible; the number of non-equivalent irreducible representations is limited by the useful formula

$$N = \sum_i n_i^2 \,, \tag{1.14}$$

where N is the order of the group and n_i the dimension of the i-th irreducible representation.

Self-representation (of a matrix group) - The irreducible representation used to define a matrix group is called sometimes self-representation.

Example
In the case of the symmetric group S_3 ($N = 3! = 6$), there are two one-dimensional and one two-dimensional non-equivalent irreducible representations.

1.2 Lie groups and Lie algebras

A Lie group combines three different mathematical structures, since it satisfies the following requirements:

1. the *group axioms* of Section 1.1;
2. the group elements form a *topological space*, so that the group is considered a special case of topological group;
3. the group elements constitute an *analytic manifold*.

As a consequence, a Lie group can be defined in different but equivalent ways. Specifically, it can be defined as a topological group with additional analytic properties, or an analytic manifold with additional group properties.

We shall give a general definition of Lie group and, for this reason, first we summarize the main concepts that are involved. For complete and detailed analyses on Lie groups we refer to the books by Cornwell[1] and Varadarajan[2] and, for more details on topological concepts, to the book by Nash and Sen[3].

---

[1] J.F. Cornwell, *Group Theory in Physics*, Vol. 1 and 2, Academic Press, 1984.

[2] V.S. Varadarajan, *Lie Groups, Lie Algebras, and their Representations*, Springer-Verlag, 1974.

[3] C. Nash, S. Sen, *Topology and Geometry for Physicists*, Academic Press, 1983.

Topological space - A topological space S is a non-empty set of elements called *points* for which there is a collection T of subsets, called *open sets*, satisfying the following conditions:

1. the empty set and the set S belong to T;
2. the union of any number of sets in T belongs to T;
3. the intersection of any finite number of sets in T belongs to T.

Hausdorff space - A Hausdorff space is a topological space S with a topology T which satisfies the *separability* axiom: any two distinct points of S belong to disjoint open subsets of T.

Cartesian product of two topological spaces - If S and S' are two topological spaces with topologies T and T' respectively, the set of pairs (P, P'), where $P \in S$ and $P' \in S'$, is defined to be the Cartesian product $S \times S'$.

Metric space - An important kind of Hausdorff space is the so-called metric space, in which one can define a *distance* function $d(P, P')$ between any two points P and P' of S. The distance or metric $d(P, P')$ is real and must satisfy the following axioms:

1. $d(P, P') = d(P', P)$;
2. $d(P, P) = 0$;
3. $d(P, P') > 0$ if $P \neq P'$;
4. $d(P, P') \leq d(P, P'') + d(P'', P')$ for any three points of S.

Examples
1. Let us consider the n-dimensional Euclidean space $\mathcal{R}^n$ and two points P and P' in $\mathcal{R}^n$ with coordinates $(x_1, x_2, ..., x_n)$ and $(x'_1, x'_2, ..., x'_n)$ respectively. With the metric defined by

$$d(P, P') = \left\{ \sum_{i=1}^{n} (x_i - x'_i)^2 \right\}^{1/2},$$
(1.15)

one can show that $\mathcal{R}^n$ is a metric space, since it satisfies the required axioms.
2. Let us consider the set $\mathcal{M}$ of all the $m \times m$ matrices $\mathbf{M}$ with complex elements and, for any two matrices $\mathbf{M}$ and $\mathbf{M}$', let us define the *distance*

$$d(\mathbf{M}, \mathbf{M}') = \left\{ \sum_{i,j=1}^{m} | M_{ij} - M'_{ij} |^2 \right\}^{1/2}.$$
(1.16)

Then one can show that the set $\mathcal{M}$ is a metric space.

Compact space - A family of open sets of the topological space S is said to be an *open covering* of S if the union of its open sets contains S. If, for every open covering of S there is always a *finite subcovering* (i.e. a union of a finite number of open sets) which contains S, the topological space S is said to be *compact*. If there exists no finite subcovering, the space S is said to be *non-compact*.

Connected space - A topological space $\mathcal{S}$ is *connected* if it is not the union of non-empty disjoint open subsets. In order to specify the notion of connectedness it is useful to give a definition of *path*. A path in $\mathcal{S}$ from the point x_0 to the point x_1 is a continuous mapping ϕ of the interval $[0, 1]$ in $\mathcal{R}$ into $\mathcal{S}$ with $\phi(0) = x_0, \phi(1) = x_1$. A *closed path or loop* is a path for which $x_0 = x_1$ and $\phi(0) = \phi(1)$. There are different kinds of loops: for instance, those which can be shrunk to a point by a continuous deformation and those for which the shrinking is not possible. Two loops are *equivalent* or *homotopic* if one can be obtained from the other by a continuous deformation. All equivalent loops can be collected in an *equivalence class*. A topological space $\mathcal{S}$ in which any loop can be shrunk to a point by continuous deformation is called *simply connected*. If there are n distinct classes of equivalence of closed paths, $\mathcal{S}$ is said to be *n-times connected*.

Examples
3. A region $\mathcal{F}$ of the Euclidean space $\mathcal{R}^n$ is compact only if it is finite; otherwise it is not compact. In fact, for any open covering there is a finite subcovering which contains $\mathcal{F}$ only if $\mathcal{F}$ is finite.
4. The space $\mathcal{R}^2$ is simply connected; however, a region of $\mathcal{R}^2$ with a "hole" is not simply connected since loops encircling the hole cannot be shrunk to a point.

Second countable space - A topological space $\mathcal{S}$ with topology $\mathcal{T}$ is said to be *second countable* if $\mathcal{T}$ contains a countable collection of open sets such that every open set of $\mathcal{T}$ is a union of sets of this collection. The topological spaces considered in the Examples 1 and 2 are second countable.

Homeomorphic mapping - Let us consider two topological spaces $\mathcal{S}$ and $\mathcal{S}'$ with topologies $\mathcal{T}$ and $\mathcal{T}'$, respectively. A mapping ϕ from $\mathcal{S}$ onto $\mathcal{S}'$ is said to be *open* if, for every open set V of $\mathcal{S}$, the set $\phi(V)$ is an open set of $\mathcal{S}'$. A mapping ϕ is *continuous* if, for every open set V' of $\mathcal{S}'$, the set $\phi^{-1}(V')$ is an open set of $\mathcal{S}$. Finally, if ϕ is a continuous and open mapping of $\mathcal{S}$ onto $\mathcal{S}'$, it is called *homeomorphic* mapping.

Locally Euclidean space - A Hausdorff topological space $\mathcal{V}$ is said to be a *locally Euclidean space of dimension* n if each point of $\mathcal{V}$ is contained in an open set which is homeomorphic to a subset of $\mathcal{R}^n$. Let V be an open set of $\mathcal{V}$ and ϕ a homeomorphic mapping of V onto a subset of $\mathcal{R}^n$. Then for each point $P \in V$ there exists a set of coordinates $(x_1, x_2, ..., x_n)$ such that $\phi(P) = (x_1, x_2, ..., x_n)$; the pair (V, ϕ) is called a *chart*.

Analytic manifold of dimension n - Let us consider a locally Euclidean space $\mathcal{V}$ of dimension n, which is second countable, and a homeomorphic mapping ϕ of an open set V onto a subset of $\mathcal{R}^n$: if, for every pair of charts (V_α, ϕ_α) and (V_β, ϕ_β) of $\mathcal{V}$ for which the intersection $V_\alpha \cap V_\beta$ is non-empty, the mapping $\phi_\beta \circ \phi_\alpha^{-1}$ is an analytic function, then $\mathcal{V}$ is an analytic manifold of dimension n. The simplest example of analytic manifold of dimension n is $\mathcal{R}^n$ itself.

We are now in the position of giving a more precise definition of a Lie group.

Lie group - A Lie group $\mathcal{G}$ of dimension n is a set of elements which satisfy the following conditions:

1. they form a *group*;
2. they form an *analytic manifold* of dimension n;
3. for any two elements a and b of $\mathcal{G}$, the mapping $\phi(a,b) = a \circ b$ of the Cartesian product $\mathcal{G} \times \mathcal{G}$ onto $\mathcal{G}$ is analytic;
4. for any element a of $\mathcal{G}$, the mapping $\phi(a) = a^{-1}$ of $\mathcal{G}$ onto $\mathcal{G}$ is analytic.

1.2.1 Linear Lie groups

The Lie groups that are important for physical applications are of the type known as *linear Lie groups*, for which a simpler definition can be given.

Let us consider a n-dimensional vector space V over the field $\mathcal{F}$ (such as the field $\mathcal{R}$ of real numbers and the field $\mathcal{C}$ of complex numbers) and the *general linear group* $GL(N, \mathcal{F})$ of $N \times N$ matrices. A Lie group $\mathcal{G}$ is said to be a *linear Lie group* if it is isomorphic to a subgroup $\mathcal{G}'$ of $GL(N, \mathcal{F})$. In particular, a *real* linear Lie group is isomorphic to a subgroup of the linear group $GL(N, \mathcal{R})$ of $N \times N$ real matrices.

A linear Lie group $\mathcal{G}$ of dimension n satisfies the following conditions:

1. $\mathcal{G}$ possesses a faithful finite-dimensional representation D. Suppose that this representation has dimension m; then the distance between two elements g and g' of $\mathcal{G}$ is given, according to Eq. (1.16), by

$$d(g,g') = \left\{ \sum_{i,j=1}^{m} | D(g)_{ij} - D(g')_{ij} |^2 \right\}^{1/2}, \qquad (1.17)$$

and the set of matrices $D(g)$ satisfies the requirement of a metric space.

2. There exists a real number $\delta > 0$ such that every element g of $\mathcal{G}$ lying in the open set V_δ, centered on the identity e and defined by $d(g,e) < \delta$, can be parametrized by n independent real parameters $(x_1, x_2, ..., x_n)$, with e corresponding to $x_1 = x_2 = ... = x_n = 0$. Then every element of V_δ corresponds to one and only one point in a n-dimensional real Euclidean space $\mathcal{R}^n$. The number n is the *dimension* of the linear Lie group.

3. There exists a real number $\epsilon > 0$ such that every point in $\mathcal{R}^n$ for which

$$\sum_{i=1}^{n} x_i^2 < \epsilon^2 \qquad (1.18)$$

corresponds to some element g in the open set V_δ defined above and the correspondence is one-to-one.

4. Let us define $D(g(x_1, x_2, ..., x_n)) \equiv D(x_1, x_2, ..., x_n)$ the representation of each generic element $g(x_1, x_2, ..., x_n)$ of $\mathcal{G}$. Each matrix element of $D(x_1, x_2, ..., x_n)$ is an analytic function of $(x_1, x_2, ..., x_n)$ for all $(x_1, x_2, ..., x_n)$ satisfying Eq. (1.18).

Before giving some examples of linear Lie groups we need a few other definitions:

Connected Lie group - A linear Lie group $\mathcal{G}$ is said to be *connected* if its topological $\mathcal{S}$ space is connected. According to the definition of connected space, $\mathcal{G}$ can be *simply connected* or *multiply connected*. In Chapter 2, we shall examine explicitly simply and doubly connected Lie groups, such as $SU(2)$ and $SO(3)$.

Center of a group - The center of a group $\mathcal{G}$ is the subgroup $\mathcal{Z}$ consisting of all the elements $g \in \mathcal{G}$ which commute with every element of $\mathcal{G}$. Then $\mathcal{Z}$ and its subgroups are Abelian; they are invariant subgroups of $\mathcal{G}$ and they are called *central invariant subgroups*.

Universal covering group - If $\mathcal{G}$ is a (multiply) connected Lie group there exist a *simply* connected group $\tilde{\mathcal{G}}$ (unique up to isomorphism) such that $\mathcal{G}$ is isomorphic to the factor group $\tilde{\mathcal{G}}/\mathcal{K}$, where $\mathcal{K}$ is a discrete central invariant subgroup of $\tilde{\mathcal{G}}$. The group $\tilde{\mathcal{G}}$ is called the *universal covering group* of $\mathcal{G}$.

Compact Lie group - A linear Lie group is said to be compact if its topological space is *compact*. A topological group which does not satisfy the above property is called *non-compact*.

Unitary representations of a Lie group - The content of the following theorems shows the great difference between compact and non-compact Lie groups.

1. If $\mathcal{G}$ is a *compact* Lie group then every representation of $\mathcal{G}$ is equivalent to a *unitary* representation;
2. If $\mathcal{G}$ is a *compact* Lie group then every reducible representation of $\mathcal{G}$ is *completely reducible*;
3. If $\mathcal{G}$ is a *non-compact* Lie group then it possesses no finite-dimensional unitary representation apart from the trivial representation in which $D(g) = 1$ for all $g \in \mathcal{G}$.

For physical applications, in the case of compact Lie group, one is interested only in *finite-dimensional* representations; instead, in the case of non-compact Lie groups, one needs also to consider *infinite-dimensional* (unitary) representations.

We list here the principal classes of groups of $N \times N$ matrices, which can be checked to be linear Lie groups:

$GL(N, C)$: *general linear group* of *complex* regular matrices M (det $M \neq 0$); its dimension is $n = 2N^2$.

$SL(N,C)$: *special linear group*, subgroup of $GL(N,C)$ with $\det M = 1$; its
   dimension is $n = 2(N^2 - 1)$.
$GL(N,R)$: *general linear group* of *real* regular matrices R ($\det R \neq 0$); dimension $n = N^2$.
$SL(N,R)$: *special linear group*, subgroup of $GL(N,R)$ with $\det R = 1$; dimension $n = N^2 - 1$.
$U(N)$: *unitary group* of complex matrices U satisfying the condition $UU^{\dagger} = U^{\dagger}U = I$, where $U^{\dagger}$ is the adjoint of U; dimension $n = N^2$.
$SU(N)$: *special unitary group*, subgroup of $U(N)$ with $\det U = 1$; dimension
   $n = N^2 - 1$.
$O(N)$: *orthogonal group* of *real* matrices O satisfying $O\tilde{O} = I$ where $\tilde{O}$ is the
   transpose of O; dimension $n = \frac{1}{2}N(N-1)$.
$SO(N)$: *special orthogonal group* or *rotation* group in N dimensions, subgroup
   of $O(N)$ with $\det O = 1$; dimension $n = \frac{1}{2}N(N-1)$.
$Sp(N)$: *symplectic group*. It is the group of the unitary $N \times N$ matrices U
   (with N even) which satisfy the condition $\tilde{U}JU = J$ ($\tilde{U}$ is the transpose
   of U and $J = \begin{pmatrix} 0 & I \\ -I & 0 \end{pmatrix}$ where I is the $\dfrac{N}{2} \times \dfrac{N}{2}$ unit matrix); dimension
   $n = \frac{1}{2}N(N+1)$.
$U(\ell, N - \ell)$: *pseudo-unitary group* of complex matrices U satisfying the condition $UgU^{\dagger} = g$, where g is a diagonal matrix with elements $g_{kk} = 1$ for
   $1 \leq k \leq \ell$ and $g_{kk} = -1$ for $\ell + 1 \leq k \leq N$. Its dimension is $n = N^2$.
$O(\ell, N - \ell)$: *pseudo-orthogonal group* of real matrices O satisfying the condition $Og\tilde{O} = g$; dimension $n = \frac{1}{2}N(N-1)$.

All the groups listed above are subgroups of $GL(N,C)$. In particular:

1. the groups $U(N)$, $SU(N)$, $O(N)$, $SO(N)$, $Sp(N)$ are compact;
2. the groups $GL(N)$, $SL(N)$, $U(\ell, N - \ell)$, $O(\ell, N - \ell)$ are not compact.

Examples
1. *The group $SO(3)$*. Its elements can be defined by the orthogonal 3×3 matrices
R satisfying

$$\tilde{R}R = I\,, \tag{1.19}$$

$$\det R = 1\,. \tag{1.20}$$

The rotation group $SO(3)$ is *compact*. In fact, its coordinate domain can be identified
with a sphere in the euclidean space $\mathcal{R}^3$, i.e. a compact domain. The rotation group
$SO(3)$ is *connected*: any two points can be connected by a continuous path. However,
not all closed paths can be shrunk to a point; in fact, the group is doubly connected.

2. *The group $O(3)$*. If one keeps only the orthogonality condition (1.19) and disregard
(1.20), one gets the larger group $O(3)$, which contains elements with both signs
of $\det R$: $\det R = \pm 1$. The group consists of two disjoint sets, corresponding to
$\det R = +1$ and $\det R = -1$. The first set coincides with the group $SO(3)$, which
is an *invariant subgroup* of $O(3)$. Then the group $O(3)$ is neither simple nor semi-
simple, while one can prove that $SO(3)$ is simple. The group $O(3)$ is *not connected*,

since it is the union of two disjoint sets. These properties are illustrated in Section 2.1.

3. *The group* $SU(2)$. The elements of the group $SU(2)$ are the 2×2 matrices u satisfying

$$uu^\dagger = u^\dagger u = I\,, \tag{1.21}$$

$$\det u = 1\,. \tag{1.22}$$

The group $SU(2)$ and all the groups of the type $SU(N)$ are simply connected. One can show that the groups $SO(3)$ and $SU(2)$ are homomorphic and that $SU(2)$ is the universal covering group of $SO(3)$. The kernel of the homomorphism is the *center* of $SU(2)$, which is the Abelian subgroup $\mathcal{Z}_2$ consisting of two elements represented by the square roots $(1, -1)$ of the identity. The group $SO(3)$ is isomorphic to the factor group $SU(2)/\mathcal{Z}_2$.

4. *The Lorentz group.* The Lorentz transformations in one dimension (say, along the x_1 axis), are characterized by a real parameter ψ (being $\cosh\psi = \gamma = (1 - \beta^2)^{-1/2}$ with $\beta = v/c$):

$$\begin{aligned} x_0' &= x_0 \cosh\psi - x_1 \sinh\psi\,, \\ x_1' &= -x_0 \sinh\psi + x_1 \cosh\psi\,. \end{aligned} \tag{1.23}$$

One can show that they form a linear Lie group, the one-dimensional Lorentz group $SO(1,1)$. This group is *non-compact*: in fact the parameter ψ varies from $-\infty$ to $+\infty$, i.e. its domain is not bounded.

1.2.2 Real Lie algebras

In the study of the Lie groups both *local* and *global* aspects are important, but most of the information on the structure of a Lie group comes from the analysis of its local properties. These properties are determined by the *real Lie algebras*; in the case of linear Lie groups, the link between Lie algebras and Lie groups is provided by the matrix exponential function.

Before going to the real Lie algebras we have to collect some relevant definitions and to state a few theorems. Also in this case, we shall not reproduce the proofs of the theorems, that the reader can find in the quoted reference[4].

Matrix exponential function - The exponential form of a $m \times m$ matrix A is given by the series

$$e^A = 1 + \sum_{k=1}^\infty \frac{1}{k!} A^k \tag{1.24}$$

which converges for any $m \times m$ matrix A. We recall that a series of $m \times m$ matrices $\sum_{k=1}^\infty A^k$ converges to a $m \times m$ matrix A only if the series of matrix elements $\sum_{k=1}^\infty (A_{rs})^k$ converges to A_{rs} for all $r, s = 1, 2, ...m$.

The matrix exponential function possesses the following properties:

---

[4] J.F. Cornwell, *Group Theory in Physics*, Vol.1 and 2, Academic Press, 1984.

1. $\left(e^A\right)^\dagger = e^{A^\dagger}$;

2. e^A is always non-singular and $\left(e^A\right)^{-1} = e^{-A}$;

3. $\det(e^A) = e^{\operatorname{tr} A}$;

4. if A and B are two $m \times m$ matrices that commute:

$$e^A e^B = e^{A+B} = e^B e^A ; \qquad (1.25)$$

5. if A and B do not commute and their entries are sufficiently small, we can write $e^A e^B = e^C$, where C is given by an infinite series

$$C = A + B + \tfrac{1}{2}[A, B] + \tfrac{1}{12}([A, [A, B]] + [B, [B, A]]) + \cdots \qquad (1.26)$$

where the successive terms contain commutators of increasingly higher order. The above equation is called *Baker-Campbell-Hausdorff* formula; a general expression can be found in ref.[5].

6. the exponential mapping $\phi(A) = e^A$ is a one-to-one continuous mapping of a small neighbourhood of the $m \times m$ zero matrix onto a small neighbourhood of the $m \times m$ unit matrix.

One-parameter subgroup of a linear Lie group - Given a linear group $\mathcal{G}$ of $m \times m$ matrices, a one-parameter subgroup $\mathcal{T}$ is a Lie subgroup of $\mathcal{G}$ which consists of the matrices $T(t)$ depending on a real parameter t such that

$$T(t)T(t') = T(t + t') \qquad (1.27)$$

for all t, t' in the interval $(-\infty, +\infty)$. $T(t)$ is a continuous and differentiable function of t. Clearly the subgroup $\mathcal{T}$ is Abelian. Eq. (1.27) for $t' = 0$ implies that $T(0)$ is the identity.

Every one-parameter subgroup of a linear Lie group of $m \times m$ matrices is formed by exponentiation

$$T(t) = e^{\omega t} \qquad (1.28)$$

where

$$\omega = \left. \frac{dT}{dt} \right|_{t=0}. \qquad (1.29)$$

In fact, taking the derivative of Eq. (1.27) with respect to t' and putting $t' = 0$, one gets $dT/dt = \omega T(t)$, from which Eq. (1.29) follows.

We can now define a real Lie algebra.

Real Lie algebra - A real Lie algebra $\mathcal{L}$ of dimension $n \geq 1$ is a real vector space of dimension n with a composition law called *Lie product* $[a, b]$ such that, for every element a, b, c of $\mathcal{L}$:

---

[5] A.A. Sagle, R.E. Walde, *An Introduction to Lie Groups and Lie Algebras*, Academic Press, 1973.

1. $[a, b] \in \mathcal{L}$;
2. $[\alpha a + \beta b, c] = \alpha [a, c] + \beta [b, c]$, with α, β real numbers;
3. $[a, b] = -[b, a]$;
4. $[a, [b, c]] + [b, [c, a]] + [c, [a, b]] = 0$.

The last relation is called *Jacoby identity*. In the case of a Lie algebra of matrices the Lie product is the *commutator*.

Abelian Lie algebra - A Lie algebra is *Abelian* if $[a, b] = 0$ for all $a, b \in \mathcal{L}$.

Subalgebra of a Lie algebra - A subalgebra $\mathcal{L}'$ of a Lie algebra $\mathcal{L}$ is a subset of elements of $\mathcal{L}$ that form a Lie algebra with the same Lie product.

Invariant subalgebra of a Lie algebra - A subalgebra $\mathcal{L}'$ of a Lie algebra $\mathcal{L}$ is said to be *invariant* if $[a, b] = 0$ for all $a \in \mathcal{L}'$ and all $b \in \mathcal{L}$.

From the definition of a linear Lie group $\mathcal{G}$ of dimension n it follows that, in the case in which $\mathcal{G}$ is a group of $m \times m$ matrices A, there is a one-to-one correspondence between the matrices A lying close to the identity and the points in $\mathcal{R}^n$ satisfying the condition (1.18). Then one can parametrize the matrices A as functions $A(x_1, x_2, ...x_n)$ of the coordinates $x_1, x_2, ...x_n$ satisfying (1.18); by assumption, the elements of $A(x_1, x_2, ...x_n)$ are analytic functions of $x_1, x_2, ...x_n$. The n matrices $a_1, a_2, ...a_n$ defined by

$$(a_r)_{ij} = \left. \frac{\partial A_{ij}}{\partial x_r} \right|_{x_1 = x_2 = ... = 0} \tag{1.30}$$

(where $i, j = 1, 2, ...m; r = 1, 2, ...n$) form a basis for a real n-dimensional vector space, which is the Lie algebra associated to the Lie group $\mathcal{G}$; the composition law is the commutator. In general, the matrices $a_1, a_2, ...a_n$ are not necessarily real, but the reality condition of a real Lie algebra $\mathcal{L}$ requires that the elements of $\mathcal{L}$ be *real linear combinations* of $a_1, a_2, ...a_n$.
In the physical applications, the quantities $a_1, a_2, ...a_n$ are usually referred to as *generators* of the Lie algebra $\mathcal{L}$. In general, they are chosen to be hermitian.

Relationship between linear Lie algebras and linear Lie groups - One can associate a real linear Lie algebra $\mathcal{L}$ of dimension n to every linear Lie group $\mathcal{G}$ of the same dimension, as specified by the following theorems.

1. Every element a of a real Lie algebra $\mathcal{L}$ of a linear Lie group $\mathcal{G}$ is associated with a one-parameter subgroup of $\mathcal{G}$ defined by $A(t) = e^{at}$ for t in the interval $(-\infty, +\infty)$.
2. Every element g of a linear Lie group $\mathcal{G}$ in some small neighbourhood of the identity e belongs to a one-parameter subgroup of $\mathcal{G}$.
3. If $\mathcal{G}$ is a compact linear Lie group, every element of a connected subgroup of $\mathcal{G}$ can be expressed in the form e^a, where a is an element of the corresponding real Lie algebra $\mathcal{L}$. In particular, if $\mathcal{G}$ is compact and connected, every element g of $\mathcal{G}$ has the form e^a, where a is an element of $\mathcal{L}$.

Examples

1. The *real Lie algebra* of $SU(N)$. Let $A(t) = e^{at}$ be a one-parameter subgroup of $SU(N)$. Since A is a $N \times N$ matrix, which satisfies the conditions $A^\dagger A = AA^\dagger = I$ and $\det A = 1$, one gets: $a^\dagger = -a$ and $tr(a) = 0$. Then the real Lie algebra of $SU(N)$ is the set of all traceless and anti-hermitian $N \times N$ matrices.

2. The *real Lie algebra* of $SL(N, R)$. The elements of the one-parameter subgroup are real $N \times N$ matrices A with $\det A = 1$. Then the real Lie algebra of $SL(N, R)$ is the set of traceless real $N \times N$ matrices.

Adjoint representation of a Lie algebra - Given a real Lie algebra $\mathcal{L}$ of dimension n and a basis $a_1, a_2, ...a_n$ for $\mathcal{L}$, we define for any $a \in \mathcal{L}$ the $n \times n$ matrix $\mathrm{ad}(a)$ by the relation

$$[a, a_s] = \sum_{p=1}^{n} \mathrm{ad}(a)_{ps} a_p \ . \tag{1.31}$$

The quantities $\mathrm{ad}(a)_{ps}$ are the entries of the set of matrices $\mathrm{ad}(a)$ which form a n-dimensional representation, called the *adjoint representation* of $\mathcal{L}$. This representation plays a key role in the analysis of semi-simple Lie algebras, as it will be shown in the next Section.

Structure constants - Let us consider the real Lie algebra $\mathcal{L}$ of dimension n and a basis $a_1, a_2, ...a_n$. Then, since $[a_r, a_s] \in \mathcal{L}$, one can write in general

$$[a_r, a_s] = \sum_{p=1}^{n} c_{rs}^p a_p \ . \tag{1.32}$$

Eqs. (1.31) and (1.32) together imply

$$\{\mathrm{ad}(a_r)\}_{ps} = c_{rs}^p \ . \tag{1.33}$$

The n^3 real number c_{rs}^p are called *structure constants* of $\mathcal{L}$ with respect to the basis $a_1, a_2, ...a_n$. The structure constants are not independent. In fact, from the relations which define the real Lie algebra it follows:

$$\begin{aligned} c_{rs}^p &= -c_{sr}^p \\ c_{pq}^s c_{rs}^t + c_{qr}^s c_{ps}^t + c_{rp}^s c_{qs}^t &= 0 \ . \end{aligned} \tag{1.34}$$

It is useful to define the $n \times n$ matrix g whose entries are expressed in terms of the structure constants:

$$g_{ij} = \sum_{\ell, k} c_{ik}^\ell c_{j\ell}^k \ . \tag{1.35}$$

Casimir operators - Let us consider the real vector space $\mathcal{V}$ of dimension n of a semi-simple Lie algebra with basis $a_1, a_2, ...a_n$ and composition law given by Eq. (1.32). One defines the second-order *Casimir operator* acting on the vector space $\mathcal{V}$ as

$$C = \sum_{i,j} g_{ij} a_i a_j \, , \qquad (1.36)$$

where g_{ij} is defined in Eq. (1.35). Making use of Eq. (1.32), one can prove that the Casimir operator commutes with all the elements of the Lie algebra:

$$[C, a_r] = 0 \, . \qquad (1.37)$$

In general, for a simple or semi-simple Lie algebra of *rank* ℓ, one can build ℓ independent Casimir operators by means of second and higher order products of the basis elements a_r; they can be used to specify the irreducible representations of the group.

1.3 Semi-simple Lie algebras and their representations

The study of semi-simple Lie algebras is very useful for physical applications, especially in the field of elementary particle theory. First we give some definitions.

Simple Lie algebra - A Lie algebra is said to be simple if it is not Abelian and it has no proper invariant Lie subalgebra.

Semi-simple Lie algebra - A Lie algebra is said to be semi-simple if it is not Abelian and it has no Abelian invariant Lie subalgebra.

Every semi-simple Lie algebra $\mathcal{L}$ is the direct sum of a set of simple Lie algebras, i.e. there exists a set of invariant simple subalgebras $\mathcal{L}_1, \mathcal{L}_2, ... \mathcal{L}_k$ $(k \geq 1)$ such that

$$\mathcal{L} = \mathcal{L}_1 \oplus \mathcal{L}_2 \oplus ... \oplus \mathcal{L}_k \, . \qquad (1.38)$$

Simple and semi-simple linear Lie group - A linear Lie group $\mathcal{G}$ is simple (semi-simple) if and only if its real Lie algebra $\mathcal{L}$ is simple (semi-simple).

Examples
1. The Lie group $SU(N)$ is simple for all $N \geq 2$.
2. The Lie group $SO(N)$ is simple for $N = 3$ and for $N \geq 5$. The group $SO(2)$ is Abelian and therefore it is not simple and the group $SO(4)$ is semi-simple, but not simple, since it is homomorphic to $SO(3) \otimes SO(3)$.

Killing form - The Killing form $B(a, b)$ corresponding to any two elements a and b of a Lie algebra $\mathcal{L}$ of dimension n is defined by the quantity

$$B(a, b) = \text{tr}\{\text{ad}(a)\text{ad}(b)\} \, , \qquad (1.39)$$

where $\text{ad}(a)$ and $\text{ad}(b)$ are the matrices of a and b in the adjoint representation of $\mathcal{L}$. Note that if $\mathcal{L}$ is a real Lie algebra, all the matrix elements of $\text{ad}(a)$ are real for each $a \in \mathcal{L}$, and in this case $B(a,b)$ is real for all $a, b \in \mathcal{L}$. The quantities B_{ij} defined by

$$B_{ij} \equiv B(a_i, a_j) = \text{tr}\{\text{ad}(a_i)\text{ad}(a_j)\}, \tag{1.40}$$

where $i, j = 1, 2, \ldots n$ and $a_1, a_2, \ldots a_n$ is a basis for $\mathcal{L}$, are seen to coincide, making use of Eq. (1.33), with the matrix elements g_{ij} given in Eq. (1.35).

The Killing form matrix (1.40) provides a criterion for determining if a Lie algebra $\mathcal{L}$ is semi-simple:

Theorem - A real Lie algebra of dimension n is semi-simple if and only if the matrix given by the Killing forms of Eq. (1.40) is non-degenerate, i.e. if and only if

$$\det\{B_{ij}\} \neq 0. \tag{1.41}$$

Equivalently, the above condition can be expressed in the form:

$$\det\{g_{ij}\} \neq 0. \tag{1.42}$$

Compact semi-simple real Lie algebra - A semi-simple real Lie algebra $\mathcal{L}$ is said to be *compact* if its Killing form is negative definite, i.e. if for any element $a \neq 0$ of $\mathcal{L}$ it is $B(a,a) < 0$. Otherwise the Lie algebra is said to be *non-compact*. A connected semi-simple Lie group is compact if and only if its corresponding Lie algebra is compact.

Cartan subalgebra - A Cartan subalgebra $\mathcal{H}$ of a semi-simple Lie algebra $\mathcal{L}$ is a subalgebra of $\mathcal{L}$ with the following properties:

1. $\mathcal{H}$ is a maximal Abelian subalgebra of $\mathcal{L}$, i.e. every subalgebra of $\mathcal{L}$ containing $\mathcal{H}$ as a proper subalgebra is not Abelian;
2. the adjoint representation $\text{ad}(h)$ of $\mathcal{H}$ is completely reducible.

Rank of a semi-simple Lie algebra - The rank of a semi-simple Lie algebra $\mathcal{L}$ is the *dimension* ℓ of the Cartan subalgebra $\mathcal{H}$. Since $\mathcal{H}$ is Abelian, the irreducible representations are one-dimensional; consequently, the matrices $\text{ad}(h_k)$ with $k = 1, 2, \ldots \ell$ must be simultaneously diagonalizable.

1.3.1 Classification of real semi-simple Lie algebras

Given a semi-simple real Lie algebra of dimension n, one can find a basis $h_1, h_2, \ldots h_\ell$; $e_1, e_2, \ldots e_{n-\ell}$, which we shall call *standard basis*, such that

$$[h_i, h_j] = 0 \tag{1.43}$$

and

$$[h_i, e_k] = r_k(h_i)e_k, \tag{1.44}$$

where the real quantities $r_k(h_i)$, (with $i = 1, 2, ... \ell$ and $k = 1, 2, ... n - \ell$) can be considered as the ℓ components of $n - \ell$ vectors $\mathbf{r}_k$, called *root vectors* or simply *roots*.

A theorem states that if $\mathbf{r}_k$ is a root, also $-\mathbf{r}_k$ is a root. Then the number of roots is even, and it is convenient to relabel the roots by $\mathbf{r}_\alpha$ and $\mathbf{r}_{-\alpha} = -\mathbf{r}_\alpha$, with $\alpha = 1, 2, ...(n - \ell)/2$ and the corresponding basis by $\mathbf{e}_{\pm\alpha}$. Accordingly, Eq. (1.44) can be rewritten in the form

$$[h_i, e_{\pm\alpha}] = r_{\pm\alpha_i} e_{\pm\alpha} \tag{1.45}$$

where we have adopted the notation $r_{\pm\alpha_i} = r_{\pm\alpha}(h_i)$.

The set of elements $e_{\pm\alpha}$ form a subspace of dimension $n - \ell$ of $\mathcal{L}$ which is called *root subspace*. It is convenient to consider the Cartan subalgebra $\mathcal{H}$ as a subspace of $\mathcal{L}$ corresponding to the *zero roots*. Altogether there are ℓ zero roots and $n - \ell$ non-zero roots.

Choosing, for the sake of convenience, the normalization

$$\sum_{\alpha_i, \alpha_j = 1}^{(n-\ell)/2} (r_{\alpha_i} r_{\alpha_j} + r_{-\alpha_i} r_{-\alpha_j}) = \delta_{ij} , \tag{1.46}$$

one can normalize the basis e_α, $e_{-\alpha}$ such that

$$[e_\alpha, e_{-\alpha}] = \sum_{i=1}^{\ell} r_{\alpha_i} h_i . \tag{1.47}$$

The other commutators are given by

$$[e_\alpha, e_\beta] = N_{\alpha\beta} \, e_{\alpha+\beta} , \tag{1.48}$$

which hold if $\mathbf{r}_\alpha + \mathbf{r}_\beta$ is a non-vanishing root; otherwise the r.h.s. of Eq. (1.48) is equal to 0. The constants $N_{\alpha\beta}$ can be directly computed once h_i and $e_{\pm\alpha}$ are known.

The roots satisfy the following useful properties:

1. *Reflection property.* If $\mathbf{r}_\alpha$ and $\mathbf{r}_\beta$ are two non-zero roots then

$$\mathbf{r}_\gamma = \mathbf{r}_\beta - 2\frac{\mathbf{r}_\alpha \cdot \mathbf{r}_\beta}{|\mathbf{r}_\alpha|^2}\mathbf{r}_\alpha \tag{1.49}$$

is also a root. In other words, the reflection of $\mathbf{r}_\beta$ through the hyperplane orthogonal to $\mathbf{r}_\alpha$ gives another root, as illustrated in Fig. 1.1.

2. *Integrality property.* If $\mathbf{r}_\alpha$ and $\mathbf{r}_\beta$ are two non-zero roots, then p and q, defined by

$$p = 2\frac{\mathbf{r}_\alpha \cdot \mathbf{r}_\beta}{|\mathbf{r}_\alpha|^2} \quad , \quad q = 2\frac{\mathbf{r}_\alpha \cdot \mathbf{r}_\beta}{|\mathbf{r}_\beta|^2} \quad , \tag{1.50}$$

are two integer numbers of the same sign. This means that the projection of one of the two roots, say $\mathbf{r}_\alpha$, on the other root $\mathbf{r}_\beta$ is a half-integral multiple of $|\mathbf{r}_\alpha|$, and viceversa.

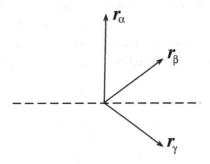

Fig. 1.1. *Graphical representation of the root reflection property.*

The above properties imply that two non-zero roots $\mathbf{r}_\alpha$ and $\mathbf{r}_\beta$ satisfy

$$\mathbf{r}_\alpha \cdot \mathbf{r}_\beta = \tfrac{1}{2}p\,|\mathbf{r}_\alpha|^2 = \tfrac{1}{2}q\,|\mathbf{r}_\beta|^2 = r_\alpha r_\beta \cos\phi\,, \qquad (1.51)$$

from which

$$\cos^2\phi = \tfrac{1}{4}pq \qquad (1.52)$$

and

$$R = \frac{r_\alpha}{r_\beta} = \sqrt{\frac{q}{p}}\,. \qquad (1.53)$$

Since p and q are integer numbers, we have only 5 possible solutions which are shown in Table 1.1 (it is sufficient to consider the angles in the interval $0 \le \phi \le \frac{\pi}{2}$).

Table 1.1. Values of ϕ and R

(p,q)	$(0,1)$	$(1,1)$	$(2,1)$	$(3,1)$	$(2,2)$
ϕ	$\dfrac{\pi}{2}$	$\dfrac{\pi}{3}$	$\dfrac{\pi}{4}$	$\dfrac{\pi}{6}$	0
R	0	1	$\sqrt{2}$	$\sqrt{3}$	1

Let us consider the case of real Lie algebras of rank $\ell = 2$. Making use of the *reflection property* shown in Fig. 1.1, one can build the so-called *root diagrams*. The number of roots, including the ℓ zero-roots, gives the dimension of the algebra.

Examples

1. Case $\phi = \dfrac{\pi}{3}$, $R = 1$. One can draw the root diagram shown in Fig 1.2 and obtain the dimension $d = 8$. It is the root diagram of the Lie algebra A_2 of the group $SU(3)$.

2. Case $\phi = \dfrac{\pi}{4}$, $R = \sqrt{2}$. The root diagram is shown in Fig 1.3; the dimension is $d = 10$ and the Lie algebra, denoted by B_2 corresponds to the group $O(5)$.

3. Case $\phi = \dfrac{\pi}{6}$, $R = \sqrt{3}$. The root diagram is shown in Fig. 1.4. The dimension is $d = 14$ and the Lie algebra corresponds to the so-called exceptional group G_2.

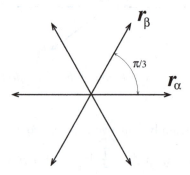

Fig. 1.2. *Root diagram of the real Lie algebra A_2 of the group $SU(3)$.*

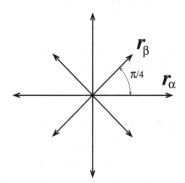

Fig. 1.3. *Root diagram of the real Lie algebra B_2 of the group $O(5)$.*

Dynkin diagrams - They are introduced to get a classification of all the semi-simple Lie algebras[6]. Taking into account Table 1.1, they are constructed as follows. For every simple root place a *dot*. Since for a simple Lie algebra the roots are of two sizes, a dark dot is made to correspond to the smaller root. Let us then consider two adjacent roots in a root diagram: the relevant

---

[6] See e.g. R. N. Cahn, *Semi-simple Lie Algebras and their Representations*, Benjamin/Cummings Publ. Co., 1984.

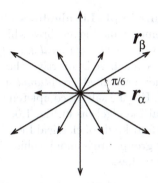

Fig. 1.4. *Root diagram of the real Lie algebra of the group G_2.*

values of the angle between them are only three, $\phi = \frac{\pi}{3}, \frac{\pi}{4}, \frac{\pi}{6}$. The Dynking diagrams for the Lie algebras of $SU(3)$, $O(5)$ and G_2 are shown in Fig. 1.5, where the single, double and triple lines indicate the three different angles.

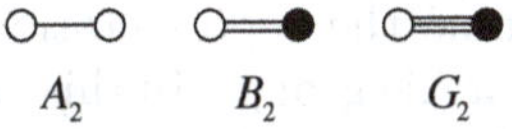

$$A_2 \qquad B_2 \qquad G_2$$

Fig. 1.5. *Dynkin diagrams of the real Lie algebras of rank $\ell = 2$.*

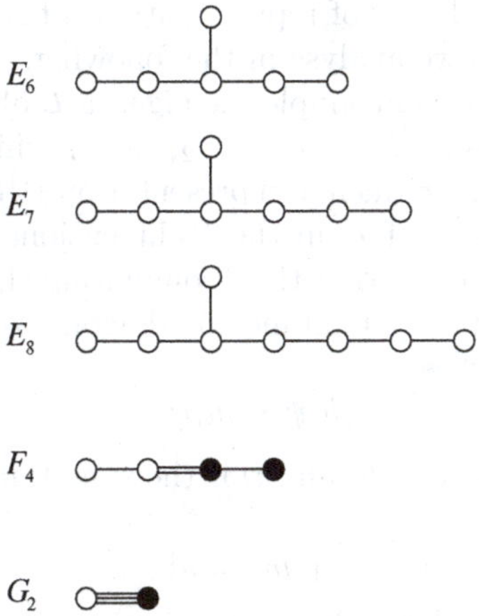

$$E_6$$
$$E_7$$
$$E_8$$
$$F_4$$
$$G_2$$

Fig. 1.6. *Dynkin diagrams of the exceptional Lie algebras.*

Classification of the semi-simple Lie algebras - It is due to Elie Cartan[7]. There are four classes of semi-simple Lie algebras which go from rank $\ell = 1$ to arbitrary large values: they are called *classical Lie algebras* and are denoted by A_ℓ, B_ℓ, C_ℓ and D_ℓ. Moreover there is a further class, which consists of only 5 kinds of Lie algebras: they are called *exceptional algebras* and are denoted by G_2, F_4, E_6, E_7 and E_8[8]. For the sake of completeness we show the Dynking diagrams of the exceptional Lie algebras in Fig. 1.6.

In Table 1.2 we list the four kinds of classical Lie algebras. From this Table it appears that some of the groups are homomorphic, since they have the same Lie algebras. Specifically, one has:

- rank $\ell = 1$: $SU(2) \sim O(3) \sim Sp(2)$;
- rank $\ell = 2$: $O(5) \sim Sp(4)$;
- rank $\ell = 3$: $SU(4) \sim O(6)$.

1.3.2 Representations of semi-simple Lie algebras and linear Lie groups

In this Subsection we examine the representations of real semi-simple Lie algebras and of *compact* linear Lie groups. In this case, all the representations can be chosen to be *unitary*. Moreover, since an important theorem states that every *reducible* representation of a semi-simple real Lie algebra is *completely reducible*, we can restrict ourselves to *unitary* and *irreducible* representations.

The key idea in the theory of representations of semi-simple Lie algebras is that of *weights*, which we analyse in the following.

Let us consider a real semi-simple Lie algebra $\mathcal{L}$ of dimension n and rank ℓ and the standard basis $h_1, h_2, ...h_\ell$; $e_1, e_2, ...e_{n-\ell}$ which satisfies Eqs. (1.43) and (1.44). Let D be an irreducible representation (IR) of $\mathcal{L}$ of dimension N and $\psi_1, \psi_2, ...\psi_N$ the basis of D in the N-dimensional vector space V. This basis is choosen in such a way that the ℓ commuting elements h_i ($i = 1, 2, ...\ell$) of the Cartan algebra $\mathcal{H}$ are simultaneouly diagonalized. Let ψ be one of the basis vector, which satisfies

$$h_i \psi = m_i \psi, \qquad (1.54)$$

where m_i is one eigenvalue of h_i and ψ is the simultaneous eigenvector of the set of eigenvalues

$$m_1, m_2, ...m_\ell. \qquad (1.55)$$

These eigenvalues can be considered the components of a vector $\mathbf{m}$ in a ℓ-dimensional space: it is this vector which is defined to be the *weight* of the representation D. The weights have interesting properties, of which we list in the following the most important ones.

---

[7] It was presented in the thesis: E. Cartan, *Sur la structure des groups de transformations finis et continues*, Paris 1894; 2nd. ed. Vuibert, Paris 1933.

[8] See e.g. R. N. Cahn, *quoted ref.*

Table 1.2. Classical Lie algebras

A			B			C			D		
$\mathcal{L}$	$\mathcal{G}$	n	$\mathcal{L}$	$\mathcal{G}$	n	$\mathcal{L}$	$\mathcal{G}$	n	$\mathcal{L}$	$\mathcal{G}$	n
A_1	$SU(2)$	3	B_1	$O(3)$	3	C_1	$Sp(2)$	3	D_1	$O(2)$	1
A_2	$SU(3)$	8	B_2	$O(5)$	10	C_2	$Sp(4)$	10	D_2	$O(4)$	6
A_3	$SU(4)$	15	B_3	$O(7)$	21	C_3	$Sp(6)$	21	D_3	$O(6)$	15
...	......	...	...	......	...	...	......	...	...	......	...
A_ℓ	$SU(\ell+1)$	$(\ell+2)\ell$	B_ℓ	$O(2\ell+1)$	$(2\ell+1)\ell$	C_ℓ	$Sp(2\ell)$	$(2\ell+1)\ell$	D_ℓ	$O(2\ell)$	$(2\ell-1)\ell$

1. Eigenvectors belonging to different weights are orthogonal. More eigenvectors may have the same weight; the number of different eigenvectors corresponding to the same weight is called *multiplicity* of the weight. If a weight belongs only to one eigenvector it is called *simple*.
2. A N-dimensional representation possesses N weights, some of which may be identical.
3. For any weight $\mathbf{m}$ and any root $\mathbf{r}_\alpha$, the quantity

$$k = 2\frac{\mathbf{m} \cdot \mathbf{r}_\alpha}{r_\alpha^2} \tag{1.56}$$

is an integer, and

$$\mathbf{m}' = \mathbf{m} - k\mathbf{r}_\alpha \tag{1.57}$$

is also a weight with the same multiplicity as $\mathbf{m}$. The situation is represented in Fig. 1.7, where $\mathbf{m}'$ is obtained by reflection of $\mathbf{m}$ in the hyperplane perpendicular to the root $\mathbf{r}_\alpha$.

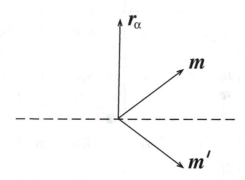

Fig. 1.7. *Graphical representation of the reflection property of the weights, according to Eq. (1.57).*

Next we give a few definitions about weights and state a few theorems.

Equivalent weights - Two weights $\mathbf{m}$ and $\mathbf{m}'$ are equivalent if one can be derived from the other by reflection. All the equivalent weights form a *class*.

Higher weight - The weight $\mathbf{m}$ is *higher* than $\mathbf{m}'$ if the first non-vanishing component of $\mathbf{m} - \mathbf{m}'$ is positive.

Highest weight of a representation - If $\mathbf{m}$ is a weight of a representation of $\mathcal{L}$, such that it is higher than every other weight of the representation, it is said to be the *highest* weight of the representation.

Theorem 1 - The highest weight of a representation is *simple* if the representation is irreducible. Two representations are equivalent if and only if they have the same highest weight.

Theorem 2 - For every irreducible representation D of a simple Lie algebra of rank ℓ there are ℓ *fundamental weights*

$$\mathbf{m}^{(1)}, \mathbf{m}^{(2)}, ... \mathbf{m}^{(\ell)}, \tag{1.58}$$

such that the highest weight of each IR is given by the linear combination

$$\mathbf{M} = \sum_{i=1}^{\ell} p_i \mathbf{m}^{(i)}, \tag{1.59}$$

where the coefficients p_i are non-negative integers. There exist ℓ IR's each of which has a fundamental weight as the highest weight; they are called *fundamental representations*. Starting from these representations, one can derive all the higher IR's. For each IR, one can draw a *weight diagram*, as shown in the following examples.

Theorem 3 - If D is an IR of a Lie algebra $\mathcal{L}$, also D^* $((D^*)_{ij} = (D_{ij})^*)$ is an IR of $\mathcal{L}$, called *complex conjugate* representation, and its weights are the *negatives* of those of D.

Theorem 4 - The weight diagram of the adjoint representation of a Lie algebra $\mathcal{L}$ coincides with the root diagram. Moreover, all the non-zero weights are simple and the number of zero-weights is equal to the rank ℓ.

In order to determine, in general, the fundamental weights of the IR's of a Lie algebra $\mathcal{L}$, as well as the weight multiplicities and dimensions of the IR's, it should be necessary to develop the machinery of the representation theory. We shall not discuss these arguments here; since we are mainly interested in the IR's of the $SU(N)$ groups, it will be sufficient to make use of the *Young tableaux*. As considered in detail in Appendix C, this method allows to determine the properties of all the IR's of $SU(N)$.

Here we limit ourselves to consider an example in the frame of the Lie algebra of $SU(3)$.

Example
Weight diagrams for the Lie algebra of $SU(3)$.
Making use of the root diagram of $SU(3)$ shown in Fig. 1.2, and of the normalization condition (1.46), we can write explicitly, in column form, three roots (the other three are the opposite vectors):

$$\mathbf{r}_{(1)} = \frac{1}{\sqrt{3}} \begin{pmatrix} 1 \\ 0 \end{pmatrix} ; \tag{1.60}$$

$$\mathbf{r}_{(2)} = \frac{1}{2\sqrt{3}} \begin{pmatrix} 1 \\ \sqrt{3} \end{pmatrix} ; \tag{1.61}$$

$$\mathbf{r}_{(3)} = \frac{1}{2\sqrt{3}} \begin{pmatrix} 1 \\ -\sqrt{3} \end{pmatrix} . \tag{1.62}$$

Then, from Eq. (1.56) applied to $\mathbf{r}_{(2)}$ and $\mathbf{r}_{(3)}$, the following relations are derived:

$$\sqrt{3}\,(m_1 + \sqrt{3}\,m_2) = p_1 \ , \quad \sqrt{3}\,(m_1 - \sqrt{3}\,m_2) = p_2 , \tag{1.63}$$

where m_1 and m_2 are the two components of the weight $\mathbf{m}$ and p_1, p_2 two integer numbers. Therefore, we can write for $\mathbf{m}$:

$$\mathbf{m} = \frac{1}{6} \begin{pmatrix} \sqrt{3} \\ 1 \end{pmatrix} p_1 + \frac{1}{6} \begin{pmatrix} \sqrt{3} \\ -1 \end{pmatrix} p_2 . \tag{1.64}$$

The pair of numbers (p_1, p_2) characterizes each IR, which we shall denote by $D(p_1, p_2)$. The dimension of the representation is given by

$$N = \tfrac{1}{2}(p_1 + 1)(p_1 + p_2 + 2)(p_2 + 1) . \tag{1.65}$$

The choices $(1,0)$ and $(0,1)$ for (p_1, p_2) correspond, respectively, to the fundamental weights

$$\mathbf{m}_{(1)} = \frac{1}{6}\begin{pmatrix} \sqrt{3} \\ 1 \end{pmatrix} \qquad , \qquad \mathbf{m}_{(2)} = \frac{1}{6}\begin{pmatrix} \sqrt{3} \\ -1 \end{pmatrix} . \tag{1.66}$$

By use of the reflection property Eq. (1.57), one can build the two corresponding weight diagrams, reported in Fig. 1.8. The highest weight of the IR $D(1,1)$ is equal

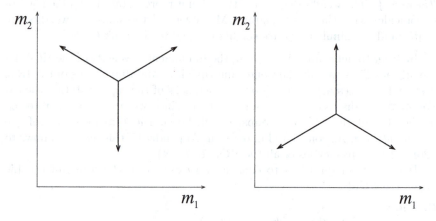

Fig. 1.8. *Weight diagrams of the IR's* 3 *and* $\bar{3}$ *of* $SU(3)$, *which correspond to the IR* $D(1,0)$ *and its conjugate* $D(0,1)$, *respectively.*

to the root given in Eq. (1.60). The weight diagram is equal to the root diagram of Fig. 1.2; the dimension is $N = 8$.

Finally, the IR $D(3,0)$ corresponds to $N = 10$. The root diagram is shown in Fig. 1.9.

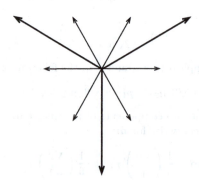

Fig. 1.9. *Weight diagrams of the IR's* 10 *of* $SU(3)$.

2

The rotation group

In this Chapter we give a short account of the main properties of the three-dimensional rotation group $SO(3)$ and of its universal covering group $SU(2)$. The group $SO(3)$ is an important subgroup of the Lorentz group, which will be considered in the next Chapter, and we think it is useful to give a separate and preliminary presentation of its properties. After a general discussion of the general characteristics of $SO(3)$ and $SU(2)$, we shall consider the corresponding Lie algebra and the irreducible representations of these groups. All the group concepts used in the following can be found in the previous Chapter.

2.1 Basic properties

The three-dimensional rotations are defined as the linear transformations of the vector $\mathbf{x} = (x_1, x_2, x_3)$

$$x_i' = \sum_j R_{ij} x_j \,, \tag{2.1}$$

which leave the square of $\mathbf{x}$ invariant:

$$x'^2 = x^2 \,. \tag{2.2}$$

Explicitly, the above condition gives

$$\sum_i x_i'^2 = \sum_{ijk} R_{ij} R_{ik} x_j x_k = \sum_j x_j^2 \,, \tag{2.3}$$

which implies

$$R_{ij} R_{ik} = \delta_{jk} \,. \tag{2.4}$$

In matrix notation Eqs. (2.1) and (2.4) can be written as

$$\mathbf{x}' = R\mathbf{x} \tag{2.5}$$

G. Costa and G. Fogli, *Symmetries and Group Theory in Particle Physics*,
Lecture Notes in Physics 823, DOI: 10.1007/978-3-642-15482-9_2,
© Springer-Verlag Berlin Heidelberg 2012

and

$$\tilde{R}R = I \,, \tag{2.6}$$

where $\tilde{R}$ is the transpose of R. Eq. (2.6) defines the *orthogonal* group $O(3)$; the matrices R are called *orthogonal* and they satisfy the condition

$$\det R = \pm 1 \,. \tag{2.7}$$

The condition

$$\det R = +1 \tag{2.8}$$

defines the *special orthogonal group* or *rotation group* $SO(3)$[1]. The corresponding transformations do not include space inversions, and can be identified with pure rotations.

A real matrix R satisfying Eqs. (2.6), (2.8) is characterized by 3 independent parameters, i.e. the dimension of the group is 3. One can choose different sets of parameters: a common parametrization, which will be considered explicitly in Section 2.4, is in terms of the three Euler angles. Another useful parametrization consists in associating to each matrix R a point of a sphere of radius π in the Euclidean space $\mathcal{R}^3$ (Fig. 2.1). For each point P inside the sphere there is a corresponding unique rotation: the direction of the vector OP individuates the axis of rotation and the lenght of OP fixes the angle ϕ ($0 \leq \phi \leq \pi$) of the rotation around the axis in counterclockwise sense. However, if $\phi = \pi$, the same rotation corresponds to the antipode P' on the surface of the sphere. We shall come back to this point later. Writing $OP = \phi\mathbf{n}$ where

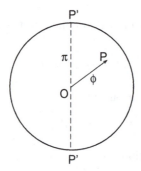

Fig. 2.1. *Parameter domain of the rotation group.*

$\mathbf{n} = (n_1, n_2, n_3)$ is a unit vector, each matrix R can be written explicitly in terms of the parameters ϕ and n_1, n_2, n_3 (only two of the n_i are independent, since $\sum n_i^2 = 1$). With the definitions $c_\phi = \cos\phi$, $s_\phi = \sin\phi$, one can write explicitly:

---

[1] For a detailed analysis see E.P. Wigner, *Group Theory and its applications to the quantum mechanics of atomic spectra*, Academic Press (1959).

$$R = \begin{pmatrix} n_1^2(1 - c_\phi) + c_\phi & n_1 n_2(1 - c_\phi) - n_3 s_\phi & n_1 n_3(1 - c_\phi) + n_2 s_\phi \\ n_1 n_2(1 - c_\phi) + n_3 s_\phi & n_2^2(1 - c_\phi) + c_\phi & n_2 n_3(1 - c_\phi) - n_1 s_\phi \\ n_1 n_3(1 - c_\phi) - n_2 s_\phi & n_2 n_3(1 - c_\phi) + n_1 s_\phi & n_3^2(1 - c_\phi) + c_\phi \end{pmatrix} .$$
(2.9)

From Eq. (2.9), one can prove that the product of two elements and the inverse element correspond to analytic functions of the parameters, i.e. the rotation group is a Lie group.

If one keeps only the orthogonality condition (2.6) and disregard (2.8), one gets the larger group $O(3)$, which contains elements with both signs, $\det R = \pm 1$. The groups consists of two disjoint sets, corresponding to $\det R = +1$ and $\det R = -1$. The first set coincides with the group $SO(3)$, which is an *invariant subgroup* of $O(3)$: in fact, if R belongs to $SO(3)$ and R' to $O(3)$, one gets

$$\det(R' R R'^{-1}) = +1 .$$
(2.10)

The group $O(3)$ is then neither simple nor semi-simple, while one can prove that $SO(3)$ is simple.

The elements with $\det R = -1$ correspond to *improper* rotations, i.e. rotations times *space inversion* I_s, where

$$I_s \mathbf{x} = -\mathbf{x} \qquad \text{i.e.} \qquad I_s = \begin{pmatrix} -1 & & \\ & -1 & \\ & & -1 \end{pmatrix} .$$
(2.11)

The element I_s and the identity I form a group $\mathcal{J}$ which is abelian and isomorphic to the permutation group $\mathcal{S}_2$. It is an invariant subgroup of $O(3)$. Each element of $O(3)$ can be written in a unique way as the product of a proper rotation times an element of $\mathcal{J}$, so that $O(3)$ is the *direct product*

$$O(3) = SO(3) \otimes \mathcal{J} .$$
(2.12)

It is important to remark that the group $SO(3)$ is *compact*; in fact its parameter domain is a sphere in the euclidean space $\mathcal{R}^3$, i.e. a compact domain. From Eq. (2.12) it follows that also the group $O(3)$ is compact, since both the disjoint sets are compact.

The rotation group $SO(3)$ is *connected*: in fact, any two points of the parameter domain can be connected by a continuous path. However, not all closed paths can be shrunk to a point. In Fig. 2.2 three closed paths are shown. Since the antipodes correspond to the same point, the path in case b) cannot be contracted to a point; instead, for case c), by moving P' on the surface, we can contract the path to a single point P. Case c) is then equivalent to case a) in which the path can be deformed to a point. We see that there are only two classes of closed paths which are distinct, so that we can say that the group $SO(3)$ is *doubly connected*. The group $O(3)$ is *not connected*, since it is the union of two disjoint sets.

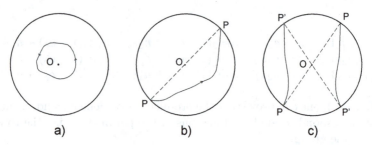

Fig. 2.2. *Different paths for SO(3).*

Since the group $SO(3)$ is not simply connected, it is important to consider its universal covering group, which is the *special unitary group $SU(2)$* of order $r = 3$. The elements of the group $SU(2)$ are the complex 2×2 matrices u satisfying

$$uu^\dagger = u^\dagger u = I \,, \tag{2.13}$$

$$\det u = 1 \,, \tag{2.14}$$

where $u^\dagger$ is the adjoint (conjugate transpose) of u. They can be written, in general, as

$$u = \begin{pmatrix} a & b \\ -b^* & a^* \end{pmatrix} \tag{2.15}$$

where a and b are complex parameters restricted by the condition

$$|a|^2 + |b|^2 = 1 \,. \tag{2.16}$$

Each matrix u is then specified by 3 real parameters. Defining

$$a = a_0 + ia_1 \,, \tag{2.17}$$

$$b = a_2 + ia_3 \,, \tag{2.18}$$

Eq. (2.16) becomes

$$a_0^2 + a_1^2 + a_2^2 + a_3^2 = 1 \,. \tag{2.19}$$

The correspondence between the matrices u and the matrices R of $SO(3)$ can be found replacing the orthogonal transformation (2.5) by

$$h' = uhu^\dagger \,, \tag{2.20}$$

where

$$h = \boldsymbol{\sigma} \cdot \mathbf{x} = \begin{pmatrix} x_3 & x_1 - ix_2 \\ x_1 + ix_2 & -x_3 \end{pmatrix} . \tag{2.21}$$

and $\boldsymbol{\sigma} = (\sigma_1, \sigma_2, \sigma_3)$ denotes the three Pauli matrices

$$\sigma_1 = \begin{pmatrix} 0 & 1 \\ 1 & 0 \end{pmatrix}, \quad \sigma_2 = \begin{pmatrix} 0 & -i \\ i & 0 \end{pmatrix}, \quad \sigma_3 = \begin{pmatrix} 1 & 0 \\ 0 & -1 \end{pmatrix}, \tag{2.22}$$

which satisfy the relation

$$\text{Tr}(\sigma_i \sigma_j) = 2\delta_{ij} . \tag{2.23}$$

Making use of eqs. (2.1), (2.20), one can express the elements of the matrix R in terms of those of the matrix u in the form:

$$R_{ij} = \tfrac{1}{2}\text{Tr}(\sigma_i u \sigma_j u^\dagger) . \tag{2.24}$$

Since this relation remains unchanged replacing u by $-u$, we see that for each matrix R of $SO(3)$ there are two corresponding matrices u and $-u$ of $SU(2)$.

The group $SU(2)$ is *compact* and *simply connected*. In fact, if we take the real parameters a_0, a_1, a_2, a_3 to characterize the group elements, we see that the parameter space, defined by Eq. (2.19), is the surface of a sphere of unit radius in a 4-dimensional euclidean space. This domain is compact and then also the group $SU(2)$ is compact. Moreover, all the closed paths on the surface can be shrunk continuously to a point, so that the group $SU(2)$ is simply connected. Since $SU(2)$ is homomorphic to $SO(3)$ and it does not contain simply connected subgroups, according to the definition given in Subsection 1.2.1, $SU(2)$ is the *universal covering group* of $SO(3)$. The kernel of the homomorphism is the invariant subgroup $\mathcal{E}(I, -I)$, and then the factor group $SU(2)/\mathcal{E}$ is isomorphic to $SO(3)$.

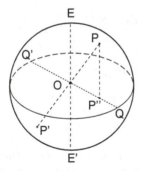

Fig. 2.3. *Parameter space for $SU(2)$ and $SO(3)$.*

The homomorphism between $SO(3)$ and $SU(2)$ can be described in a pictorial way, as shown in Fig. 2.3. The sphere has to be thought of as a 4-dimensional sphere, and the circle which divides it into two hemispheres as a three-dimensional sphere. The points E, E' correspond to the elements I and $-I$ of $SU(2)$. In general, two antipodes, such as P and P', correspond to a pair of elements u and $-u$. Since both u and $-u$ correspond to the same element of $SO(3)$, the parameter space of this group is defined only by the surface of one hemisphere; this can be projected into a three-dimensional sphere, and we see that two opposite points Q, Q' on the surface (on the circle in Fig. 2.3) correspond now to the same element of the group.

2.2 Infinitesimal transformations and Lie algebras of the rotation group

In order to build the Lie algebra of the rotation group, we consider the infinitesimal transformations of $SO(3)$ and $SU(2)$ in a neighborhood of the unit element. There is a *one-to-one correspondence* between the infinitesimal transformations of $SO(3)$ and $SU(2)$, so that the two groups are *locally isomorphic*. Therefore, the groups $SO(3)$ and $SU(2)$ have the same Lie algebra. We can build a basis of the Lie algebra in the following way. We can start from the three independent elements R_1, R_2, R_3 of $SO(3)$ corresponding to the rotations through an angle ϕ around the axis x_1, x_2, x_3 respectively. From Eq. (2.9) we get

$$R_1 = \begin{pmatrix} 1 & 0 & 0 \\ 0 & \cos\phi & -\sin\phi \\ 0 & \sin\phi & \cos\phi \end{pmatrix}, \quad R_2 = \begin{pmatrix} \cos\phi & 0 & \sin\phi \\ 0 & 1 & 0 \\ -\sin\phi & 0 & \cos\phi \end{pmatrix}, \quad R_3 = \begin{pmatrix} \cos\phi & -\sin\phi & 0 \\ \sin\phi & \cos\phi & 0 \\ 0 & 0 & 1 \end{pmatrix}$$
(2.25)

and, according to Eq. (1.30), we obtain the *generators* of the Lie algebra. For the sake of convenience, we use the definition:

$$J_k = i\frac{dR_k(\phi)}{d\phi}\bigg|_{\phi=0} \qquad (k = 1, 2, 3) , \tag{2.26}$$

so that the three generators are given by

$$J_1 = \begin{pmatrix} 0 & 0 & 0 \\ 0 & 0 & -i \\ 0 & i & 0 \end{pmatrix}, \quad J_2 = \begin{pmatrix} 0 & 0 & i \\ 0 & 0 & 0 \\ -i & 0 & 0 \end{pmatrix}, \quad J_3 = \begin{pmatrix} 0 & -i & 0 \\ i & 0 & 0 \\ 0 & 0 & 0 \end{pmatrix}. \tag{2.27}$$

We can take J_1, J_2, J_3 as the basis elements of the Lie algebra; making use of the relation between linear Lie algebras and connected Lie groups (see Subsection 1.2.2), the three rotations (2.25) can be written in the form:

$$R_1 = e^{-i\phi J_1}, \quad R_2 = e^{-i\phi J_2}, \quad R_3 = e^{-i\phi J_3} . \tag{2.28}$$

In general, a rotation through an angle ϕ about the direction **n** is represented by

$$R = e^{-i\phi \, \mathbf{J} \cdot \mathbf{n}} . \tag{2.29}$$

We have considered the specific case of a three-dimensional representation for the rotations R_i and the generators J_i; in general, one can consider J_i as hermitian operators and the R_i as unitary operators in a n-dimensional linear vector space. One can check that J_1, J_2, J_3 satisfy the commutation relations

$$[J_i, J_j] = i\epsilon_{ijk}J_k, \tag{2.30}$$

which show that the algebra has *rank* 1. The structure constants are given by the antisymmetric tensor ϵ_{ijk}, and Eq. (1.35) reduces to

$$g_{ij} = -\delta_{ij}. \tag{2.31}$$

Since the condition (1.42) is satisfied, the algebra is simple and the Casimir operator (1.36) becomes, with a change of sign,

$$C = J^2 = J_1^2 + J_2^2 + J_3^2. \tag{2.32}$$

The above relations show that the generators J_k have the properties of the *angular momentum operators*[2].

2.3 Irreducible representations of $SO(3)$ and $SU(2)$

We saw that the group $SO(3)$ can be defined in terms of the orthogonal transformations given in Eq. (2.1) in a 3-dimensional Euclidean space. Similarly, the group $SU(2)$ can be defined in terms of the unitary transformations in a 2-dimensional complex linear space

$$\xi'^i = \sum_j u_{ij}\xi^j. \tag{2.33}$$

This equation defines the self-representation of the group. Starting from this representation, one can build, by reduction of *direct products*, the higher irreducible representations (IR's). A convenient procedure consists in building, in terms of the basic vectors, higher tensors, which are then decomposed into *irreducible tensors*. These are taken as the bases of irreducible representations; in fact, their transformation properties define completely the representations (for details see Appendix B).

However, starting from the basic vector $\mathbf{x} = (x_1, x_2, x_3)$, i.e. from the three-dimensional representation defined by Eq. (2.1), one does not get all the irreducible representations of $SO(3)$, but only the so-called *tensorial* IR's which correspond to integer values of the angular momentum j. Instead, all the IR's can be easily obtained considering the universal covering group $SU(2)$. The basis of the self-representation consists, in this case, of two-component vectors, usually called *spinors*[3], such as

---

[2] We recall that the eigenvalues of J^2 are given by $j(j+1)$; see e.g. W. Greiner, *Quantum Mechanics, An Introduction*, Springer-Verlag (1989).

[3] Strictly speaking, one should call the basis vectors ξ "spinors" with respect to $SO(3)$ and "vectors" with respect to $SU(2)$.

$$\xi = \begin{pmatrix} \xi^1 \\ \xi^2 \end{pmatrix} , \tag{2.34}$$

which transforms according to (2.33), or in compact notation

$$\xi' = u\xi . \tag{2.35}$$

We call ξ *controvariant* spinor of rank 1. In order to introduce a scalar product, it is useful to define the *covariant* spinor η of rank 1 and components η_i, which transforms according to

$$\eta' = \eta u^{-1} = \eta u^\dagger , \tag{2.36}$$

so that

$$\eta\xi = \eta'\xi' = \sum_i \eta_i \xi^i . \tag{2.37}$$

In terms of the components η_i:

$$\eta_i{}' = \sum_j u^\dagger_{ji}\eta_j = \sum_j u^*_{ij}\eta_j . \tag{2.38}$$

Taking the complex conjugate of (2.33)

$$\xi'^{*i} = \sum_j u^*_{ij}\xi^{*j} , \tag{2.39}$$

we see that the component ξ_i transform like ξ^{*i}, i.e.

$$\xi^{*i} \equiv \xi_i . \tag{2.40}$$

The representation u^* is called the *conjugate representation*; the two IR's u and u^* are equivalent. One can check, using the explicit expression (2.15) for u, that u and u^* are related by a similarity transformation

$$u^* = SuS^{-1} , \tag{2.41}$$

with

$$S = \begin{pmatrix} 0 & 1 \\ -1 & 0 \end{pmatrix} . \tag{2.42}$$

We see also that the spinor

$$\bar{\xi} = S^{-1}\xi^* = \begin{pmatrix} -\xi_2 \\ \xi_1 \end{pmatrix} \tag{2.43}$$

transforms in the same way as ξ. Starting from ξ^i and ξ_i we can build all the higher irreducible tensors, whose transformation properties define all the IR's of SU(2). Besides the tensorial representations, one obtains also the *spinorial* representations, corresponding to half-integer values of the angular momentum j.

We consider here only a simple example. The four-component tensor

$$\zeta_i^j = \xi^j \xi_i \tag{2.44}$$

can be splitted into a scalar quantity

$$\mathrm{Tr}\{\zeta\} = \sum_i \xi^i \xi_i \tag{2.45}$$

and a traceless tensor

$$\hat{\zeta}_i^j = \xi^j \xi_i - \tfrac{1}{2}\delta_i^j \sum_k \xi^k \xi_k \ , \tag{2.46}$$

where δ_i^j is the Kronecker symbol.

The tensor $\hat{\zeta}_i^j$ is not further reducible. It is equivalent to the 3-vector $\mathbf{x}$; in fact, writing it as a 2×2 matrix $\hat{\zeta}$, it can be identified with the matrix h defined in (2.21)

$$\hat{\zeta} = \boldsymbol{\sigma} \cdot \mathbf{x} = \begin{pmatrix} x_3 & x_1 - ix_2 \\ x_1 + ix_2 & -x_3 \end{pmatrix} . \tag{2.47}$$

Its transformation properties, according to those of ξ, η, are given by

$$\hat{\zeta}' = u\hat{\zeta}u^\dagger . \tag{2.48}$$

Let us consider the specific case

$$u = \begin{pmatrix} e^{-\frac{1}{2}i\phi} & 0 \\ 0 & e^{\frac{1}{2}i\phi} \end{pmatrix} . \tag{2.49}$$

Using for $\hat{\zeta}$ the expression (2.47), we get from (2.49):

$$\begin{aligned} x_1' &= \cos\phi\, x_1 - \sin\phi\, x_2 \ , \\ x_2' &= \sin\phi\, x_1 + \cos\phi\, x_2 \ , \\ x_3' &= x_3 \ . \end{aligned} \tag{2.50}$$

The matrix (2.49) shows how a spinor is transformed under a rotation through an angle ϕ and it corresponds to the 3-dimensional rotation

$$R_3 = \begin{pmatrix} \cos\phi & -\sin\phi & 0 \\ \sin\phi & \cos\phi & 0 \\ 0 & 0 & 1 \end{pmatrix} \tag{2.51}$$

given in eq. (2.25). In particular, taking $\phi = 2\pi$, we get $\xi' = -\xi$, i.e. *spinors change sign* under a rotation of 2π about a given axis. The angles ϕ and $\phi + 2\pi$ correspond to the same rotation, i.e. to the same element of $SO(3)$; on the other hand, the two angles correspond to two different elements of $SU(2)$, i.e.

to the 2×2 unitary matrices u and $-u$. Then to each matrix R of $SO(3)$ there correspond two elements of $SU(2)$; for this reason, often the matrices u and $-u$ are said to constitute a "double-valued" representation[4] of $SO(3)$.

In general, we can distinguish two kinds of IR's

$$D(u) = +D(-u) \,, \tag{2.52}$$
$$D(u) = -D(-u) \,, \tag{2.53}$$

which are called *even* and *odd*, respectively. The direct product decomposition $u \otimes u \otimes ... \otimes u$, where the self-representation u appears n times, gives rises to even and odd IR's according to whether n is even or odd. The even IR's are the *tensorial* representations of $SO(3)$, the odd IR's are the *spinorial* representations of $SO(3)$.

The IR's are usually labelled by the eigenvalues of the squared angular momentum operator J^2, i.e. the Casimir operator given in Eq. (2.32), which are given by $j(j+1)$, with j integer or half integer. An IR of $SO(3)$ or $SU(2)$ is simply denoted by $D^{(j)}$; its dimension is equal to $2j+1$. Even and odd IR's correspond to *integer* and *half-integer* j, respectively.

The basis of the $D^{(j)}$ representation consists of $(2j+1)$ elements, which correspond to the eigenstates of J^2 and J_3; it is convenient to adopt the usual notation $|\, j, m >$ specified by[5]

$$\begin{aligned} J^2 \,|\, j, m > &= j(j+1) \,|\, j, m > \,, \\ J_3 \,|\, j, m > &= m \,|\, j, m > \,. \end{aligned} \tag{2.54}$$

Finally, we want to mention the IR's of $O(3)$. We have seen that the group $O(3)$ can be written as the direct product

$$O(3) = SO(3) \otimes \mathcal{J} \,, \tag{2.12}$$

where $\mathcal{J}$ consists of I and I_s. Since $I_s^2 = I$, the element I_s is represented by

$$D(I_s) = \pm I \,. \tag{2.55}$$

According to (2.12), we can classify the IR's of $O(3)$ in terms of those of $SO(3)$, namely the $D^{(j)}$'s. In the case of *integer* j, we can have two kinds of IR's of $O(3)$, according to the two possibilities

$$D^{(j)}(I_s R) = +D^{(j)}(R) \,, \tag{2.56}$$
$$D^{(j)}(I_s R) = -D^{(j)}(R) \,. \tag{2.57}$$

---

[4] See e.g. J.F. Cornwell, *Group Theory in Physics*, Vol. 1 and 2, Academic Press (1984); M. Hamermesh, *Group Theory and its Applications to Physical Problems*, Addison-Wesley (1962).

[5] M.E. Rose, *Elementary Theory of Angular Momentum*, John Wiley and Sons (1957).

Only the second possibility corresponds to *faithful* representations, since the improper rotations $I_s R$ are distinguished from the proper rotations R. We denote the two kinds of IR's (2.56), (2.57) by $D^{(j+)}$, $D^{(j-)}$, respectively. The bases of these IR's are called *tensors* and *pseudotensors*; in general, one calls tensors (scalar, vector, etc.) the basis of $D^{(0+)}$, $D^{(1-)}$, $D^{(2+)}$, ..., and pseudotensors (pseudoscalar, axial vector, etc.) the basis of $D^{(0-)}$, $D^{(1+)}$, $D^{(2-)}$, etc.

In the case of half-integer j, since the IR's of $SO(3)$ are double-valued, i.e. each element R is represented by $\pm D^{(j)}(R)$ also for the improper element, one gets

$$I_s R \to D^{(j)}(I_s R) = \pm D^{(j)}(R) . \tag{2.58}$$

Then, for each half-integer j, there is one double-valued IR of $O(3)$.

2.4 Matrix representations of the rotation operators

For the applications in many sectors of physics one needs the explicit expressions of the rotation matrices in an arbitrary representation. Following the notation established in the literature, it is useful to specify a rotation R in terms of the so-called Euler angles α, β, γ. For their definition, we consider a fixed coordinate system (x, y, z). Any rotation R can be regarded as the result of three successive rotations, as indicated in Fig. 2.4.

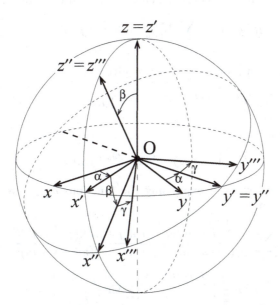

Fig. 2.4. *Sequence of rotations that define the three Euler angles α, β and γ. The planes in which the rotations take place are also indicated.*

1. Rotation $R_\alpha \equiv R_\alpha(z)$ through an angle α $(0 \leq \alpha < 2\pi)$ about the z-axis, which carries the coordinate axes (x, y, z) into $(x', y', z' = z)$;
2. Rotation $R'_\beta \equiv R_\beta(y')$ through an angle β $(0 \leq \beta \leq \pi)$ about the y'-axis, which carries the system (x', y', z') into $(x'', y'' = y', z'')$;
3. Rotation $R''_\gamma \equiv R_\gamma(z'')$ through an angle γ $(0 \leq \gamma < 2\pi)$ about the z''-axis, which carries the system (x'', y'', z'') into $(x''', y''', z''' = z'')$.

The three rotations can be written in the form

$$R_\alpha = e^{-i\alpha J_z}, \quad R'_\beta = e^{-i\beta J_{y'}}, \quad R''_\gamma = e^{-i\gamma J_{z''}}, \tag{2.59}$$

where J_z, $J_{y'}$ and $J_{z''}$ are the components of $\mathbf{J}$ along the z, y' and z'' axes. The complete rotation R is then given by:

$$R(\alpha, \beta.\gamma) = R''_\gamma R'_\beta R_\alpha = e^{-i\gamma J_{z''}} e^{-i\beta J_{y'}} e^{-i\alpha J_z}. \tag{2.60}$$

We leave as an exercise the proof that the three rotations can be carried out in the *same* coordinate system if the order of the three rotations in inverted, i.e. Eq. (2.60) can be replaced by

$$R(\alpha, \beta, \gamma) = R_\alpha R_\beta R_\gamma = e^{-i\alpha J_z} e^{-i\beta J_y} e^{-i\gamma J_z}. \tag{2.61}$$

The procedure for determining the rotation matrices, i.e. the matrix representations of the rotation operator R, is straightforward, even if the relative formulae may appear to be rather involved. One starts from the basis $|j, m>$ given in Eq. (2.54) and considers the effect of a rotation R on it:

$$R \, | \, j, m >= \sum_{m'} D^{(j)}_{m'm}(\alpha, \beta, \gamma) \, | \, j, m' > . \tag{2.62}$$

An element of the rotation matrix $D^{(j)}$ is given by

$$D^{(j)}_{m'm}(\alpha, \beta, \gamma) = <j, m' \, | \, e^{-i\alpha J_z} e^{-i\beta J_y} e^{-i\gamma J_z} \, | \, j, m >= e^{-i\alpha m'} d^{(j)}_{m'm}(\beta) e^{-i\gamma m}, \tag{2.63}$$

where one defines

$$d^{(j)}_{m'm}(\beta) = <j, m' \, | \, e^{-i\beta J_y} \, | \, j, m > . \tag{2.64}$$

There are different ways of expressing the functions $d^j_{m'm}$; we report here Wigner's expression[6]:

$$d^{(j)}_{m'm}(\beta) = \sum_s \frac{(-)^s [(j+m)!(j-m)!(j+m')!(j-m')!]^{1/2}}{s!(j-s-m')!(j+m-s)!(m'+s-m)!} \times$$
$$\times \left(\cos\frac{\beta}{2}\right)^{2j+m-m'-2s} \left(-\sin\frac{\beta}{2}\right)^{m'-m+2s}, \tag{2.65}$$

[6] See e.g. M.E. Rose, *Elementary Theory of Angular Momentum*, John Wiley and Sons (1957).

where the sum is over the values of the integer s for which the factorial arguments are equal or greater than zero.

It is interesting to note that, in the case of integral angular momentum, the d-functions are connected to the well-known spherical harmonics by the relation:

$$d_{m,0}^{\ell}(\theta) = \sqrt{\frac{4\pi}{2\ell+1}} Y_{\ell}^{m}(\theta, \phi) e^{-im\phi} , \qquad (2.66)$$

where θ and ϕ are the angles of the spherical coordinates. In fact, $Y_{\ell}^{m}(\theta, \phi)$ represents the eigenfunction corresponding to the state $\mid j, m >$ of a particle with orbital angular momentum $j = \ell$.

In Appendix A we collect the explicit expressions of the spherical harmonics and the d-functions for the lowest momentum cases.

2.5 Addition of angular momenta and Clebsch-Gordan coefficients

An important application of the IR's of the rotation group is related to the addition of angular momenta and the construction of the relevant orthonormal bases[7].

We start from the IR's $D^{(j_1)}$ and $D^{(j_2)}$ and the direct product decomposition

$$D^{(j_1)} \otimes D^{(j_2)} = D^{(j_1+j_2)} \oplus D^{(j_1+j_2-1)} \oplus .... \oplus D^{(|j_1-j_2|)} . \qquad (2.67)$$

The IR's on the r.h.s. correspond to the different values of total angular momenta obtained by the quantum addition rule

$$\mathbf{J} = \mathbf{J_1} + \mathbf{J_2} , \qquad (2.68)$$

with $|J_1 - J_2| \leq J \leq J_1 + J_2$. From the commutation properties of angular momentum operators, one finds that the eigenvectors

$$\mid j_1, j_2; m_1, m_2 > \equiv \mid j_1, m_1 > \otimes \mid j_2, m_2 > \qquad (2.69)$$

constitute an orthogonal basis for the direct product representation, while the eigenvectors

$$\mid j_1, j_2; j, m > \qquad (2.70)$$

are the bases of the IR's on the r.h.s. of eq. (2.67). One can pass from one basis to the other by a unitary transformation, which can be written in the form

---

[7] For a detailed treatment of this subject see e.g. W. Greiner, *Quantum Mechanics, An Introduction*, Springer-Verlag (1989) and M. Hamermesh, *Group Theory and its Applications to Physical Problems*, Addison-Wesley (1962).

$$| j_1, j_2; j, m > = \sum_{m_1, m_2} C(j_1, j_2, j; m_1, m_2, m) \, | j_1, j_2; m_1, m_2 > \ . \qquad (2.71)$$

The elements of the transformation matrix are called *Clebsh-Gordan coefficients* (or simply *C-coefficients*), defined by

$$C(j_1, j_2, j; m_1, m_2, m) = < j_1, j_2; m_1, m_2 \, | \, j_1, j_2; j, m > \ , \qquad (2.72)$$

where $m = m_1 + m_2$. With the standard phase convention the C-coefficients are real and they satisfy the orthogonality relation (replacing m_2 by $m - m_1$):

$$\sum_{m_1} C(j_1, j_2, j; m_1, m - m_1) C(j_1, j_2, j'; m_1, m - m_1) = \delta_{jj'} \ . \qquad (2.73)$$

Moreover, the transformation (2.71) is *orthogonal*, and the inverse transformation can be easily obtained:

$$| j_1, j_2; m_1, m_2 > = \sum_{j, m} C(j_1, j_2, j; m_1, m - m_1) \, | j_1, j_2; j, m > \ . \qquad (2.74)$$

We report the values of the Clebsh-Gordan coefficients for the lowest values of j_1 and j_2 in Appendix A, while for other cases and for a general formula we refer to specific texbooks[8].

In connection with the C-coefficients it is convenient to quote without proof the *Wigner-Eckart theorem* which deals with matrix elements of tensor operators. An irreducible tensor operator of rank J is defined as a set of $(2J+1)$ functions T_{JM} (where $M = -J, -J+1, ..., J-1, J$) which transform under the $(2J+1)$ dimensional representations of the rotation group in the following way:

$$R T_{JM} R^{-1} = \sum_{M'} D^J_{M'M}(\alpha, \beta, \gamma) T_{JM'} \ . \qquad (2.75)$$

The Wigner-Eckart theorem states that the dependence of the matrix element $< j', m' \, | \, T_{JM} \, | \, j, m >$ on the quantum number M, M' is entirely contained in the C-coefficients:

$$< j', m' \, | \, T_{JM} \, | \, j, m > = C(j, J, j'; m, M, m') < j' \, | \, T_J \, | \, j > \ . \qquad (2.76)$$

We note that the C-coefficient vanishes unless $m' = M + m$, so that one has $C(j, J, j'; m, M, m') = C(j, J, j'; m, m' - m)$. The matrix element on the r.h.s. of the above equation is called reduced matrix element.

---

[8] See e.g. D.R. Lichtenberg, *Unitary Symmetry and Elementary Particles*, Academic Press (1970); M.E. Rose, *Elementary Theory of Angular Momentum*, John Wiley and Sons (1957).

Problems

2.1. Give the derivation of Eq. (2.24) and write explicitly the matrix R in terms of the elements of the u matrix.

2.2. Consider the Schrödinger equation $H \mid \psi> = E \mid \psi>$ in which the Hamiltonian H is invariant under rotations. Show that the angular momentum J commutes with H and then it is conserved.

2.3. The πN scattering shows a strong resonance at the kinetic energy about 200 MeV; it occurs in the P-wave ($\ell = 1$) with total angular momentum $J = \frac{3}{2}$. Determine the angular distribution of the final state.

2.4. Prove the equivalence of the two expressions for a general rotation R given in Eqs. (2.60) and (2.61).

2.5. Consider the eigenstates $\mid \frac{1}{2}, \pm \frac{1}{2}>$ of a particle of spin $\frac{1}{2}$ and spin components $\pm \frac{1}{2}$ along the z-axis. Derive the corresponding eigenstates with spin components along the y-axis by a rotation about the x-axis.

3

The homogeneous Lorentz group

In this Chapter we consider the general properties of the homogeneous Lorentz transformations. We concentrate here on their group theoretical aspects, which give insight into the central role played by special relativity in the description of the elementary particle physics. The restricted Lorentz group (which does not contain space and time inversions), denoted in the following by $\mathcal{L}_+^\uparrow$, is analysed in greater detail. Specifically, we derive the Lie algebra of the group from its infinitesimal transformations, we introduce the group of unimodular complex 2×2 matrices which is homomorphic to $\mathcal{L}_+^\uparrow$ being its universal covering group; finally, we discuss the properties of the finite-dimensional irreducible representations. All the group concepts used in the following can be found in Chapters 1 and 2, where they are illustrated mainly in terms of the rotation group, which is a subgroup of $\mathcal{L}_+^\uparrow$. Keeping in mind the examples given for the rotation group, one should be able to proceed with little effort through the material of this Chapter.

3.1 Basic properties

The most general Lorentz transformations are given in terms of the coordinate transformations connecting any two inertial frames of reference.

We neglect here space-time translations, and consider the general *homogeneous Lorentz transformations*. Following the usual notation of Special Relativity[1], these transformations are defined by

$$x'^\mu = \Lambda^\mu{}_\nu x^\nu , \tag{3.1}$$

with the condition

$$(x')^2 = (x)^2 , \tag{3.2}$$

and are expressed in terms of the *controvariant* four-vector x of components

---

[1] See e.g. N.K. Glendenning, *Special and General Relativity*, Springer (2007).

G. Costa and G. Fogli, *Symmetries and Group Theory in Particle Physics*, Lecture Notes in Physics 823, DOI: 10.1007/978-3-642-15482-9_3, © Springer-Verlag Berlin Heidelberg 2012

$$x^\mu = (x^0, x^1, x^2, x^3) = (ct, x, y, z) ,\tag{3.3}$$

where c is the speed of light.

In Eq. (3.1), the sum from 0 to 3 over the repeated index ν is implied; this convention will be always used in the following. The square of a four-vector is defined by

$$(x)^2 = g_{\mu\nu} x^\mu x^\nu = x^\mu x_\mu = (x^0)^2 - (\mathbf{x})^2 ,\tag{3.4}$$

where the *metric tensor* $g_{\mu\nu} = g^{\mu\nu}$ can be written as the 4×4 matrix

$$g = \{g_{\mu\nu}\} = \begin{pmatrix} 1 & 0 & 0 & 0 \\ 0 & -1 & 0 & 0 \\ 0 & 0 & -1 & 0 \\ 0 & 0 & 0 & -1 \end{pmatrix}\tag{3.5}$$

and $x_\mu = (x^0, -x^1, -x^2, -x^3)$ is the *covariant* four-vector.

The condition (3.2) can be written explicitly in terms of (3.1), as

$$g_{\mu\nu} x'^\mu x'^\nu = g_{\mu\nu} \Lambda^\mu{}_\rho \Lambda^\nu{}_\sigma x^\rho x^\sigma = g_{\rho\sigma} x^\rho x^\sigma ,\tag{3.6}$$

which implies

$$\Lambda^\mu{}_\rho g_{\mu\nu} \Lambda^\nu{}_\sigma = g_{\rho\sigma} ,\tag{3.7}$$

or in matrix form

$$\tilde{\Lambda} g \Lambda = g .\tag{3.8}$$

From (3.8) one gets

$$(\det \tilde{\Lambda})(\det \Lambda) = (\det \Lambda)^2 = 1 ,\tag{3.9}$$

i.e.

$$\det \Lambda = \pm 1 .\tag{3.10}$$

Moreover, if we take the $\rho = \sigma = 0$ component of (3.7), we get

$$(\Lambda^0{}_0)^2 - \sum_{i=1}^{3} (\Lambda^i{}_0)^2 = 1 ,\tag{3.11}$$

which implies

$$(\Lambda^0{}_0)^2 \geq 1 ,\tag{3.12}$$

i.e.

$$\Lambda^0{}_0 \geq 1 \qquad \text{or} \qquad \Lambda^0{}_0 \leq -1 .\tag{3.13}$$

The homogeneous Lorentz transformations are defined by (3.1) as *linear* transformations which leave invariant the quadratic form (3.4); they are completely characterized by the real 4×4 matrices Λ which satisfy the condition (3.8). In the following we show that the matrices Λ form a group, i.e. the homogeneous Lorentz group, denoted by $\mathcal{L}$, and specify its main properties.

1. The set $\mathcal{L}$ of the matrices Λ, which characterize the homogeneous Lorentz transformations, is a *group*. In fact, the properties given in Section 1.1 are satisfied:
   - if Λ_1 and Λ_2 are elements of $\mathcal{L}$, we can define the product $\Lambda_3 = \Lambda_1\Lambda_2$ which is an element Λ_3 of $\mathcal{L}$:

$$(\widetilde{\Lambda_1\Lambda_2})g(\Lambda_1\Lambda_2) = \tilde{\Lambda}_2\tilde{\Lambda}_1 g\Lambda_1\Lambda_2 = \tilde{\Lambda}_2 g\Lambda_2 = g , \qquad (3.14)$$

   i.e.

$$\tilde{\Lambda}_3 g\Lambda_3 = g . \qquad (3.15)$$

   - Clearly, the product is associative.
   - The *identity* is the unit matrix I, which belongs to $\mathcal{L}$:

$$\tilde{I}gI = g . \qquad (3.16)$$

   - Eq. (3.10) assures that for each element Λ there is the *inverse* Λ^{-1}; from (3.8) we get

$$\tilde{\Lambda}^{-1}g\Lambda^{-1} = g , \qquad (3.17)$$

   i.e. Λ^{-1} belongs to $\mathcal{L}$.
2. The group $\mathcal{L}$ is isomorphic to the pseudo-orthogonal group $O(1,3)$ defined in Subsection 1.2.1, i.e. the group of rotations in a four-dimensional Minkowski space; then $\mathcal{L}$ is a *linear Lie group of dimension* $n = 6$. In fact, the matrix Λ depends upon 16 real parameters (the transformations preserve the reality properties of the four-vector x^μ), but they have to satisfy Eq. (3.8), i.e. 10 independent conditions, so that there are only 6 independent real parameters.
3. The group $\mathcal{L}$ is *not connected*. From Eqs. (3.10) and (3.13) we see that the group consists of 4 *disjoint sets*, which are called *components*, corresponding to the 4 possibilities presented in Table 3.1. This situation corresponds to the fact that $\mathcal{L}$ contains 3 *discrete* transformations, which are defined as follows:
   - *space inversion* I_s

$$I_s x^\mu = g_{\mu\nu}x^\nu , \qquad (3.18)$$

   - *time inversion* I_t

$$I_t x^\mu = -g_{\mu\nu}x^\nu , \qquad (3.19)$$

   - *space-time inversion* $I_{st} = I_s I_t = I_t I_s$

$$I_{st} x^\mu = -x^\mu . \qquad (3.20)$$

Together with the identity I, the discrete elements I_s, I_t, I_{st} form an *Abelian subgroup* $\mathcal{I}$ of $\mathcal{L}$. This subgroup is not invariant (compare with the situation of the group $O(3)$ and its Abelian subgroup I, I_s in Section 2.1). In fact, while the relation

$$I_{st}\Lambda = \Lambda I_{st} \qquad (3.21)$$

holds for any Λ, the analogous relation does not hold for I_s and I_t. The group is not connected since it is the union of disjoint sets; however, each component is *connected* since any element can be connected by a continuous path to any other element in the same component. Among the four components, only $\mathcal{L}_+^\uparrow$ is a subgroup of $\mathcal{L}$: it is called the *proper orthochronous* or *restricted* Lorentz group. From Table 3.1 we see that, taking two arbitrary elements Λ of $\mathcal{L}$ and Λ' of $\mathcal{L}_+^\uparrow$, $\Lambda\Lambda'\Lambda^{-1}$ belongs to $\mathcal{L}_+^\uparrow$, so that $\mathcal{L}_+^\uparrow$ is an *invariant subgroup* of $\mathcal{L}$. Then the group $\mathcal{L}$ is isomorphic to the *semi-direct product* $\mathcal{L}_+^\uparrow \circledS \mathcal{I}$. The other components can be identified with the *cosets* of $\mathcal{L}_+^\uparrow$ in $\mathcal{L}$, which are: $I_s\mathcal{L}_+^\uparrow$, $I_t\mathcal{L}_+^\uparrow$, $I_{st}\mathcal{L}_+^\uparrow$. This show that any element of $\mathcal{L}$ can be obtained by applying a discrete operation to an element of $\mathcal{L}_+^\uparrow$.

Table 3.1. Properties of the four components of the Lorentz group $\mathcal{L}$.

Component	det Λ	$\Lambda^0{}_0$	Discrete transformation
$\mathcal{L}_+^\uparrow$	$+1$	$\geq +1$	I
$\mathcal{L}_-^\uparrow$	-1	$\geq +1$	$I_s = g$
$\mathcal{L}_+^\downarrow$	$+1$	≤ -1	$I_t = -g$
$\mathcal{L}_-^\downarrow$	-1	≤ -1	$I_{st} = -I$

Other *subgroups* of $\mathcal{L}$ can be obtained by combining $\mathcal{L}_+^\uparrow$ with another component; the elements of $\mathcal{L}_+^\uparrow$ and $\mathcal{L}_+^\downarrow$ form the subgroup $\mathcal{L}_+$, which is the *proper Lorentz group*; the elements of $\mathcal{L}_+^\uparrow$ and $\mathcal{L}_-^\uparrow$ form the *orthochronous Lorentz group* $\mathcal{L}^\uparrow$.

3.2 The proper orthochronous Lorentz group $\mathcal{L}_+^\uparrow$

First we consider some specific examples of Lorentz transformations which are important in practice and which will be used in the following.

1. *Space rotations*

A rotation by an angle ϕ around the first axis x^1 is represented by the matrix

$$R_1 = \begin{pmatrix} 1 & 0 & 0 & 0 \\ 0 & 1 & 0 & 0 \\ 0 & 0 & \cos\phi & -\sin\phi \\ 0 & 0 & \sin\phi & \cos\phi \end{pmatrix} \qquad (3.22)$$

Rotations around the axes x^2 and x^3 are given by analogous matrices R_2 and R_3, respectively (compare with Eq. (2.25)).

A general rotation by an angle ϕ around a fixed arbitrary direction is represented by a matrix

$$R = \left(\begin{array}{c|c} 1 & 0 \\ \hline 0 & R_n \end{array} \right) \tag{3.23}$$

where R_n is an element of the three dimensional group $SO(3)$ given explicitly in Eq. (2.9). It is then clear that the rotation group is a subgroup of $\mathcal{L}_+^\uparrow$.

2.*Pure Lorentz transformations*
A so-called pure Lorentz transformation with velocity $\mathbf{v}$ in the direction of the axis x^1 is given by

$$L_1 = \begin{pmatrix} \gamma & -\beta\gamma & 0 & 0 \\ -\beta\gamma & \gamma & 0 & 0 \\ 0 & 0 & 1 & 0 \\ 0 & 0 & 0 & 1 \end{pmatrix} \tag{3.24}$$

where $\beta = v$ $(c = 1)$, $\gamma = (1 - \beta^2)^{-\frac{1}{2}}$. It is useful to re-write L_1 in terms of a parameter ψ defined by

$$\gamma = \cosh\psi , \tag{3.25}$$

so that (3.24) becomes

$$L_1 = \begin{pmatrix} \cosh\psi & -\sinh\psi & 0 & 0 \\ -\sinh\psi & \cosh\psi & 0 & 0 \\ 0 & 0 & 1 & 0 \\ 0 & 0 & 0 & 1 \end{pmatrix} . \tag{3.26}$$

Compared with (3.22), L_1 looks like a rotation through an imaginary angle $\phi = i\psi$ in the (x^0, ix^1) plane. Similarly, we can write two matrices L_2, L_3 which represent pure Lorentz transformations along the axis x^2, x^3, respectively.

In general, a pure Lorentz transformation corresponding to a velocity $\mathbf{v}$ is represented by the matrix L with matrix elements

$$L^\mu{}_\nu = \left(\begin{array}{c|c} \cosh\psi & n_j \sinh\psi \\ \hline -n^i \sinh\psi & \delta^i{}_j - n^i n_j(\cosh\psi - 1) \end{array} \right) = \left(\begin{array}{c|c} \gamma & \beta_j\gamma \\ \hline -\beta^i\gamma & \delta^i{}_j - \frac{\beta^i\beta_j}{\beta^2}(\gamma - 1) \end{array} \right) \tag{3.27}$$

where $\mathbf{n} = (n^1, n^2, n^3)$ is the unit vector along $\boldsymbol{\beta} = \mathbf{v}/c$ and $\gamma = \cosh\psi$. It can be obtained from L_3 by combining it with the rotation R_n which takes the axis x^3 along $\mathbf{n}$:

$$L = R_n L_3 R_n^{-1} . \tag{3.28}$$

The pure Lorentz transformations, in general, do not form a group; only the one-dimensional Lorentz transformations, such as (3.26), form a group (see example 4 in Subsection 1.2.1).

Next we consider some general properties of the group $\mathcal{L}_+^\uparrow$. At the end of this Section, it will be shown that a generic transformation Λ of $\mathcal{L}_+^\uparrow$ can be obtained as the product

$$\Lambda = LR \tag{3.29}$$

of a pure Lorentz transformation L times a pure rotation R. We can associate 3 of the 6 parameters characterizing Λ to a rotation, and the other 3 to a pure Lorentz transformation. The parameter space corresponding to rotations can be taken as a three-dimensional sphere of radius π (see Fig. 2.1). The parameter space corresponding to the transformations L can be taken as a *hyperboloid* in a four-dimensional euclidean space (see Fig. 3.1): in fact, the quantity

$$(x)^2 = (x^0)^2 - (\mathbf{x})^2 = \text{const} \tag{3.30}$$

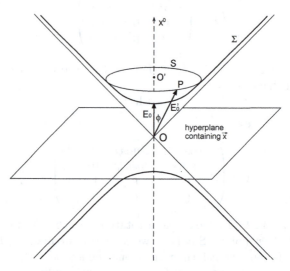

Fig. 3.1. *Parameter domain of the pure Lorentz transformations.*

is invariant under a transformation L. The hyperboloid consists of two branches inside the light cone, but only the upper one (Σ) is needed if one consider $x^0 > 0$ (the transformations of $\mathcal{L}_+^\uparrow$ cannot change the sign of x^0). A pure Lorentz transformation L takes the unit four-vector E_0 into E_0'; it is uniquely defined by the point P. For a given ϕ, P moves on a sphere S (a circle in Fig. 3.1), and so one needs two more parameters to fix its position on S. In the Minkowski space, the angle ϕ becomes imaginary, and it corresponds to the parameter ψ in Eq. (3.27).

From the topological properties of the parameter space, we can infer the following properties for the group $\mathcal{L}_+^\uparrow$:

- the group $\mathcal{L}_+^\uparrow$ is *not compact*. In fact, the parameter space Σ is a hyperboloid, i.e. a domain which is not compact;

- the group $\mathcal{L}_+^\uparrow$ is *doubly connected*. The subset of the L transformations is simply connected: in fact, any closed path on Σ can be shrunked to a point. However, the subset of rotations, which form the subgroup $SO(3)$, is doubly connected (see Section 2.1), and so is the group $\mathcal{L}_+^\uparrow$ itself.

The fact that $\mathcal{L}_+^\uparrow$ is doubly connected leads us to look for its *universal covering group*, which is simply connected and homomorphic to $\mathcal{L}_+^\uparrow$. The situation is similar to what occurs for the groups $SO(3)$ and $SU(2)$. In analogy to what shown in Section 2.1 (see, in particular, Eqs. (2.20) and (2.21)), we associate to each four-vector $x = (x^0, x^1, x^2, x^3)$ a 2×2 hermitian matrix

$$X = \sigma_\mu x^\mu = \begin{pmatrix} x^0 + x^3 & x^1 - ix^2 \\ x^1 + ix^2 & x^0 - x^3 \end{pmatrix} , \tag{3.31}$$

where

$$\sigma_\mu = (\sigma_0, \boldsymbol{\sigma}) \tag{3.32}$$

stands for the set of the unit matrix and the three Pauli matrices (2.22):

$$\sigma_0 = \begin{pmatrix} 1 & 0 \\ 0 & 1 \end{pmatrix} , \quad \sigma_1 = \begin{pmatrix} 0 & 1 \\ 1 & 0 \end{pmatrix} , \quad \sigma_2 = \begin{pmatrix} 0 & -i \\ i & 0 \end{pmatrix} , \quad \sigma_3 = \begin{pmatrix} 1 & 0 \\ 0 & -1 \end{pmatrix} . \tag{3.33}$$

It is convenient to introduce also the set of matrices

$$\sigma^\mu = \underline{\sigma}_\mu = (\sigma_0, -\boldsymbol{\sigma}) ; \tag{3.34}$$

obviously this definition implies $\underline{\sigma}^\mu = \sigma_\mu$. One gets easily the relation

$$\mathrm{Tr}(\underline{\sigma}_\mu \sigma_\nu) = 2g_{\mu\nu} , \tag{3.35}$$

which allows to express x^μ as

$$x^\mu = \tfrac{1}{2}\mathrm{Tr}(\underline{\sigma}^\mu X) . \tag{3.36}$$

The determinant of X is given by:

$$\det X = (x^0)^2 - (\mathbf{x})^2 = (x)^2 . \tag{3.37}$$

Let us introduce a transformation of X through a unimodular complex matrix A

$$X' = A X A^\dagger , \tag{3.38}$$

where

$$A = \begin{pmatrix} \alpha & \beta \\ \gamma & \delta \end{pmatrix} , \tag{3.39}$$

with the condition

$$\det A = \alpha\delta - \beta\gamma = 1 . \tag{3.40}$$

From (3.38) we get, since $\det A = 1$,

$$(x')^2 = \det X' = \det X = (x)^2 . \tag{3.41}$$

This shows that Eq. (3.38) describes a linear transformation of x^μ which leaves $(x)^2$ invariant, so that it corresponds to a Lorentz transformation Λ. The connection between the matrices A and Λ is given by

$$\Lambda^\mu_{\ \nu} = \tfrac{1}{2} \text{Tr}(\underline{\sigma}^\mu A \sigma_\nu A^\dagger) . \tag{3.42}$$

The matrices A given by (3.39), (3.40) form a group denoted by $SL(2,C)$ (following the notation of Subsection 1.2.1, the symbol $L(2,C)$ refers to *linear* transformations in a two-dimensional complex space; S stands for *special* and refers to the unimodularity condition). It is clear from Eq. (3.42) that for each matrix A there is a corresponding Lorentz transformation Λ; viceversa, for a given Lorentz transformation Λ there are two corresponding matrices A and $-A$. In particular, both matrices I and $-I$ in $SL(2,C)$ correspond to the identity element of $\mathcal{L}^\uparrow_+$.

The correspondence (3.42) is preserved under multiplication, which proves the *homomorphism* between the group $\mathcal{L}^\uparrow_+$ and $SL(2,C)$. Moreover, it can be shown that the matrices Λ given by (3.42) satisfy the conditions $\det\Lambda = +1$, $\Lambda^0_{\ 0} \geq +1$. This proves that the group $SL(2,C)$ is homomorphic to the proper orthochronous Lorentz group $\mathcal{L}^\uparrow_+$. The proof of these statements is left as an exercise.

It is instructive to write the matrices A of $SL(2,C)$ in exponential form

$$A = e^S , \tag{3.43}$$

where the 2×2 matrix S has to satisfy the condition

$$\text{Tr}\, S = 0 , \tag{3.44}$$

since $\det A = 1$. There are 6 independent 2×2 traceless matrices: we can choose the 3 (hermitian) Pauli matrices σ_k and the 3 anti-hermitian matrices $i\sigma_k$. In general, the matrix S can be written as the sum $S = S_1 + S_2$, where

$$S_1 = -\tfrac{1}{2} i\phi \, \boldsymbol{\sigma} \cdot \mathbf{n} \qquad \text{and} \qquad S_2 = -\tfrac{1}{2} \psi \boldsymbol{\sigma} \cdot \boldsymbol{\nu} , \tag{3.45}$$

ϕ and ψ being real parameters with $\mathbf{n}$ and $\boldsymbol{\nu}$ unit (real) vectors. From Eq. (3.43) we get two kinds of matrices A:

$$U = e^{iS_1} = \cos \tfrac{1}{2}\phi - i\boldsymbol{\sigma} \cdot \mathbf{n} \sin \tfrac{1}{2}\phi , \tag{3.46}$$

$$H = e^{iS_2} = \cosh \tfrac{1}{2}\psi - \boldsymbol{\sigma} \cdot \boldsymbol{\nu} \sinh \tfrac{1}{2}\psi . \tag{3.47}$$

It is easy to check that the matrices U are *unitary*, while the matrices H are *hermitian*. Moreover, one can show that the matrix U given in (3.46) represents a *rotation* through an angle ϕ about the $\mathbf{n}$ direction (in particular, taking $\mathbf{n} = (1,0,0)$ one gets (3.22)), and that the matrix H given in (3.47) represents a *pure Lorentz transformation* L with $\boldsymbol{\beta} = \boldsymbol{\nu} \tanh \psi$. All this can be shown by deducing explicitly the corresponding transformations of x^μ from Eq. (3.38).

The above relations show that a general matrix A of $SL(2,C)$ can be written as the product

$$A = HU .\tag{3.48}$$

Due to the homomorphism between $SL(2,C)$ and $\mathcal{L}_+^\uparrow$, Eq. (3.48) proves the decomposition of a Lorentz transformation Λ into the product LR (see Eq. (3.29)).

Since the matrices U are unitary and unimodular, they form the group $SU(2)$ considered in Section 2.1, which is a subgroup of $SL(2,C)$. Instead, the subset of the matrices H is not a group (the product of two hermitian matrices is not, in general, hermitian), in agreement with what stated about pure Lorentz transformations.

Since the group $SU(2)$ is simply connected, and so is the set of pure Lorentz transformations, we can conclude from (3.48) that also the group $SL(2,C)$ is *simply connected*; one can convince oneself that it is the *universal covering group* of $\mathcal{L}_+^\uparrow$.

3.3 Lie algebra of the group $\mathcal{L}_+^\uparrow$

We examine here the infinitesimal transformations of the restricted Lorentz group $\mathcal{L}_+^\uparrow$, and the *generators* of the Lie algebra.

First, we consider the subgroup of rotations and the three independent elements $R_1(\phi)$, $R_2(\phi)$, $R_3(\phi)$ (see (3.23) and (2.25)). The generators of these rotations are defined, according to (2.26), by

$$J_k = i\frac{\partial R_k}{\partial \phi}\bigg|_{\phi=0} ,\tag{3.49}$$

from which we obtain:

$$J_1 = \begin{pmatrix} 0 & 0 & 0 & 0 \\ 0 & 0 & 0 & 0 \\ 0 & 0 & 0 & -i \\ 0 & 0 & i & 0 \end{pmatrix}, \quad J_2 = \begin{pmatrix} 0 & 0 & 0 & 0 \\ 0 & 0 & 0 & i \\ 0 & 0 & 0 & 0 \\ 0 & -i & 0 & 0 \end{pmatrix}, \quad J_3 = \begin{pmatrix} 0 & 0 & 0 & 0 \\ 0 & 0 & -i & 0 \\ 0 & i & 0 & 0 \\ 0 & 0 & 0 & 0 \end{pmatrix}.\tag{3.50}$$

An infinitesimal rotation about a direction $\mathbf{n}$ can be written as

$$R = I - i\delta\phi\,\mathbf{J}\cdot\mathbf{n}\tag{3.51}$$

where $\mathbf{J} = (J_1, J_2, J_3)$, and can be obtained directly from (3.23) with R given by (2.9). A finite rotation can be written in terms of $\mathbf{J}$ by exponentiation (see Eq. (2.29)):

$$R = e^{-i\phi\mathbf{J}\cdot\mathbf{n}} .\tag{3.52}$$

Similarly, from $L_1(\psi)$, $L_2(\psi)$, $L_3(\psi)$ (see (3.26) and (3.27)), we obtain the generators of the pure Lorentz transformations:

$$K_\ell = i\frac{\partial L_\ell}{\partial \psi}\bigg|_{\psi=0} , \qquad (3.53)$$

i.e.

$$K_1 = \begin{pmatrix} 0 & -i & 0 & 0 \\ -i & 0 & 0 & 0 \\ 0 & 0 & 0 & 0 \\ 0 & 0 & 0 & 0 \end{pmatrix}, K_2 = \begin{pmatrix} 0 & 0 & -i & 0 \\ 0 & 0 & 0 & 0 \\ -i & 0 & 0 & 0 \\ 0 & 0 & 0 & 0 \end{pmatrix}, K_3 = \begin{pmatrix} 0 & 0 & 0 & -i \\ 0 & 0 & 0 & 0 \\ 0 & 0 & 0 & 0 \\ -i & 0 & 0 & 0 \end{pmatrix}. \quad (3.54)$$

An infinitesimal pure Lorentz transformation L is written as

$$L = I - i\delta\psi\, \mathbf{K} \cdot \boldsymbol{\nu} , \qquad (3.55)$$

where $\mathbf{K} = (K_1, K_2, K_3)$, $\boldsymbol{\nu} = \mathbf{v}/|\mathbf{v}|$, and can be obtained directly from (3.27). Finite pure Lorentz transformations can be obtained by exponentiation in the form

$$L = e^{-i\psi\mathbf{K}\cdot\boldsymbol{\nu}} . \qquad (3.56)$$

Then a general transformation of $\mathcal{L}_+^\uparrow$ can be written, according to (3.29), (3.52), (3.56) as

$$\Lambda = e^{-i(\phi\mathbf{J}\cdot\mathbf{n} + \psi\mathbf{K}\cdot\boldsymbol{\nu})} . \qquad (3.57)$$

One can check, from the explicit expressions given above, that J_i, K_i satisfy the following commutation relations

$$\begin{aligned} [J_i , J_j] &= i\epsilon_{ijk} J_k , \\ [K_i, K_j] &= i\epsilon_{ijk} J_k , \\ [J_i , K_j] &= i\epsilon_{ijk} K_k , \end{aligned} \qquad (3.58)$$

where ϵ_{ijk} is the completely antisymmetric tensor.

The above commutators define the *Lie algebra* of the groups $\mathcal{L}_+^\uparrow$ and $SL(2, C)$. The quantities $J_k(k = 1, 2, 3)$ and $K_\ell(\ell = 1, 2, 3)$ form the basis of the Lie algebra; we shall refer to them as *generators* of the Lie algebra of $\mathcal{L}_+^\uparrow$. The relations (3.57) and (3.58), that we derived making use of the 4-dimensional representation, hold in the general case. We stress the fact that, while the generators J_i are hermitian (they represent dynamical variables, i.e. the components of the angular momentum), the generators K_i (see (3.54)) are *anti-hermitian*.

Eqs. (3.58) show that the Lie algebra has *rank 2*. In terms of the structure constants ϵ_{ijk}, one can check that the Lie algebra is at least semi-simple; in fact, it is *simple*, and so is the group $\mathcal{L}_+^\uparrow$.

By defining the antisymmetric tensor $M_{\mu\nu}$ with components

$$(M_{12}, M_{23}, M_{31}) = (J_3, J_1, J_2) \; ,$$
$$(M_{01}, M_{02}, M_{03}) = (K_1, K_2, K_3) \; , \tag{3.59}$$

the commutators (3.58) can be re-written in a covariant form

$$[M_{\lambda\rho}, M_{\mu\nu}] = -i(g_{\lambda\mu}M_{\rho\nu} + g_{\rho\nu}M_{\lambda\mu} - g_{\lambda\nu}M_{\rho\mu} - g_{\rho\mu}M_{\lambda\nu}) \; . \tag{3.60}$$

From (3.57) and (3.59) it follows that a general transformation of $\mathcal{L}_+^\uparrow$ can be written as

$$\Lambda = e^{-\frac{1}{2}i\omega^{\mu\nu}M_{\mu\nu}} \; , \tag{3.61}$$

where $\omega^{\mu\nu}$ is a real antisymmetric matrix.

In terms of $M_{\mu\nu}$, one can obtain the following *Casimir operators*

$$\frac{1}{2}M^{\mu\nu}M_{\mu\nu} = \mathbf{J}^2 - \mathbf{K}^2 \; ,$$
$$\frac{1}{2}\epsilon^{\mu\nu\sigma\tau}M_{\mu\nu}M_{\sigma\tau} = -\mathbf{J} \cdot \mathbf{K} \; , \tag{3.62}$$

where $\epsilon^{\mu\nu\sigma\tau} = \epsilon_{\mu\nu\sigma\tau}$ is the completely antisymmetric tensor in four dimensions. In fact, it can be shown that they commute with all the generators $M_{\mu\nu}$.

For the classification of the irreducible representations of $\mathcal{L}_+^\uparrow$, it is useful to introduce the linear combinations of J_i, K_i

$$M_i = \frac{1}{2}(J_i + iK_i) \; ,$$
$$N_i = \frac{1}{2}(J_i - iK_i) \; , \tag{3.63}$$

which are *hermitian*. Eqs. (3.58) give in terms of M_i, N_i

$$[M_i, M_j] = i\epsilon_{ijk}M_k \; ,$$
$$[N_i, N_j] = i\epsilon_{ijk}N_k \; ,$$
$$[M_i, N_j] = 0 \; . \tag{3.64}$$

Moreover one has

$$[M^2, M_j] = 0 \; ,$$
$$[N^2, N_j] = 0 \; , \tag{3.65}$$

where

$$M^2 = \sum_i M_i^2 \; , \qquad N^2 = \sum_i N_i^2 \; . \tag{3.66}$$

Since (3.63) are not *real* linear combinations of the basic elements, the commutators (3.64) do not define the real Lie algebra of $\mathcal{L}_+^\uparrow$. However, due to the analogy with angular momentum, it is convenient to use the eigenvalues of the hermitian operators M^2, M_3, N^2, N_3 to label the elements of the irreducible representations.

3.4 Irreducible representations of the group $\mathcal{L}_+^\uparrow$

We are interested here in the *finite-dimensional* irreducible representations
(IR's) of $\mathcal{L}_+^\uparrow$. As stated in Section 3.2, since $\mathcal{L}_+^\uparrow$ is a *non-compact* group, its
finite-dimensional IR's *cannot be unitary*. This can be seen explicitly from
Eq. (3.57): since the generators K_i are not hermitian, the matrices Λ are, in
general, not unitary.

The IR's of $\mathcal{L}_+^\uparrow$ are usually labelled by the eigenvalues of the Casimir
operators M^2, N^2 given by (3.66). From (3.64), (3.65) we see that M_i, N_i
behave as the components of two angular momenta $\mathbf{M}$, $\mathbf{N}$. Their eigenvalues
are then given in terms of two numbers j, j', which can take independently
zero or positive integral or half-integral values. To each pairs of values (j, j')
there correspond $(2j + 1) \cdot (2j' + 1)$ eigenstates, which can be taken as the
basis of the IR's of $\mathcal{L}_+^\uparrow$ of dimension $(2j + 1) \cdot (2j' + 1)$. Each element Λ of $\mathcal{L}_+^\uparrow$
is then represented by $D^{(j,j')}(\Lambda)$.

If one restricts oneself to the subgroup of rotations, the representations
are no longer irreducible, and they can be decomposed in terms of the IR's of
$SO(3)$ as follows:

$$D^{(j,j')}(R) = D^{(j)}(R) \otimes D^{(j')}(R) = D^{(j+j')}(R) \oplus ... \oplus D^{(|j-j'|)}(R) \ . \quad (3.67)$$

As in the case of the group $SO(3)$, we have for $\mathcal{L}_+^\uparrow$ two kinds of IR's:
the *tensorial representations* ($j + j'$ = integer) and the *spinor representations*
($j + j'$ = half-integer). The spinor representations, as discussed in Section 2.3
for $SO(3)$, are *double-valued*. This feature is due to the fact that the group
$\mathcal{L}_+^\uparrow$ is *doubly connected*. All its IR's can be found by looking for the IR's of its
universal covering group $SL(2, C)$.

We shall examine in more detail the lowest spinor representations. There
are two IR's of dimension 2, which are *not equivalent*: $D^{(\frac{1}{2}, 0)}$, $D^{(0, \frac{1}{2})}$. In order
to understand this point, we go to the group $SL(2, C)$ and to Eq. (3.42) which
gives the connection between the matrices Λ and A. It is clear that the two
matrices $+A$, $-A$ correspond to the same matrix Λ. The homomorphism between $SL(2, C)$ and $\mathcal{L}_+^\uparrow$ (two-to-one correspondence) explains the occurrence
of *both signs*; this is the analogue of what seen in Section 2.3 for the $SU(2)$
and $SO(3)$ groups. However, while for the rotation group a representation
and its complex conjugate are equivalent, in the present case A and A^* are
not equivalent. In fact, since the matrices A are, in general, *not unitary*, one
cannot find a similarity transformation relating A to A^*.

Therefore, the matrices A and A^* constitute two non equivalent irreducible
representations of $\mathcal{L}_+^\uparrow$, acting on two different two-dimensional vector spaces.
We have then *two non equivalent bases*, and, in general, two kinds of (controvariant) *spinors*[2], which shall be denoted by ξ and ξ^* and transform ac-

---

[2] In analogy with the footnote related to eq. (2.34), one should call the basis ξ
"spinor" with respect to $\mathcal{L}_+^\uparrow$ and "vector" with respect to $SL(2, C)$.

cording to[3]

$$\xi' = A\xi , \tag{3.68}$$

$$\xi^{*\prime} = A^*\xi^* . \tag{3.69}$$

We can show that the two IR's defined by (3.68), (3.69) correspond to the two IR's $D^{(\frac{1}{2},0)}$, $D^{(0,\frac{1}{2})}$ of $\mathcal{L}_+^\uparrow$. In fact we remember that a general 2×2 matrix A can be written, according to (3.43), (3.45) as

$$A = e^{-\frac{1}{2}i(\phi\boldsymbol{\sigma}\cdot\mathbf{n}-i\psi\boldsymbol{\sigma}\cdot\boldsymbol{\nu})} . \tag{3.70}$$

Comparison with (3.57) shows that $\frac{1}{2}\sigma_i$, $-\frac{1}{2}i\sigma_i$ provide a 2×2 representation for the generators J_i, K_i, i.e.

$$J_i \to \tfrac{1}{2}\sigma_i , \qquad\qquad K_i \to -\tfrac{1}{2}i\sigma_i , \tag{3.71}$$

and, from (3.63), for M_i, N_i,

$$M_i \to \tfrac{1}{2}\sigma_i , \qquad\qquad N_i \to 0 . \tag{3.72}$$

The representation given by (3.70) is then equivalent to $D^{(\frac{1}{2},0)}(\Lambda)$.

Similarly, taking the conjugate A^* of (3.70), one gets the representation

$$J_i \to -\tfrac{1}{2}\sigma_i^* , \qquad\qquad K_i \to -\tfrac{1}{2}i\sigma_i^* . \tag{3.73}$$

However, since

$$\sigma_2\sigma_i^*\sigma_2 = -\sigma_i , \tag{3.74}$$

the representation (3.73) is *equivalent* to

$$J_i \to \tfrac{1}{2}\sigma_i , \qquad\qquad K_i \to \tfrac{1}{2}i\sigma_i , \tag{3.75}$$

which give

$$M_i \to 0 , \qquad\qquad N_i \to \tfrac{1}{2}\sigma_i . \tag{3.76}$$

This shows that the representation provided by A^* is equivalent to $D^{(0,\frac{1}{2})}(\Lambda)$.

It is useful to introduce covariant spinors, which transform according to[4] (compare with (2.36))

---

[3] Written in terms of the (controvariant) components $\xi = \begin{pmatrix} \xi^1 \\ \xi^2 \end{pmatrix}$, $\xi^* = \begin{pmatrix} \xi^{\dot{1}} \\ \xi^{\dot{2}} \end{pmatrix}$,

Eqs. (3.68), (3.69), following the standard notation, become

$$\xi'^\alpha = A^\alpha{}_\beta\xi^\beta , \qquad\qquad \xi'^{\dot\alpha} = A^*{}^{\dot\alpha}{}_{\dot\beta}\xi^{\dot\beta} .$$

[4] In terms of the (covariant) components $\eta = (\eta_1 \ \eta_2)$, $\eta^* = (\eta_{\dot{1}} \ \eta_{\dot{2}})$, Eqs. (3.77), (3.78) become:

$$\eta'_\alpha = \eta_\beta(A^{-1})^\beta{}_\alpha , \qquad\qquad \eta'_{\dot\alpha} = \eta_{\dot\beta}(A^{-1})^{\dot\beta}{}_{\dot\alpha} .$$

$$\eta' = \eta A^{-1} , \tag{3.77}$$
$$\eta^{*\prime} = \eta^*(A^*)^{-1} , \tag{3.78}$$

and are defined in such a way that the products $\eta\xi$, $\eta^*\xi^*$ are invariant (scalar under $\mathcal{L}_+^\uparrow$).

It is easy to see that a matrix of $SL(2,C)$ exists which relates A and $\tilde{A}^{-1}$ (A and $\tilde{A}^{-1}$ correspond, in fact, to equivalent representations) according to

$$\tilde{A}^{-1} = CAC^{-1} , \tag{3.79}$$

that is

$$C = \tilde{A}CA . \tag{3.80}$$

The above relation is satisfied by any antisymmetric matrix C, in particular by

$$C = i\sigma_2 = \begin{pmatrix} 0 & 1 \\ -1 & 0 \end{pmatrix} . \tag{3.81}$$

Eq. (3.77) can be re-written, using (3.81), as

$$\tilde{\eta}' = \tilde{A}^{-1}\tilde{\eta} = CAC^{-1}\tilde{\eta} , \tag{3.82}$$

which, compared with eq. (3.68), gives

$$\tilde{\eta} = C\xi . \tag{3.83}$$

This shows that the matrix C transforms controvariant into covariant spinors, and viceversa. Since $C = C^*$, the same relations hold between A^*, $(A^\dagger)^{-1}$ and ξ^*, η^*:

$$C = A^\dagger CA^*, \tag{3.84}$$
$$\tilde{\eta}^* = C\xi^* . \tag{3.85}$$

In conclusion, we have examined the bases of the two lowest spinor representations, $D^{(\frac{1}{2},0)}$, $D^{(0,\frac{1}{2})}$. In terms of these bases, one can build not only all the higher spinor representations, but also all the tensorial representations. With this respect, the spinor IR's are more fundamental than the tensor IR's.

Let us consider, as an example, the four-dimensional representation. Its basis can be taken as the four vector x^μ, which transforms according to Eq. (3.1). We saw that this equation can be replaced by

$$X' = AXA^\dagger , \tag{3.38}$$

where X represents the four vector written as a 2×2 matrix (see Eq. (3.31)). We consider now the quantity $\xi\xi^\dagger$; its transformation properties follow from Eqs. (3.68), (3.69) :

$$\xi'\xi'^\dagger = A\xi\xi^\dagger A^\dagger . \tag{3.86}$$

Then $\xi\xi^\dagger$ transforms in the same way of the four vector X, so it can be taken as the basis of the four-dimensional IR. It is denoted by $D^{(\frac{1}{2},\frac{1}{2})}$, which is consistent with the fact that its basis contains both kind of spinors ξ, ξ^*; in fact, one can write:

$$D^{(\frac{1}{2},\frac{1}{2})} = D^{(\frac{1}{2},0)} \oplus D^{(0,\frac{1}{2})} . \tag{3.87}$$

In general, one can obtain higher IR's by decomposition of direct products; it can be shown that the following relation holds

$$D^{(j_1,j_2)} \otimes D^{(j_1',j_2')} = D^{(j_1+j_1',j_2+j_2')} \oplus D^{(j_1+j_1'-1,j_2+j_2')} \oplus \ldots \oplus D^{(|j_1-j_1'|,|j_2-j_2'|)} . \tag{3.88}$$

which is the analogue of the relation (2.67), valid for the rotation group.

Finally, we mention a very important point, on which we shall come back in Chapter 7. According to Eq. (3.67), we see that an IR of $\mathcal{L}_+^\uparrow$ contains, in general, several IR's of $SO(3)$. We know that each element of the basis of $D^{(j)}$ describes one of the $(2j+1)$ states of a particle with spin J. If we want to keep this correspondence between IR's and states with *definite spin* also in the case of $\mathcal{L}_+^\uparrow$, we have to introduce *supplementary conditions* which reduce the number of independent basis elements.

Going back to (3.67), if we want the basis of $D^{(j,j')}$ to describe a unique value of spin (we choose for it the highest value, since the lower ones can be described by lower IR's), we have to keep only $2(j+j')+1$ elements out of $(2j+1) \cdot (2j'+1)$, so that these can be grouped together to form the basis of $D^{(j+j')}$ in $SO(3)$. The number of conditions is then given by $4jj'$. For instance, in the simple case

$$D^{(\frac{1}{2},\frac{1}{2})}(R) = D^{(1)}(R) \oplus D^{(0)}(R) , \tag{3.89}$$

we see that a spin 1 particle is described by a four-dimensional basis; so that one needs a supplementary condition (called in this case Lorentz condition) to leave only 3 independent elements which describes the 3 different spin 1 states.

3.5 Irreducible representations of the complete Lorentz group

It was pointed out in Section 3.1 that the complete homogeneous Lorentz group $\mathcal{L}$, corresponding to the general homogeneous Lorentz transformations, can be obtained from the group $\mathcal{L}_+^\uparrow$ by inclusion of the three inversion I_s, I_t, I_{st}.

We shall not consider here all kinds of finite-dimensional IR's of $\mathcal{L}$, but limit ourselves to specific cases, which are important for physical applications and which will be used in the following.

First, we examine the transformation properties of the spinors ξ and ξ^*, bases of the two-dimensional IR's $D^{(\frac{1}{2},0)}$ and $D^{(0,\frac{1}{2})}$ of $\mathcal{L}_+^\uparrow$, under the inversion operations. From this analysis, one should be able to determine the behaviour of the higher IR's of $\mathcal{L}_+^\uparrow$ under the discrete operations, and, therefore, to build the IR's of $\mathcal{L}$.

We shall follow a heuristic approach, starting from the transformation properties of the matrix (3.31), which we rewrite here

$$X = \sigma_\mu x^\mu = \begin{pmatrix} x^0 + x^3 & x^1 - ix^2 \\ x^1 + ix^2 & x^0 - x^3 \end{pmatrix} . \tag{3.31}$$

One can check that, under the operations I_s, I_t, I_{st}, the matrix X transforms as follows

$$X \xrightarrow{I_s} X' = -CX^*C^{-1} , \tag{3.90}$$

$$X \xrightarrow{I_t} X' = CX^*C^{-1} , \tag{3.91}$$

$$X \xrightarrow{I_{st}} X' = -X , \tag{3.92}$$

where C is the matrix defined by eq. (3.81).

We assume that the quantity $\xi\xi^\dagger$, which behaves as the matrix X under a transformation of $SL(2,C)$, has also the same transformation properties under the discrete operations. Similarly for the quantity $\tilde{\eta}\eta^*$. From the above equations we can then extract the behaviour of ξ, η (and ξ^*, η^*) under I_s, I_t, I_{st}, aside from phase ambiguities.

We give here one of the possible choices[5]

$$\begin{cases} \xi \xrightarrow{I_s} i\eta^\dagger , \\ \eta^\dagger \xrightarrow{I_s} i\xi , \end{cases} \tag{3.93}$$

$$\begin{cases} \xi \xrightarrow{I_t} \eta^\dagger , \\ \eta^\dagger \xrightarrow{I_t} -\xi , \end{cases} \tag{3.94}$$

$$\begin{cases} \xi \xrightarrow{I_{st}} i\xi , \\ \eta^\dagger \xrightarrow{I_{st}} -i\eta^\dagger . \end{cases} \tag{3.95}$$

The above transformations show that it is not possible to build scalar and pseudoscalar quantities using only either ξ, η or ξ^*, η^*, but both kinds of spinors are needed. In fact, one can easily check that the quantities

$$\eta^*\xi^* + \eta\xi , \tag{3.96}$$

$$\eta^*\xi^* - \eta\xi , \tag{3.97}$$

are. respectively, *scalar* and *pseudoscalar* under space inversion. Making use, together with the matrices $\sigma_\mu = (\sigma_0, \boldsymbol{\sigma})$, of the conjugate matrices $\sigma^\mu = \underline{\sigma}_\mu = C\sigma_\mu^* C^{-1} = (\sigma_0, -\boldsymbol{\sigma})$ introduced in Eq. (3.34), one can show that the quantity

---

[5] Only one of the possible choices is given here. For a more complete discussion see: A.J. Macfarlane, Jour. of Math. Phys. **3** (1962) 1116, and references therein.

$$\xi^\dagger \underline{\sigma}_\mu \tilde{\eta} \pm \eta \sigma_\mu \xi^* \tag{3.98}$$

behaves, according to the $\pm$ sign, as a *vector* or a *pseudovector*.

It is useful to give a more general classification of the IR's of the orthochronous Lorentz group $\mathcal{L}^\uparrow$, which contains I_s and the subgroup $O(3)$. We shall denote by $\mathcal{D}^{(j_1,j_2)}$ the IR's of $\mathcal{L}^\uparrow$ to distinguish them from those of $\mathcal{L}_+^\uparrow$, denoted by $D^{(j_1,j_2)}$. The following cases may occur:

a) $j_1 = j_2$. Each IR of $\mathcal{L}^\uparrow$ corresponds to a definite IR of $\mathcal{L}_+^\uparrow$. In analogy with $O(3)$, (see Section 2.3), there are two kinds of IR's: $\mathcal{D}^{(j,j,+)}$ and $\mathcal{D}^{(j,j,-)}$; however, in this case both are faithful. Their difference is understood going to the rotation subgroup $O(3)$: in terms of the IR's of $O(3)$ they are decomposed as follows

$$\begin{aligned}
\mathcal{D}^{(j,j,+)} &= \mathcal{D}^{(0,+)} \oplus \mathcal{D}^{(1,-)} \oplus \dots \oplus \mathcal{D}^{(2j,\pm)} \;, \\
\mathcal{D}^{(j,j,-)} &= \mathcal{D}^{(0,-)} \oplus \mathcal{D}^{(1,+)} \oplus \dots \oplus \mathcal{D}^{(2j,\mp)} \;.
\end{aligned} \tag{3.99}$$

In the last term, the upper or lower sign is to be taken according to $2j$ being even or odd.

b) $j_1 \neq j_2$, $j_1 + j_2 =$ integer. Each IR of $\mathcal{L}^\uparrow$ contains two IR of $\mathcal{L}_+^\uparrow$, which can be decomposed as

$$\mathcal{D}^{(j_1,j_2)} = D^{(j_1,j_2)} \oplus D^{(j_2,j_1)} = \sum_{j=|j_1-j_2|}^{j_1+j_2} \mathcal{D}^{(j,+)} \oplus \mathcal{D}^{(j,-)} \;. \tag{3.100}$$

c) $j_1 \neq j_2$, $j_1 + j_2 =$ half integer. Each IR of $\mathcal{L}^\uparrow$ contains, also in this case, two IR of $\mathcal{L}_+^\uparrow$, and it is reduced according to

$$\mathcal{D}^{(j_1,j_2)} = D^{(j_1,j_2)} \oplus D^{(j_2,j_1)} = \sum_{j=|j_1-j_2|}^{j_1+j_2} \mathcal{D}^{(j)} \;. \tag{3.101}$$

One can understand the meaning of the above decomposition, by noting that the infinitesimal generators J_i, K_i transform under space inversion as follows

$$\begin{aligned}
I_s J_i I_s &= J_i \;, \\
I_s K_i I_s &= -K_i \;.
\end{aligned} \tag{3.102}$$

Correspondingly, according to Eq. (3.63) M_i, N_i are transformed into each other, so that the two IR's $D^{(j_1,j_2)}$, $D^{(j_2,j_1)}$ are interchanged under space inversion. This fact explains why, for $j_i \neq j_2$ both $D^{(j_1,j_2)}$ and $D^{(j_2,j_1)}$ are contained in the IR $\mathcal{D}^{(j_1,j_2)}$ of $\mathcal{L}^\uparrow$. In particular, as indicated by (3.93), both lowest spinor representations of $\mathcal{L}_+^\uparrow$ enter into the lowest IR of $\mathcal{L}^\uparrow$, according to

$$\mathcal{D}^{(\frac{1}{2},0)} = D^{(\frac{1}{2},0)} \oplus D^{(0,\frac{1}{2})} \;. \tag{3.103}$$

Problems

3.1. Derive the generic pure Lorentz transformation (3.27) making use of the relation (3.28).

3.2. Let us denote by $\Lambda(A)$ the matrix Λ of $\mathcal{L}_+^\uparrow$ corresponding to a matrix A of $SL(2, C)$. Check that the following relation holds $\Lambda(A)\Lambda(B) = \Lambda(AB)$, i.e. that the correspondence is preserved under multiplication.

3.3. Making use of (3.42), express a generic matrix Λ in terms of the elements of the matrix A given by (3.39). Show that the matrix $\Lambda(A)$ so obtained satisfies the condition $\det \Lambda(A) = +1$.

3.4. Prove that the element $\Lambda^0{}_0$ of the matrix $\Lambda(A)$ satisfies the condition $\Lambda^0{}_0(A) \geq +1$.

3.5. Show that the unitary transformation U given by (3.46) corresponds to the rotation matrix R given by (2.9), (3.23).

3.6. Show that the hermitian matrix H given by (3.47) corresponds to the pure Lorentz transformation L given by (3.27).

3.7. A generic infinitesimal Lorentz transformation can be written in the form $\Lambda^\rho{}_\sigma = g^\rho{}_\sigma + \delta\omega^\rho{}_\sigma$ with $\delta\omega^\rho{}_\sigma$ real, infinitesimal and antisymmetric: $\delta\omega_{\rho\sigma} = -\delta\omega_{\sigma\rho}$. Making use of this form and of (3.61) determine the matrix elements of the generators $M_{\mu\nu}$ and verify (3.59).

3.8. Consider the group $SO(4)$, i.e. the group of proper rotations in a four-dimensional euclidean space, and its Lie algebra. Discuss its properties and compare with those of the group $\mathcal{L}_+^\uparrow$.

3.9. Write explicitly the transformations of the spinors ξ and ξ^* under: a) a rotation about the axis x^3, b) a pure Lorentz transformation along the axis x^3.

3.10. Given a tensor $T_{\mu\nu}$, its *dual* is defined by $T_{\mu\nu}^D = \frac{1}{2}\epsilon_{\mu\nu\sigma\tau}T^{\sigma\tau}$; a tensor is *selfdual* if $T_{\mu\nu}^D = T_{\mu\nu}$ and *anti-selfdual* if $T_{\mu\nu}^D = -T_{\mu\nu}$. Show that any antisymmetric tensor $A_{\mu\nu} = -A_{\nu\mu}$ can be decomposed in the sum of a selfdual and an anti-selfdual tensor and that both are irreducible under $\mathcal{L}_+^\uparrow$. Consider explicitly the case of the electromagnetic field tensor $F_{\mu\nu}$.

3.11. Show that the bases of the IR's obtained by the decomposition

$$D^{(\frac{1}{2},\frac{1}{2})} \otimes D^{(\frac{1}{2},\frac{1}{2})} = D^{(1,1)} \oplus D^{(1,0)} \oplus D^{(0,1)} \oplus D^{(0,0)}$$

are a traceless symmetric tensor, a selfdual and anti-selfdual anti-symmetric tensor and a scalar, respectively.

4

The Poincaré transformations

In this Chapter, we examine the general properties of the Poincaré group, i.e. the group of inhomogeneous Lorentz transformations, which are space-time coordinate transformations between any two inertial frames of reference. Then we consider those unitary irreducible representations of this group which are the most suitable for the description of the quantum mechanical states of one or more particles.

It is well known from special relativity that all inertial frames of reference are completely equivalent for the description of the physical phenomena; in fact, the physical laws are the same in all inertial frame or, in other words, they are *invariant* under the Poincaré transformations. The importance of the Poincaré group is related to this invariance principle. We refer to Section 9.2 for a brief discussion on *invariance principles*, symmetries and conservation laws.

4.1 Group properties

The inhomogeneous Lorentz transformations, or Poincaré transformations, connect the space-time coordinates of any two frames of reference whose relative velocity is constant.

A general Poincaré transformation can be written, as a generalization of a homogeneous Lorentz transformation (3.1), in the form

$$x'^{\mu} = \Lambda^{\mu}{}_{\nu}x^{\nu} + a^{\mu} , \tag{4.1}$$

where a^{μ} ($\mu = 0, 1, 2, 3$) stand for the components of a vector in $\mathcal{R}^4$. We shall denote the above transformation by (a, Λ).

It is easy to see that the transformations (a, Λ) form a group: the so-called *Poincaré group*, denoted in the following by $\mathcal{P}$. In particular, the application of two successive transformations (a_1, Λ_1), (a_2, Λ_2) gives

$$x' = \Lambda_1 x + a_1 , \qquad x'' = \Lambda_2 x' + a_2 = \Lambda_2 \Lambda_1 x + \Lambda_2 a_1 + a_2 , \tag{4.2}$$

G. Costa and G. Fogli, *Symmetries and Group Theory in Particle Physics*,
Lecture Notes in Physics 823, DOI: 10.1007/978-3-642-15482-9_4,
© Springer-Verlag Berlin Heidelberg 2012

so that the product of the two transformations is

$$(a_2, \Lambda_2)(a_1 \Lambda_1) = (a_2 + \Lambda_2 a_1, \Lambda_2 \Lambda_1) . \tag{4.3}$$

Clearly the unit element is $(0, I)$; then from Eq. (4.3) one gets the inverse element

$$(a, \Lambda)^{-1} = (-\Lambda^{-1} a, \Lambda^{-1}) . \tag{4.4}$$

The group $\mathcal{P}$ is characterized by 10 parameters: the 6 parameters of the homogeneous Lorentz group plus the 4 parameters a^μ.

Obviously, the Poincaré group contains, as a subgroup, the homogeneous Lorentz group $\mathcal{L}$ of the transformations $(0, \Lambda)$; it contains also the subgroup $\mathcal{S}$ of the four dimensional translations (a, I), which is an Abelian invariant subgroup of $\mathcal{P}$. Any Poincaré transformation can be written in a unique way as the product of a pure translation and a homogeneous Lorentz transformation

$$(a, \Lambda) = (a, I)(0, \Lambda) , \tag{4.5}$$
$$(a, \Lambda) = (0, \Lambda)(\Lambda^{-1} a, I) . \tag{4.6}$$

This shows that the Poincaré group is the semi-direct product of the two subgroups:

$$\mathcal{P} = \mathcal{S} \circledS \mathcal{L} . \tag{4.7}$$

We note that the order of the transformations in Eqs. (4.5), (4.6) is important, since the translations do not commute with the Lorentz transformations. From the above relations one gets easily the useful result

$$(0, \Lambda^{-1})(a, I)(0, \Lambda) = (\Lambda^{-1} a, I) . \tag{4.8}$$

We know that the group $\mathcal{L}$ consists of four disjoint components $\mathcal{L}_+^\uparrow$, $\mathcal{L}_-^\uparrow$, $\mathcal{L}_+^\downarrow$, $\mathcal{L}_-^\downarrow$, corresponding to the possible choices of the signs of $\det \Lambda$ and $\Lambda^0{}_0$ listed in Table 3.1. Similarly, the group $\mathcal{P}$ consists of four disjoint components $\mathcal{P}_+^\uparrow$, $\mathcal{P}_-^\uparrow$, $\mathcal{P}_+^\downarrow$, $\mathcal{P}_-^\downarrow$, each of which contains the corresponding component of the subgroup $\mathcal{L}$.

In this Chapter, we shall limit ourselves to the transformations of $\mathcal{P}_+^\uparrow = \mathcal{S} \circledS \mathcal{L}_+^\uparrow$, which form the proper orthochronous inhomogeneous Lorentz group. We briefly mention the topological properties of $\mathcal{P}_+^\uparrow$, which are easily derived from those of its subgroups:

a) The group $\mathcal{P}_+^\uparrow$ is *non-compact*; in fact, it contains the subgroups $\mathcal{L}_+^\uparrow$ and $\mathcal{S}$ which are both non-compact (the translation group $\mathcal{S}$ is clearly non-compact, since $\mathcal{R}^4$ is not compact);

b) The group $\mathcal{P}_+^\uparrow$ is *doubly-connected*: it has the same connectedness of its subgroup $\mathcal{L}_+^\uparrow$ (specifically of $SO(3)$). Its *universal covering group* is defined by the transformations (a, A), where A is an element of $SL(2, C)$;

c) The group $\mathcal{P}_+^\uparrow$ is neither simple nor semi-simple, in fact, the translation group $\mathcal{S}$ is an invariant subgroup of $\mathcal{P}_+^\uparrow$.

Next we consider the Lie algebra of the group $\mathcal{P}_+^\uparrow$; its basic elements are the infinitesimal generators of $\mathcal{L}_+^\uparrow$ and $\mathcal{S}$.

In general, an infinitesimal translation is given by

$$(\delta a, I) = I - i\delta a_\mu P^\mu , \tag{4.9}$$

introducing four operators P^μ, which are the infinitesimal generators of the translations. A finite translation is given by exponentiation, as

$$(a, I) = e^{-ia_\mu P^\mu} . \tag{4.10}$$

The commutation relations of the infinitesimal generators are easily obtained giving a specific representation of the Lie algebra of $\mathcal{P}_+^\uparrow$. For this purpose it is convenient to write (a, Λ) as a 5×5 matrix

$$\begin{pmatrix} \Lambda & a \\ 0 & 1 \end{pmatrix} , \tag{4.11}$$

where the four vector a is written as a column matrix. Eq. (4.2) is then obtained by matrix product, and Eq. (4.1) can be considered as the transformation of a vector

$$y = \begin{pmatrix} x \\ 1 \end{pmatrix} \tag{4.12}$$

in a five-dimensional space. In fact, it can be re-written as

$$\begin{pmatrix} x' \\ 1 \end{pmatrix} = \begin{pmatrix} \Lambda & a \\ 0 & 1 \end{pmatrix} \begin{pmatrix} x \\ 1 \end{pmatrix} = \begin{pmatrix} \Lambda x + a \\ 1 \end{pmatrix} . \tag{4.13}$$

Expressing in similar way the infinitesimal transformation (4.9), one can obtain explicitly

$$P_0 = \begin{pmatrix} 0 & 0 & 0 & 0 & i \\ 0 & 0 & 0 & 0 & 0 \\ 0 & 0 & 0 & 0 & 0 \\ 0 & 0 & 0 & 0 & 0 \\ 0 & 0 & 0 & 0 & 0 \end{pmatrix} , \qquad P_1 = \begin{pmatrix} 0 & 0 & 0 & 0 & 0 \\ 0 & 0 & 0 & 0 & i \\ 0 & 0 & 0 & 0 & 0 \\ 0 & 0 & 0 & 0 & 0 \\ 0 & 0 & 0 & 0 & 0 \end{pmatrix} , \tag{4.14}$$

and similarly for P_2, P_3.

In the same representation, the generators J_i and K_i are replaced by 5×5 matrices, obtained from (3.50) and (3.54) by adding a fifth row and a fifth column of zeros.

One can check that the following commutation relations hold:

$$\begin{aligned}
[P_\mu, P_\nu] &= 0 , \\
[J_i , P_0] &= 0 , \\
[J_i , P_i] &= 0 , \\
[J_i , P_j] &= i\epsilon_{ijk}P_k , \\
[K_i, P_0] &= -iP_i , \\
[K_i, P_i] &= -iP_0 , \\
[K_i, P_j] &= 0 .
\end{aligned} \tag{4.15}$$

In terms of the antisymmetric tensor $M_{\mu\nu}$ given by (3.59), the above commutators become:

$$[P_\mu, P_\nu] \;=\; 0\,, \tag{4.16}$$

$$[M_{\mu\nu}, P_\rho] = -i(g_{\mu\rho}P_\nu - g_{\nu\rho}P_\mu)\,, \tag{4.17}$$

and, together with the commutators

$$[M_{\lambda\rho}, M_{\mu\nu}] \;=\; -i(g_{\lambda\mu}M_{\rho\nu} + g_{\rho\nu}M_{\lambda\mu} - g_{\lambda\nu}M_{\rho\mu} - g_{\rho\mu}M_{\lambda\nu})\,, \tag{3.60}$$

they define the Lie algebra of $\mathcal{P}_+^\uparrow$.

One can check that the operator

$$P^2 = P_\mu P^\mu \tag{4.18}$$

commutes with all the generators P_μ, $M_{\mu\nu}$ of the Lie algebra; hence P^2 is an invariant operator under the Poincaré transformations.

Let us introduce the operator

$$W_\mu = \epsilon_{\mu\nu\sigma\tau} M^{\nu\sigma} P^\tau\,; \tag{4.19}$$

it follows immediately that

$$W_\mu P^\mu = 0\,. \tag{4.20}$$

Moreover, making use of Eqs. (3.60), (4.17), one can obtain the following commutators

$$[P_\mu, W_\nu] \;=\; 0\,,$$
$$[M_{\mu\nu}, W_\sigma] = -i(g_{\nu\sigma}W_\mu - g_{\mu\sigma}W_\nu)\,, \tag{4.21}$$
$$[W_\mu, W_\nu] \;=\; i\epsilon_{\mu\nu\sigma\tau} W^\sigma P^\tau\,.$$

Finally, one can prove that the operator

$$W^2 = W_\mu W^\mu \tag{4.22}$$

commutes with all the P_μ and $M_{\mu\nu}$ and hence it is an invariant operator.

The eigenvalues of the invariant operators P^2 and W^2 are used to label the IR's of the group; for some classes of IR's one can have additional invariants.

4.2 Unitary representations of the proper orthochronous Poincaré group

In the previous Section, we have exhibited a 5-dimensional representation of the proper orthochronous Poincaré group $\mathcal{P}_+^\uparrow$, which is not unitary, since both the generators P_μ and K_i are expressed by non-hermitian matrices. In fact, as pointed out in Subsection 1.2.1, since the group $\mathcal{P}$ and its subgroup $\mathcal{P}_+^\uparrow$ are not compact, no finite-dimensional unitary IR's exists.

We recall from quantum mechanics[1] the well known fact that the infinitesimal generators of translations P_μ can be identified with the *energy-momentum operators*. Moreover, the infinitesimal generators $M_{\mu\nu}$ can be identified with the components of the *angular momentum tensor*. For physical applications, we are interested in those IR's in which the operators P_μ and $M_{\mu\nu}$ are *hermitian*, since they correspond to dynamical variables, i.e. in the *unitary*, and hence infinite-dimensional, IR's of the Poincaré group $\mathcal{P}$. Here we are mainly concerned with the IR's of the restricted group $\mathcal{P}_+^\uparrow$.

Before going to the IR's of $\mathcal{P}_+^\uparrow$, we want to review briefly some general properties of the invariance principles for a quantum mechanical system. Each state of the system is represented by a vector $|\Phi\rangle$ in a Hilbert space. The state vectors are normalized to unity, to allow the probability interpretation; two vectors that differ only by a phase represent the same physical state. In fact, the results of experiments are expressed in terms of transition probabilities

$$|\langle\Psi|\Phi\rangle|^2 \tag{4.23}$$

between the states $|\Phi\rangle$ and $|\Psi\rangle$, which depend only on the absolute value of the scalar product between the two states. This shows that each vector in the set $e^{i\alpha}|\Phi\rangle$, where α is a real number, corresponds to the same physical state. The collection of all the state vectors of the form $e^{i\alpha}|\Phi\rangle$ is called a *ray*, which we denote simply by Φ.

An *invariance principle*, or *symmetry operation*, of a physical system is a one-to-one correspondence between two rays Φ and Φ' (and similarly between Ψ and Ψ'), representing physically realizable states, such that the transition probabilities remain unchanged:

$$|\langle\Psi'|\Phi'\rangle|^2 = |\langle\Psi|\Phi\rangle|^2 . \tag{4.24}$$

An important theorem due to Wigner[2] states that, if $\Phi \to \Phi'$ is a symmetry operation of a physical theory, there exists a *linear unitary* or *antilinear unitary*[3] (antiunitary) operator U, determined up to a phase, such that:

$$\Phi' = U\Phi . \tag{4.25}$$

---

[1] See e.g. S. Weinberg, *The Quantum Theory of Fields, Vol. I, Foundations*, Cambridge University Press (1995).

[2] See e.g. F.R. Halpern, *Special Relativity and Quantum Mechanics*, Prentice-Hall (1968).

[3] An antilinear unitary operator U is defined by

$$\langle U\Psi|U\Phi\rangle = \langle\Psi|\Phi\rangle^*,$$
$$U|\alpha\Psi + \beta\Phi\rangle = \alpha^*U|\Psi\rangle + \beta^*U|\Phi\rangle .$$

In particular, if a physical system is invariant under the transformations of the Poincaré group, to each element (a, Λ) of $\mathcal{P}$ there corresponds a linear operator $U(a, \Lambda)$ which transforms the state Φ into

$$\Phi' = U(a, \Lambda)\, \Phi \,, \tag{4.26}$$

leaving Eq. (4.24) invariant. The operator $U(a, \Lambda)$ is determined up to a phase; however, for those Poincaré transformations which are connected with the identity, such as the transformations of the subgroup $\mathcal{P}_+^\uparrow$, it can be shown that it is possible to choose a phase convention such that the product of any two elements of $\mathcal{P}_+^\uparrow$ is always given by[4]:

$$U(a_1, \Lambda_1)U(a_2, \Lambda_2) = \pm U(a_3, \Lambda_3) \,. \tag{4.27}$$

The sign ambiguity is related to the fact that the group $\mathcal{P}$ is doubly connected, and it can have double-valued representations.

We stress the fact that the operators $U(a, \Lambda)$ of $\mathcal{P}_+^\uparrow$ are *unitary*. In fact, for any element (a, Λ) of $\mathcal{P}_+^\uparrow$ one can find an element (a', Λ') such that

$$(a, \Lambda) = (a', \Lambda')(a', \Lambda') \,. \tag{4.28}$$

Then, from Eq. (4.27):

$$U(a, \Lambda) = \pm U(a', \Lambda')U(a', \Lambda') \,. \tag{4.29}$$

According to Wigner's theorem, $U(a', \Lambda')$ could be either unitary or antiunitary, but in both cases, since the square of an antiunitary operator is unitary, $U(a, \Lambda)$ can only be unitary. Since $U(a, \Lambda)$ is an arbitrary element of $\mathcal{P}_+^\uparrow$, it follows that all the linear operator $U(a, \Lambda)$ are unitary.

We have thus obtained a representation of the group $\mathcal{P}_+^\uparrow$ in the Hilbert space of the state vectors $|\Phi>$ in terms of the unitary operators $U(a, \Lambda)$.

The unitary representation realized by the operators $U(a, \Lambda)$ is, of course, *infinite-dimensional*, since the Hilbert space of the state vectors $|\Phi>$ is infinite-dimensional, but it is, in general, reducible. To obtain the *unitary irreducible representations*, one has to find the invariant subspaces, i.e. invariant under the transformations (a, Λ).

To make this point clear, let us consider a given set of vectors $|\Phi>$ in the Hilbert space. Since the P_μ commute among themselves (see eq. (4.16)), it would be convenient to start from the vectors $|p, \zeta>$, labelled by the eigenvalues p_μ of P_μ and by other, for the moment unspecified, quantum number ζ:

$$P_\mu|p, \zeta> = p_\mu|p, \zeta> \,. \tag{4.30}$$

Now we are facing the following problem: the vectors $|p, \zeta>$ corresponding to particular points p_μ in the momentum space are not normalizable, so that they

---

[4] F.R. Halpern, quoted ref.

do not belong to the Hilbert space. One can obtain normalizable vectors by convenient linear superposition of $|p, \zeta >$[5]. However, for the sake of simplicity, we shall adopt the normalization in terms of the Dirac delta function, widely used in the physics literature[6] and, with the inclusion of the convenient factor $2p_0$, define:

$$<p', \zeta'|p, \zeta> = 2p_0 \, \delta^3(\mathbf{p}' - \mathbf{p})\delta_{\zeta'\zeta} \, . \tag{4.31}$$

For a pure translation, from Eq. (4.10), we get

$$U(a, I) |p, \zeta> = e^{-ia_\mu p^\mu}|p, \zeta> \, , \tag{4.32}$$

while, for a homogeneous Lorentz transformation $(0, \Lambda)$, we obtain

$$U(0, \Lambda) |p, \zeta> = \sum_{\zeta'} Q_{\zeta'\zeta}|p', \zeta'> \, , \tag{4.33}$$

where Q is a unitary matrix and

$$p'^\mu = \Lambda^\mu_{\ \nu}p^\nu \, . \tag{4.34}$$

Eq. (4.34) is obtained from eq. (4.30) and from the relation

$$U^{-1}(0, \Lambda)P^\mu U(0, \Lambda) = \Lambda^\mu_{\ \nu}P^\nu \, , \tag{4.35}$$

which can be derived from Eqs. (4.8) and (4.9). This shows that the four operators P^μ transform as the components of a four-vector.

Once more we see that $P^2 = P_\mu P^\mu$ is invariant; in fact, we know that P^2 is one of the Casimir operators of the Poincaré group. Its eigenvalues p^2, when $p^2 = m^2 > 0$, correspond to the squared total energy in the c. m. system. If the physical system is a particle, m is the *mass of the particle*.

The subspace spanned by the vectors $|p, \zeta >$ with a given value of p^2 is then *invariant* under the transformations (a, Λ), so that the representation $U(0, \Lambda)$ given by (4.33) is, in general, reducible and can be decomposed into an (infinite) direct sum of representations corresponding to different eigenvalues m^2.

To reduce further the representation, we need to consider also the operator W_μ (Eq. (4.19)), whose square W^2 is the other Casimir operator of the Poincaré group. Since the components W_μ commute with P_μ but not among themselves (Eq. (4.21)), we can diagonalize W^2 and *one* component, say W_3. Among the quantum numbers ζ labelling the eigenstates $|p, \zeta>$ one can then single out the eigenvalues w_3 of W_3. For the sake of simplicity, we shall neglect in the following the other unspecified quantum numbers which should be added to make the set of observables complete.

---

[5] See, e.g., A. Messiah, *Quantum Mechanics*, North Holland (1962).
[6] See e.g. M.E. Peskin, D.V. Schroeder, *An Introduction to Quantum Field Theory*, Addison-Wesley (1995); S. Weinberg, *The Quantum Theory of Fields, Vol. I, Foundations*, Cambridge University Press (1995)

Let us first consider those representations for which $p^2 = m^2 > 0$. In that case $p_0/|p_0|$, i.e. the sign of the energy, is also an invariant of the group $\mathcal{P}_+^\uparrow$. In the four-momentum space, the eigenstates of $p^2 = m^2 > 0$ with $p_0 > 0$, which correspond to physical states, are represented by the points in the upper branch of the hyperboloid shown in Fig. 4.1. Under a transformation of $\mathcal{P}_+^\uparrow$ the representative point moves on the *same* branch of the hyperboloid.

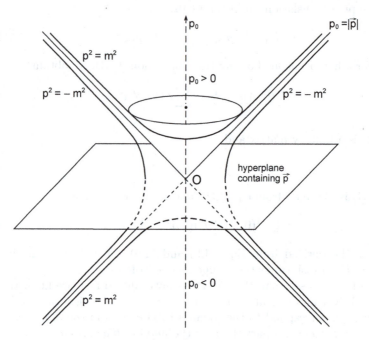

Fig. 4.1. *Hyperboloid $p^2 = m^2$ in the four-momentum space.*

An irreducible representation is characterized by the eigenvalues of P^2, W^2 and the sign of $p_0/|p_0|$; an IR with $p_0 > 0$ is then spanned by the vectors $|p, w_3 >$ with $p^2 = m^2 = $ const., and $p_0 = +\sqrt{\mathbf{p}^2 + m^2}$.

The physical meaning of W^μ is made clear by considering explicitly its components. Using the definitions (3.59), one gets from (4.19)

$$W_0 = \mathbf{P} \cdot \mathbf{J} , \tag{4.36}$$

$$\mathbf{W} = P_0 \mathbf{J} - \mathbf{P} \times \mathbf{K} . \tag{4.37}$$

It is convenient to go to the rest frame, in which $\mathbf{p} = 0$. In this frame Eqs. (4.36), (4.37) are equivalent to

$$W^\mu = m(0, J^1, J^2, J^3) , \tag{4.38}$$

i.e. W^μ reduces essentially to the components of the total angular momentum $\mathbf{J}$, which is the *spin* in the case of a particle. A rest frame can then be specified by $|\tilde{p}, j_3 >$, where $\tilde{p} = (m, \mathbf{0})$. According to Eq. (4.38), one gets

$$
\begin{aligned}
W_3 |\tilde{p}, j_3> &= m j_3 |\tilde{p}, j_3> , \\
(W)^2 |\tilde{p}, j_3> &= -m^2 j(j+1) |\tilde{p}, j_3> .
\end{aligned}
\tag{4.39}
$$

If $p^2 = 0$, one cannot apply the above arguments, since one cannot go to the rest frame. Since we are interested in the physical case of a massless particle, we take also $w^2 = 0$ ($w^2 \neq 0$ would represent continuous spin values, as we will see in Section 5.3). However, these two null quantum numbers are not sufficient to characterize an IR. Taking into account also Eq. (4.20) we have: $p^2 = 0$, $w^2 = 0$ and $w_\mu p^\mu = 0$. These conditions are satisfied only if the two four-vectors w_μ and p_μ are parallel: $w_\mu = \lambda p_\mu$. In terms of operators we can write[7]:

$$
W_\mu = \lambda P_\mu ,
\tag{4.40}
$$

where λ is an *invariant*. The physical interpretation of λ becomes evident making use of Eq. (4.36) and of the relation $\mathbf{p}^2 = p_0^2$:

$$
\lambda = \frac{W_0}{P_0} = \frac{\mathbf{P} \cdot \mathbf{J}}{P_0} = \frac{\mathbf{P} \cdot \mathbf{J}}{|\mathbf{P}|} .
\tag{4.41}
$$

This shows that λ represents the component of the spin along the direction of motion, which is usually called *helicity*. In the case $p^2 = 0$, we label the state vector with the eigenvalues of λ; i.e. by $|p, \lambda>$.

The above examples show that the operator W_μ describes the polarization states of both massive and massless particles; it represent the covariant generalization of the spin, and it is called, in fact, *covariant spin*.

We have considered only two kinds of IR's of $\mathcal{P}_+^\uparrow$ of physical interest. To summarize and complete the preceding discussion, we list in the following all the IR's of $\mathcal{P}_+^\uparrow$[8]:

a) $p^2 = m^2 > 0$.
   For each pair of values of p^2 and w^2, there are two IR's, one for each value of $p_0/|p_0|$. The vector basis of an IR is given by $|p, w_3>$; in the rest frame it reduces to the $(2j+1)$-dimensional basis of the IR of the rotation group, which corresponds to the $(2j + 1)$ states of polarization.

b) $p^2 = 0$, $p_\mu \neq 0$, $w^2 = 0$.
   Each IR is characterized by the sign of p_0 and by the value (integer or half-integer) of the helicity λ. For a given momentum p_μ and given $\lambda \neq 0$, there exist two independent states (they belong to two independent IR's), which correspond to the two different helicities $\pm \lambda$.

---

[7] Usually we denote operators by capital letters (P_μ, W_μ...) and the corresponding eigenvalues by small letters (p_μ, w_μ...).

[8] I.M. Gel'fand, R.A. Minlos, Z.Ya. Shapiro, *Representations of the Rotations and Lorentz Groups and Their Applications*, Pergamon Press (1963); M.A. Naimark, *Linear Representations of the Lorentz Group*, Pergamon Press (1964).

c) $p^2 = 0$, $p_\mu \neq 0$, $w^2 < 0$.

The IR's are infinite-dimensional also in the spin variable and would correspond to particles with continuous spin.

d) $p^2 = 0$, $p_\mu = 0$, $w^2 < 0$.

The unitary IR's coincide with those of the homogeneous Lorentz group, and they will not be considered further (they provide a powerful tool for the study of the analytic properties of the forward scattering amplitude in the angular momentum variable).

e) $p^2 < 0$.

These IR's correspond to space-like vectors p_μ, so they cannot be used in the description of physical states (they can be useful, however, in the study of the analytic properties of the scattering amplitude).

Problems

4.1. Show that the Poincaré transformations (a, Λ) form a group and give the explicit proof that the translation group S is an invariant subgroup.

4.2. Prove that P^2 and W^2 are invariant operators of the group $\mathcal{P}_+^\uparrow$.

4.3. Derive the transformation properties of the generators P_μ under $\mathcal{L}_+^\uparrow$ and the commutation relations (4.17), making use of the general form

$$U(a, \Lambda) = e^{-ia_\mu P^\mu} e^{-\frac{1}{2}i\omega_{\mu\nu} M^{\mu\nu}}$$

of the unitary IR's of $\mathcal{P}_+^\uparrow$.

4.4. Derive the transformation properties of the generators $M^{\mu\nu}$ under $\mathcal{L}_+^\uparrow$ and the commutation relations (3.60) starting from the general form

$$U(0, \Lambda) = e^{-\frac{1}{2}i\omega_{\mu\nu} M^{\mu\nu}}$$

of the unitary IR's of $\mathcal{L}_+^\uparrow$.

4.5. Making use of the Lorentz transformation which brings the state $|p, \zeta>$ at rest, show that the commutators $[W_\mu, W_\nu] = i\epsilon_{\mu\nu\sigma\tau} W^\sigma P^\tau$ reduce to $[J_i, J_j] = i\epsilon_{ijk} J_k$.

4.6. Show that the Lorentz transformations of the form

$$U(0, \Lambda) = e^{-in_\mu W^\mu} ,$$

where n is an arbitrary four-vector, leave the eigenvalues p_μ of P_μ invariant.

One particle and two particle states

This Chapter contains a detailed analysis of one-particle physical states, described in terms of the unitary IR's of the restricted Poincaré group $\mathcal{P}_+^\uparrow$ corresponding to $p^2 = m^2 > 0$ and $p^2 = 0$, $p^\mu \neq 0$. It will be shown that these IR's can be obtained from the IR's of a subgroup of $\mathcal{P}_+^\uparrow$, which is the three-dimensional rotation group $SO(3)$ in the case $p^2 = m^2 > 0$, and the abelian group $SO(2)$ of rotations about an axis in the case $p^2 = 0$, $p^\mu \neq 0$. In other words, once we know how the states of a particle transform under these rotations, we know also their transformation properties under $\mathcal{P}_+^\uparrow$. This important point will be briefly discussed in more general terms, introducing the concept of *little group*. Finally, it will be shown how the two-particle states can be constructed in terms of one-particle states as IR's of $\mathcal{P}_+^\uparrow$.

5.1 The little group

As we have seen in the previous Chapter, one needs, in general, other quantum numbers, besides the eigenvalues p_μ of P_μ, to fully specify the eigenstates $|p, \zeta\rangle$ of a physical system. The set of eigenvectors $|p, \zeta\rangle$ of P_μ relative to the *same* eigenvalue p form a Hilbert space H_p (improper in the sense specified in connection with Eq. (4.30)), which is a subspace of the total space H_{p^2}.

We can show that the spaces H_p corresponding to the four-vector p with the *same value* of p^2 are all *isomorphic* to each other (i.e. they are related by a one-to-one mapping).

Let us consider a fixed four-vector $\bar{p}$ and a transformation of $\mathcal{L}_+^\uparrow$, denoted in the following by $L_{p\bar{p}}$, which brings $\bar{p}$ into p:

$$L_{p\bar{p}}\, \bar{p} = p \,. \tag{5.1}$$

The inverse transformation is performed by $L_{p\bar{p}}^{-1}$, which obviously belongs to $\mathcal{L}_+^\uparrow$. A unitary representation of $\mathcal{P}_+^\uparrow$ is defined, in general, by[1]

---

[1] For the sake of simplicity we shall often write for short $U(\Lambda)$ instead of $U(0, \Lambda)$.

G. Costa and G. Fogli, *Symmetries and Group Theory in Particle Physics*,
Lecture Notes in Physics 823, DOI: 10.1007/978-3-642-15482-9_5,
© Springer-Verlag Berlin Heidelberg 2012

$$U(\Lambda)|p,\zeta> = \sum_{\zeta'} Q_{\zeta'\zeta}|p' = \Lambda p,\zeta'> , \tag{5.2}$$

while the transformation (5.1) is defined in H by

$$U(L_{p\bar{p}})|\bar{p},\zeta> = |p,\zeta> . \tag{5.3}$$

There is some arbitrariness in Eq. (5.1); in fact, the transformation $L_{p\bar{p}}$ is not uniquely defined, since the same result is obtained by multiplying $L_{p\bar{p}}$ on the right by a rotation about $\bar{p}$. We can fix $L_{p\bar{p}}$ in such a way that $U(L_{p\bar{p}})$ leaves the quantum number ζ unchanged.

The specific transformation (5.3) is usually called *Wigner boost*. Since also $U(L_{p\bar{p}})^{-1} = U(L_{p\bar{p}}^{-1})$ is defined, we see that the mapping between $|\bar{p},\zeta>$ and $|p,\zeta>$ is one-to-one.

The structure of the Hilbert space H of the state vectors allows one to obtain in a simple fashion the transformation properties of $|p,\zeta>$ under $\mathcal{P}_+^\uparrow$. From Eqs. (5.2), (5.3) we get

$$U(\Lambda)|p,\zeta> = U(\Lambda)U(L_{p\bar{p}})|\bar{p},\zeta> = \sum_{\zeta'} U(L_{p'\bar{p}})Q_{\zeta'\zeta}|\bar{p},\zeta'> , \tag{5.4}$$

so that

$$U(L_{p'\bar{p}}^{-1})U(\Lambda)U(L_{p\bar{p}})|\bar{p},\zeta> = U(L_{p'\bar{p}}^{-1}\Lambda L_{p\bar{p}})|\bar{p},\zeta> = \sum_{\zeta'} Q_{\zeta'\zeta}|\bar{p},\zeta'> . \tag{5.5}$$

We see that the matrix Q corresponds to a unitary representation $D(R)$ of the operation of $\mathcal{L}_+^\uparrow$:

$$\mathcal{R} = L_{p'\bar{p}}^{-1}\Lambda L_{p\bar{p}} . \tag{5.6}$$

One can check that this operation leaves the four-vector $\bar{p}$ invariant; this is shown graphycally in Fig. 5.1 in the case of a time-like vector $\bar{p}$.

The set of transformations $\mathcal{R}$ which leave the four-vector $\bar{p}$ invariant form a subgroup of $\mathcal{L}_+^\uparrow$, which is called *little group*; the operation $\mathcal{R}$ is usually referred to as *Wigner rotation*.

Eq.(5.5) can be re-written as

$$U(\mathcal{R})|\bar{p},\zeta> = \sum_{\zeta'} D_{\zeta'\zeta}(\mathcal{R})|\bar{p},\zeta'> , \tag{5.7}$$

and it provides a representation $D(\mathcal{R})$ of the little group in the subspace $H_{\bar{p}}$, which is called *little Hilbert space*.

Eq. (5.4) becomes

$$U(\Lambda)|p,\zeta> = \sum_{\zeta'} D_{\zeta'\zeta}(\mathcal{R})U(L_{p'\bar{p}})|\bar{p},\zeta'> = \sum_{\zeta'} D_{\zeta'\zeta}(\mathcal{R})|p',\zeta'> . \tag{5.8}$$

Including a translation, from Eqs. (4.6) and (4.32) one gets finally

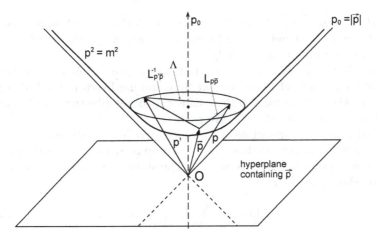

Fig. 5.1. *A Wigner rotation in the four-momentum space. The three points $\bar{p}$, p and p' lie on the hyperboloid $p^2 = m^2$.*

$$U(a, \Lambda)|p, \zeta> = e^{-ip'\cdot a} \sum_{\zeta'} D_{\zeta'\zeta}(\mathcal{R})|p', \zeta'> , \qquad (5.9)$$

which provides a unitary IR of $\mathcal{P}_+^\uparrow$ expressed explicitly in terms of the unitary representation $D(\mathcal{R})$ of the little group.

Since the little Hilbert spaces H_p corresponding to the four-vector p with the same value of p^2 are all isomorphic, the IR's $D(\mathcal{R})$ corresponding to any p obtained from $\bar{p}$ by a Wigner boost (5.1) are all *equivalent*. It is easy to find that the same little group corresponds to each class of four-vectors p, i.e. time-like, space-like, light-like and null, so that for each class one can make a convenient choice of a "standard vector" $\bar{p}$. In the following, we shall examine the cases of time-like and light-like $\bar{p}$. In the case of a null vector $\bar{p} = 0$, the little group coincides with $\mathcal{L}_+^\uparrow$ itself.

5.2 States of a massive particle

As shown in the previous Chapter, the states of a free massive particle are classified according to the IR's of the Poincaré group corresponding to $p^2 = m^2 > 0$. In the following, we shall consider only positive energy states, which will be denoted by $|m, s; \mathbf{p}, \sigma>$, where s is the spin of the particle and σ its projection along the x^3-axis, defined in the rest frame. They are normalized according to (compare with Eq. (4.31))

$$<m, s; \mathbf{p}', \sigma'|m, s; \mathbf{p}, \sigma> = 2p^0 \delta(\mathbf{p} - \mathbf{p}')\delta_{\sigma\sigma'} , \qquad (5.10)$$

with $p^0 = +\sqrt{\mathbf{p}^2 + m^2}$. In the following, we shall drop, whenever possible without causing ambiguities, the labels (m, σ) which specify the IR, and denote a state simply by $|\mathbf{p}, \sigma>$, or $|p, \sigma>$.

Since, in the present case, p is a time-like vector, we can go to the rest frame of the particle, choosing $\tilde{p} = (m, \mathbf{0})$ as standard vector. The little group relative to the four-vector $\tilde{p}$ is then sufficient to describe the transformation properties of the states of a massive particle.

Clearly, the transformations of $\mathcal{L}_+^\uparrow$ which leave $\tilde{p} = (m, \mathbf{0})$ invariant are the 3-dimensional rotations, so that the little group is $SO(3)$. (The little group of the covering group $SL(2, C)$ is $SU(2)$). The state $|\tilde{p}, \sigma>$ transforms under a pure rotation R according to

$$U(R)|\tilde{p}, \sigma> = \sum_{\sigma'} D^{(s)}_{\sigma'\sigma}(R)|\tilde{p}, \sigma'> , \qquad (5.11)$$

where $D^{(s)}(R)$ is the $(2s + 1)$-dimensional IR of $SO(3)$.

The Wigner boost $L_{p\tilde{p}}$ is given by the pure Lorentz transformation L_p along $\mathbf{p}$, called simply *boost*, which brings $\tilde{p}$ into p. In fact any other state of the particle, belonging to an IR of $\mathcal{P}_+^\uparrow$, can be obtained from a rest state by a boost L_p

$$|p, \sigma> = U(L_p)|\tilde{p}, \sigma> \qquad (5.12)$$

since, as specified in connection with Eq. (5.3), L_p can be fixed in such a way to leave the quantum numbers σ invariant.

The generic element of the little group follows then from Eq. (5.6), which in this case becomes

$$\mathcal{R}_{p'p} = L_{p'}^{-1} \Lambda L_p , \qquad (5.13)$$

where $p' = \Lambda p$. The transformation properties of a generic states $|p, \sigma>$ under $\mathcal{P}_+^\uparrow$ are obtained by applying Eq. (5.9) to the present case:

$$U(a, \Lambda)|p, \sigma> = e^{-ip' \cdot a} \sum_{\sigma'} D^{(s)}_{\sigma'\sigma}(\mathcal{R})|p', \sigma'> . \qquad (5.14)$$

The above equation provides a unitary IR of $\mathcal{P}_+^\uparrow$.

In conclusion, it appears quite natural to define a particle as a *system with definite mass and spin*, and identify its states with specific unitary IR's of the Poincaré group.

In the previous Chapter, we have introduced the *helicity* for massless particles. The helicity operator

$$\frac{\mathbf{P} \cdot \mathbf{J}}{|\mathbf{P}|} . \qquad (5.15)$$

can also be used in the case of massive particles: it represents the component of the spin along the direction of motion. Also the states of a massive particle can then be labelled by the helicity eigenvalues. While Lorentz transformations mix different helicity states, rotations leave helicity invariant, as one can check by evaluating the commutators of the operator (5.15) with the infinitesimal generators of the Lorentz transformations. This makes the use of helicity states very convenient. In the following, we shall examine this alternative description for massive one-particle states.

Let us consider a helicity state $|\breve{p}, \lambda>$ with $\breve{p} = (p^0, 0, 0, p^3 = |\mathbf{p}|)$ of definite momentum $\mathbf{p}$ along the x^3-axis; this state can be related to a rest state $|\tilde{p}, \sigma>$, where the spin component σ along x^3 coincides with λ, through a boost $L_3(p)$ along the x^3-direction

$$|\breve{p}, \lambda> = U(L_3(p))|\tilde{p}, \lambda> . \tag{5.16}$$

In general, a helicity state of a particle moving with momentum $\mathbf{p}$ along an arbitrary direction (θ, ϕ) with respect to the x^3-axis can be defined by

$$|p, \lambda> = U(R_\mathbf{p})|\breve{p}, \lambda> = U(R_\mathbf{p})U(L_3(p))|\tilde{p}, \lambda> . \tag{5.17}$$

where $R_\mathbf{p}$ represents the rotation which brings $\breve{p}$ into $p = (p^0, \mathbf{p})$.

The above relation allows one to take $\tilde{p}$ as standard vector, so that the little group is, also in this case, the rotation group $SO(3)$. From the same relation it appears that a generic Wigner boost is given now by

$$L_{p\tilde{p}} = R_\mathbf{p}L_3(p) . \tag{5.18}$$

The elements of the little group are obtained from Eq. (5.6) in the form

$$\mathcal{R}^{(\lambda)} = L_3^{-1}(p')R_{\mathbf{p}'}^{-1}\Lambda R_\mathbf{p}L_3(p) , \tag{5.19}$$

and the transformation properties of the state $|p, \lambda>$ under $\mathcal{P}_+^\uparrow$ are then given by

$$U(a, \Lambda)|p, \lambda> = e^{-ip' \cdot a} \sum_{\lambda'} D_{\lambda'\lambda}^{(s)}(\mathcal{R}^{(\lambda)})|p', \lambda'> . \tag{5.20}$$

It is useful to express the helicity states $|p, \lambda>$ in terms of the spin component states $|p, \sigma>$. This can be easily obtained from the relation

$$R_\mathbf{p}L_3(p) = L_p R_\mathbf{p} , \tag{5.21}$$

which is the analogue of Eq. (3.28). In fact, applying it to a rest state $|\tilde{p}, \lambda>$, and making use of Eqs. (5.17), (5.11) and (5.12), one gets

$$|p, \lambda> = \sum_\sigma D_{\sigma\lambda}^{(s)}(R_\mathbf{p})|p, \sigma> . \tag{5.22}$$

Moreover, Eq. (5.21) allows one to re-write Eq. (5.19) in the form

$$\mathcal{R}^{(\lambda)} = R_{\mathbf{p}'}^{-1}L_{p'}^{-1}\Lambda L_p R_\mathbf{p} = R_{\mathbf{p}'}^{-1}\mathcal{R}^{(\sigma)} R_\mathbf{p} , \tag{5.23}$$

where $\mathcal{R}^{(\sigma)} \equiv \mathcal{R}_{p'p}$ is the Wigner rotation given by (5.13).

The above results can be summarized as follows:

- The states of a particle with mass m and spin s are completely characterized by the momentum $\mathbf{p}$ and either the spin component σ along an abitrary fixed direction or the helicity λ;
- the rotation group is sufficient to describe the behaviour of a particle state under a Poincaré transformation in both formalisms.

5.3 States of a massless particle

The description in terms of helicity states becomes essential for massles particles. In fact, we have seen that the zero-mass states are classified according to the IR's of the Poincaré group characterized by $\lambda = \mathbf{P} \cdot \mathbf{J}/|\mathbf{P}|$.

It is instructive also in this case to consider the relevant little group. Since $\bar{p}$ is now a light-like vector, we can choose it in the form $\breve{p} = (p^0, o, o, p^3)$ with $|p^3| = p^0$. It is then clear that spatial rotations around the x^3-axis leave $\breve{p}$ invariant; however, they are not the most general transformations of the little group.

In the present case, it is easier to look for the Lie algebra of the little group. In the basis $|\breve{p}, \lambda>$ of the eigenstates of a massless particle of momentum $\breve{p}$ and helicity λ, one has $W_0 = W_3$ (see Eq. (4.20)) and, from Eq. (4.21), one gets

$$
\begin{aligned}
[W_1, W_2] &= 0\,, \\
[W_0/P_0, W_1] &= iW_2\,, \\
[W_0/P_0, W_2] &= -iW_1\,.
\end{aligned}
\tag{5.24}
$$

The above commutators define the Lie algebra of the two-dimensional *Euclidean group*, i.e. the group of rotations and translations in a plane. This is easily recognized by extracting from (4.15) the commutators of P_1, P_2 and J_3, which are, respectively, the infinitesimal generators of translations and rotations in the x^1-x^2 plane:

$$
\begin{aligned}
[P_1, P_2] &= 0\,, \\
[J_3, P_1] &= iP_2\,, \\
[J_3, P_2] &= -iP_1\,.
\end{aligned}
\tag{5.25}
$$

By the formal correspondence

$$
W_1 \to P_1\,, \qquad W_2 \to P_2\,, \qquad W_0/P_0 \to J_3\,,
\tag{5.26}
$$

the Lie algebras defined by (5.24) and (5.25) coincide.

Since the four-vector w_μ and p_μ are orthogonal (Eq. (4.20)), and p_μ is light-like, w_μ can be either light-like or space-like. In the particular basis chosen, we get

$$
w^2 = -w_1^2 - w_2^2\,.
\tag{5.27}
$$

We disregard the case $w^2 < 0$ since it corresponds to a continuous spectrum of representations, each state being characterized by the eigenvalues of W_1 and W_2. These representations seem to have no physical meaning, since they would describe massless particles with continuous spin. We are then left with the case $w^2 = 0$, in which w_μ and p_μ are *collinear*. From Eqs. (4.40), (4.41) we see that the correspondence (5.26) holds effectively. However, since $W_1 = W_2 = 0$, only $W_0/P_0 = \lambda$ has a *non-trivial* representation in the given basis. Since only the infinitesimal generator W_0/P_0, equivalent to J_3, survives, the little group reduces to the rotation group in a plane, i.e. the abelian group $SO(2)$ homomorphic to $U(1)$.

All the unitary IR's are one-dimensional and are given by $e^{-i\lambda\alpha}$, where λ is a fixed parameter corresponding to an eigenvalue of helicity and α is the angle of rotation. Since we require the IR's of $\mathcal{P}_+^\uparrow$ to be at most double-valued, the parameter λ can have *only integer or half-integer* values. Both positive and negative values of λ are allowed, and the values $+\lambda$ and $-\lambda$ correspond to two inequivalent representations.

One can derive explicitly the transformation properties of the helicity states of a massless particle under $\mathcal{P}_+^\uparrow$, following a procedure analogous to that used in the massive case. Eq. (5.16) is now replaced by

$$|\breve{p}, \lambda> = U(L_{\breve{p}\bar{p}})|\bar{p}, \lambda> , \qquad (5.28)$$

where $\bar{p}$ is a generic standard light-like vector, and a generic helicity state is obtained by

$$|p, \lambda> = U(R_\mathbf{p})|\breve{p}, \lambda> = U(R_\mathbf{p})U(L_{\breve{p}\bar{p}})|\bar{p}, \lambda> , \qquad (5.29)$$

where $R_\mathbf{p}$ is also here the rotation which takes $\breve{p}$ into p. The Wigner rotation is then given by

$$\mathcal{R} = L_{\breve{p}'\bar{p}}^{-1} R_{\mathbf{p}'}^{-1} \Lambda R_\mathbf{p} L_{\breve{p}\bar{p}} = R_{\mathbf{p}'}^{-1} L_{p'\bar{p}}^{-1} \Lambda L_{p\bar{p}} R_\mathbf{p} , \qquad (5.30)$$

where the second equality follows from Eq. (5.21). The transformation properties of a state $|p, \lambda>$ under $\mathcal{P}_+^\uparrow$ are of the type given in Eq. (5.20); in the present case, however, the helicity λ remains invariant, so that only the diagonal term survives in the sum, and one can simply write

$$U(a, \Lambda)|p, \lambda> = e^{-ip'\cdot a} e^{-i\lambda\alpha(p,p')}|p', \lambda> , \qquad (5.31)$$

where $\alpha(p, p')$ represents the rotation angle around the direction of motion, depending on the initial and final momenta.

Finally, let us conclude with the following analysis, which should clarify the physical implications of the above considerations about massive and massless particles, and justifies the fact that a massless particle can have only two states of polarizations with $\lambda = \pm s$. Let us consider a state $|p, \lambda = s>$ of a particle of mass m, moving along the x^3-axis and having the spin parallel to the direction of motion. According to Eq. (5.16) this state can be obtained from the rest state $|\tilde{p}, \lambda>$

$$|p, \lambda> = U(L_3(-v))|\tilde{p}, \lambda> , \tag{5.32}$$

where we have denoted the boost $L_3(p)$ as $L_3(-v)$ to indicate more explicitly the velocity $v = |\mathbf{p}|/p^0$ relative to the Lorentz transformation.

If a second Lorentz transformation of velocity $-v'$ along the x^2-axis is applied, we obtain a state

$$U(L_2(-v'))|p, \lambda> = U(L_2(-v'))U(L_3(-v))|\tilde{p}, \lambda> , \tag{5.33}$$

which represent a particle moving vith velocity $\mathbf{V}$ along a direction in the (x^2, x^3) plane making an angle θ with the x^3-axis. We cannot say, however, if the spin is still parallel to the direction of motion. One can express V and θ in terms of v, v' by means of the well-known formulae of relativistic kinematics

$$\begin{aligned} V &= (v^2 + v'^2 - v^2 v'^2)^{\frac{1}{2}} , \\ \tan \theta &= \frac{v'}{v(1 - v'^2)^{\frac{1}{2}}} . \end{aligned} \tag{5.34}$$

On the other hand, the little group matrix corresponding to the Lorentz transformation $L_2(-v')$ is, according to Eq. (5.19),

$$L_3^{-1}(-V)R_1^{-1}(\theta)L_2(-v')L_3(-v) = R_1^{-1}(\epsilon) , \tag{5.35}$$

where

$$\tan \epsilon = \frac{v'}{v}(1 - v^2)^{\frac{1}{2}} , \tag{5.36}$$

so that

$$L_2(-v')L_3(-v)R_1(\epsilon) = R_1(\theta)L_3(-V) . \tag{5.37}$$

Making use of Eq. (5.33), from the above relation we get

$$U(L_2(-v'))U(L_3(-v))U(R_1(\epsilon))|\tilde{p}, \lambda = \sigma> = |p', \lambda = \sigma> . \tag{5.38}$$

We remark that we have obtained, in this way, a state of a particle moving with velocity $\mathbf{V}$ and spin parallel to the direction of motion. Comparing Eq. (5.33) with (5.37) we see that the only difference is that the initial state $|\tilde{p}, \lambda>$ is replaced by the rest state $U(R_1(\epsilon))|\tilde{p}, \lambda>$. The rotation through ϵ is just what we need to align the spin to the linear momentum, and therefore ϵ can be interpreted as the angle between the spin direction in the state (5.33) and $\mathbf{V}$. The situation is illustrated in Fig. 5.2 a) and b) for the states defined by Eqs. (5.33) and (5.38), respectively (the vectors $\mathbf{n}_i$ and $\mathbf{n}_f$ indicate the spin directions in the initial and final states).

The above analysis and in particular Eq. (5.36) provide clear conclusions about the behaviour of massive and massless particles. If we start from the

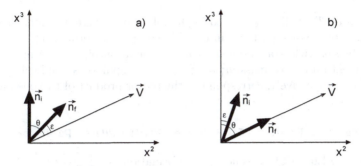

Fig. 5.2. *Effects of the Lorentz transformation (5.33) on the momentum and spin directions of a state $|p, \lambda = s>$.*

state $|p, \lambda >$ of Eq. (5.32), representing the particle moving with velocity v along the x^3-axis, then the transformation (5.33) shows how the state appears in a Lorentz frame in motion with velocity $-v'$ in the x^2 direction.

For a massive particle all values between 0 and θ are allowed for the angle ϵ: in particular, if v and v' are small with respect to 1 (velocity of light), then Eqs. (5.34), (5.36) give $\epsilon \approx \theta$, so that the spin direction is left almost unchanged by $L_2(-v')$. If, conversely, v is close to 1, Eq. (5.36) gives $\epsilon \simeq 0$, i.e. the spin remains almost parallel to the direction of motion.

For a massless particle, only the value $\epsilon = 0$ occurs: in this case the Lorentz transformation maintains the *parallelism* between spin and linear momentum. The analogous result can be obtained for the case in which the spin is *antiparallel* to the direction of motion. We can conclude that for a massless particle there are, in principle, two states of polarization, which transform into each other by space inversion. Both states are expected to occur in nature for the same particle, if the theory is invariant under space inversion; otherwise, only one state may exist.

5.4 States of two particles

We have considered up to now only states of one particle, which transform according to IR's of the restricted Poincaré group $\mathcal{P}_+^\uparrow$. The combination of two (or more) particles leads to a system which transforms, in general, according to a *reducible* representation of $\mathcal{P}_+^\uparrow$. The simplest description is given by decomposing the reducible representation into its irreducible components. The problem is analogous to the one encountered in the non relativistic quantum mechanics, when one combines two systems of angular momentum J_1 and J_2. The combined system contains all the values of angular momentum between $|J_1 - J_2|$ and $J_1 + J_2$, and its states are decomposed, accordingly, by means of the Clebsh-Gordan formula for the rotation group. The analogous results for the group $\mathcal{P}_+^\uparrow$ will be analyzed in the following, where we limit ourselves to the case of massive particles in the spin component formalism.

The two-particle states obtained by combination of the two particles of four-momentum p_1, p_2, mass $m_1^2 = p_1^2$, $m_2^2 = p_2^2$, spin s_1, s_2 and spin projection σ_1, σ_2, respectively, correspond to the tensor product of two one-particle states:

$$|m_1, m_2, s_1, s_2; \mathbf{p}_1, \mathbf{p}_2, \sigma_1, \sigma_2> = |m_1, s_1; \mathbf{p}_1, \sigma_1> |m_2, s_2; \mathbf{p}_2, \sigma_2> . \quad (5.39)$$

According to Eq. (5.14), a general transformation of $\mathcal{P}_+^\uparrow$ gives

$$U\,(a, \Lambda)|p, \sigma> = e^{-ia(p_1' + p_2')} .$$
$$\cdot \sum_{\sigma_1' \sigma_2'} D_{\sigma_1' \sigma_1}^{(s_1)} (L_{p_1'}^{-1} \Lambda L_{p_1}) D_{\sigma_2' \sigma_2}^{(s_2)} (L_{p_2'}^{-1} \Lambda L_{p_2})|p_1', s_1, \sigma_1'; p_2', s_2, \sigma_2'> . \quad (5.40)$$

where we have rewritten $|m_i, s_i; \mathbf{p}_i, \sigma_i>$ as $|p_i, s_i, \sigma_i>$.

One sees from Eq. (5.40) that the translation operator $U(a)$ is already diagonal; the two-particle state corresponds to the total momentum

$$P = p_1 + p_2 . \quad (5.41)$$

The total mass P^2 has a minimum value equal to $(m_1 + m_2)^2$; it may range from this value to ∞.

For a fixed four-vector P, one can parametrize p_1, p_2 by the "relative barycentric momentum" four-vector

$$q = q_{12} = -q_{21} = \tfrac{1}{2}\left(p_1 - p_2 - \frac{m_1^2 - m_2^2}{P^2}P\right) = p_1 - \frac{p_1 \cdot P}{P^2}P = -p_2 + \frac{p_2 \cdot P}{P^2}P . \quad (5.42)$$

The four-vector q lies in the plane defined by p_1 and p_2 and orthogonal to P:

$$q \cdot P = 0 , \quad (5.43)$$

so that one can specify q in terms of two polar angles and its modulus. Its square is given by

$$q^2 = -\frac{\lambda(m_1^2, m_2^2, P^2)}{4P^2} . \quad (5.44)$$

where

$$\lambda(Z_1, Z_2, Z_3) = Z_1^2 + Z_2^2 + Z_3^2 - 2(Z_1 Z_2 + Z_1 Z_3 + Z_2 Z_3) . \quad (5.45)$$

From Eq. (5.42) one gets the inverse formulae

$$p_1 = \frac{P^2 + m_1^2 - m_2^2}{2P^2}P + q , \qquad p_2 = \frac{P^2 - m_1^2 + m_2^2}{2P^2}P - q . \quad (5.46)$$

5.5 The ℓ-s coupling scheme

Let us go now to the rest frame of the compound system defined by $\mathbf{P} = 0$. The state (5.39) can then be specified by

$$|\tilde{P}, \mathbf{q}; s_1, \sigma_1; s_2, \sigma_2> , \qquad (5.47)$$

where $\tilde{P} = (E, \mathbf{0})$, E being the total c.m. energy, and $q = (0, \mathbf{q})$. From this we can project out the states of well-defined orbital angular momentum

$$|\tilde{P}, \ell, \mu; s_1, \sigma_1; s_2, \sigma_2> = \int d\hat{\mathbf{q}} Y_\ell^\mu(\hat{\mathbf{q}}) |\tilde{P}, \mathbf{q}; s_1, \sigma_1; s_2, \sigma_2> , \qquad (5.48)$$

where the $Y_\ell^\mu(\hat{\mathbf{q}})$ are the usual spherical harmonics, and $\hat{\mathbf{q}} = \mathbf{q}/|\mathbf{q}|$.

We have now to deal with three angular momenta (ℓ, s_1, s_2) and there are different coupling schemes that we can use. We follow here the ℓ-s *coupling*, in which we first couple the two spins s_1, s_2 to give a total spin s, and then s with ℓ to give the total angular momentum J. Using the Clebsch-Gordan coefficients, we get from Eq. (5.48):

$$|\tilde{P}, J, j; \ell, s; s_1, s_2> = \sum_{\mu\sigma} \sum_{\sigma_1\sigma_2} < s_1, s_2, s; \sigma_1, \sigma_2, \sigma> < s, \ell, J; \sigma, \mu, j> \cdot$$
$$\cdot |\tilde{P}, \ell, \mu; s_1, \sigma_1; s_2, \sigma_2> =$$
$$\qquad (5.49)$$
$$= \sum_{\mu\sigma} \sum_{\sigma_1\sigma_2} < s_1, s_2, s; \sigma_1, \sigma_2, \sigma> < s, \ell, J; \sigma, \mu, j> \cdot$$
$$\cdot \int d\hat{\mathbf{q}} Y_\ell^\mu(\hat{\mathbf{q}}) |\tilde{P}, \mathbf{q}; s_1, \sigma_1; s_2, \sigma_2> .$$

The application of a rotation R to the above state gives:

$$U(R)|\tilde{P}, J, j; \ell, s; s_1, s_2> = \sum_{\mu\sigma} \sum_{\sigma_1\sigma_1'} \sum_{\sigma_2\sigma_2'} \int d\hat{\mathbf{q}}' Y_\ell^\mu(\hat{\mathbf{q}}') \cdot$$
$$\cdot < s_1, s_2, s; \sigma_1, \sigma_2, \sigma> < s, \ell, J; \sigma, \mu, j> \cdot \quad (5.50)$$
$$\cdot D_{\sigma_1'\sigma_1}^{(s_1)}(R) D_{\sigma_2'\sigma_2}^{(s_2)}(R) |\tilde{P}, \mathbf{q}'; s_1, \sigma_1'; s_2, \sigma_2'> ,$$

where $\mathbf{q}' = R\mathbf{q}$.

Making use of the propertes of the Clebsch-Gordan coefficients and of the rotation matrices, Eq. (5.50) can be re-written as[2]

$$U(R)|\tilde{P}, J, j; \ell, s; s_1, s_2> = \sum_{j'} \sum_{\mu'\sigma'} \sum_{\sigma_1'\sigma_2'} D_{j'j}^{(J)}(R) \int d\hat{\mathbf{q}}' Y_\ell^{\mu'}(\hat{\mathbf{q}}') \cdot$$
$$\cdot < s_1, s_2, s; \sigma_1, \sigma_2, \sigma> < s, \ell, J; \sigma, \mu, j> \cdot$$
$$\cdot |\tilde{P}, \mathbf{q}'; s_1, \sigma_1'; s_2, \sigma_2'> = \qquad (5.51)$$
$$= \sum_{j'} D_{j'j}^{(J)}(R) |\tilde{P}, J, j'; \ell, s; s_1, s_2> ,$$

---

[2] See F.R. Halpern, *Special Relativity and Quantum Mechanics*, Prentice-Hall (1968).

showing that the states $|\tilde{P}, J, j'; \ell, s; s_1, s_2>$ transform according to the IR's of the rotation group.

Following a procedure similar to the one used in the derivation of Eq. (5.14), we get finally for a transformation of $\mathcal{P}_+^\uparrow$

$$U(a, \Lambda)|P, J, j; \ell, s; s_1, s_2> = e^{-iaP'} \sum_{j'} D_{j'j}^{(J)}(L_{P'}^{-1} \Lambda L_P)|\tilde{P}, J, j'; \ell, s; s_1, s_2> ,$$

(5.52)

which shows that the above states correspond to IR's of the group $\mathcal{P}_+^\uparrow$.

Problems

5.1. Show that the Wigner rotations $\mathcal{R}$ defined by (5.6) form a group.

5.2. Show explicitly the isomorphism between little groups corresponding to different standard vectors with the same value of p^2, finding the equivalence relation between their elements.

5.3. Show that the little group of a space-like vector $\bar{p}$ is the group $SO(1, 2)$ of Lorentz transformations in a 3-dimensional space-time.

5.4. Evaluate explicitly the Wigner rotation $\mathcal{R}_{p'p}$ given by Eq. (5.13) in the case in which Λ is a pure Lorentz transformation. What is the meaning of $\mathcal{R}$ in the limit $p^0 \gg m$, $p'^0 \gg m$? What is the non-relativistic limit?

5.5. Show that if the generic Lorentz transformation of Eq. (5.13) is a rotation R, then the corresponding element of the little group is R.

5.6. Find the transformation properties of the helicity states $|p, \lambda>$ starting from the transformation properties of the $|p, \sigma>$ states and taking into account the relation between them (Eq. (5.22)).

5.7. Show that the one-particle state $|\mathbf{p}, \lambda> = |m, s; \mathbf{p}, \lambda>$ transforms under a pure rotation R according to

$$U(R)|\mathbf{p}, \lambda> = e^{-i\lambda\alpha(\mathbf{p},\mathbf{p}')}|\mathbf{p}, \lambda> ,$$

where $\mathbf{p}' = R\mathbf{p}$. Find the explicit expression for $\alpha(\mathbf{p}, \mathbf{p}')$.

5.8. Show that, when Λ is a pure rotation R, the angle $\alpha(p', p)$ of Eq. (5.31) turns out to be the same as in the case of a massive particle.

5.9. Show directly that the little group of a light-like vector is the two-dimensional Euclidean group.

5.10. Find $\alpha(p', p)$ of Eq. (5.31) in the case in which Λ is a pure Lorentz transformation.

5.11. Derive explicitly the little group matrix $R_1(\epsilon)$ of Eq. (5.35).

6

Discrete operations

In this Chapter we shall consider in more detail the properties of the discrete operations I_s and I_t (space inversion and time reversal) which, as seen previously, belong to the complete Lorentz and Poincaré groups $\mathcal{L}$ and $\mathcal{P}$. The transformation properties of the one- and two-particle states under these operations will be analyzed, and their physical implications will be discussed.

6.1 Space inversion

In Chapters 3 and 4, we have considered the transformations of the restricted groups $\mathcal{L}_+^\uparrow$ and $\mathcal{P}_+^\uparrow$. We include now the space inversion operator I_s, defined by Eq. (3.18), and consider the larger groups $\mathcal{L}^\uparrow$ and $\mathcal{P}^\uparrow$, i.e. the orthochronous groups.

In the four-dimensional IR of $\mathcal{L}^\uparrow$, I_s is given by:

$$
I_s = \begin{pmatrix} 1 & 0 & 0 & 0 \\ 0 & -1 & 0 & 0 \\ 0 & 0 & -1 & 0 \\ 0 & 0 & 0 & -1 \end{pmatrix} , \tag{6.1}
$$

which coincides, as noticed already in Section 3.1, with the metric tensor g.

It is then easy to compute the commutation relations of I_s with the generators J_i, K_i and P_μ of the Poincaré group[1]:

---

[1] The commutation relations with P_0, P_i are obtained in the 5-dimensional IR of $\mathcal{P}$, using Eq. (4.14) and replacing (6.1) by the 5×5 matrix

$$
I_s = \begin{pmatrix} g & 0 \\ 0 & 1 \end{pmatrix} .
$$

G. Costa and G. Fogli, *Symmetries and Group Theory in Particle Physics*,
Lecture Notes in Physics 823, DOI: 10.1007/978-3-642-15482-9_6,
© Springer-Verlag Berlin Heidelberg 2012

$$[I_s, J_i] = 0 \,,$$
$$\{I_s, K_i\} = 0 \,,$$
$$[I_s, P_0] = 0 \,,$$ (6.2)
$$\{I_s, P_i\} = 0 \,.$$

The fact that the J_i commute with I_s, i.e. that they are invariant under space inversion, means that $\mathbf{J}$ behaves as an *axial* vector; on the other hand, the K_i and P_i anticommute with I_s, i.e. they change their sign, so that $\mathbf{K}$ and $\mathbf{P}$ transform as *polar* vectors; finally, P_0 behaves as a *scalar*. This behaviour under I_s is what one expects on physical ground.

The above relations show that I_s commutes with any rotation, but not, in general, with any transformation of $\mathcal{L}^\uparrow$ and $\mathcal{P}^\uparrow$. We know, however, that each transformation of the component $\mathcal{L}_-^\uparrow$ of $\mathcal{L}^\uparrow$ can be obtained as the product of a corresponding transformation of $\mathcal{L}_+^\uparrow$ by I_s.

In Section 3.5 we have already analysed the transformation properties under I_s of the bases of the IR's of $\mathcal{L}_+^\uparrow$. Here, we consider the effect of space inversion on one-particle states.

Let us denote by P the linear operator in the Hilbert space, corresponding to I_s: $P = U(I_s)$. From the Wigner theorem, P could be either a unitary or an antiunitary operator. The first choice will appear to be the one consistent with Eqs. (6.2).

We know that the unitary or antiunitary operator $U(I_x)$ corresponding in the Hilbert space to a discrete operation I_x, with $I_x^2 = I$, is defined up to a phase. Unless $U(I_x)$ is unitary, it is not possible to fix the phase in such a way to get $U^2(I_x) = I$. However, one can show that the following relations hold in general[2]

$$U(a, \Lambda)U(I_x) = U(I_x)U(I_x a, I_x \Lambda I_x) \,,$$ (6.3)

$$U(I_x)U(a, \Lambda) = U(I_x a, I_x \Lambda I_x)U(I_x) \,.$$ (6.4)

They are, obviously, equivalent if $U^2(I_x) = I$.

In particular, taking $I_x = I_s$ and $\Lambda = I$, and applying Eq. (6.3) to a one-particle state $|p, \sigma>$, one gets:

$$U(a, I)P|p, \sigma> = P\,U(\underline{a}, I)|p, \sigma>$$ (6.5)

where $\underline{a} = (a^0, -\mathbf{a})$. If P is unitary, since $p\underline{a} = \underline{p}a$, making use of Eq. (4.32) (with $p_0 > 0$), Eq. (6.5) becomes:

$$U(a, I)P|p, \sigma> = e^{-i\underline{p}a}P|p, \sigma> \,,$$ (6.6)

and, if P is antiunitary,

$$U(a, I)P|p, \sigma> = e^{i\underline{p}a}P|p, \sigma> \,.$$ (6.7)

---

[2] See F.R. Halpern, *Special Relativity and Quantum Mechanics*, Prentice-Hall (1968).

If P is a symmetry operation of the theory, $P|p, \sigma>$ must represent a physical state, and its energy must be positive. Since $p = (p^0, -\mathbf{p})$, this condition is satisfied only by Eq. (6.6), and then P has to be *unitary*. In this case one can choose the arbitrary phase of P such that:

$$P^2 = I . \tag{6.8}$$

Since P is unitary, Eq. (6.8) makes it hermitian

$$P = P^\dagger , \tag{6.9}$$

and its eigenvalues η are

$$\eta = \pm 1 . \tag{6.10}$$

The operator P is therefore related to an *observable* called *parity*, with eigenvalues ± 1.

Let us now apply P to a state of a massive particle with spin s. We consider first a rest state; since I_s commute with J_i, the spin eigenstates remain unchanged and one gets independently of σ

$$P|\tilde{p}, \sigma> = \eta|\tilde{p}, \sigma> , \tag{6.11}$$

where the factor $\eta = \pm 1$ is called *intrinsic parity* of the particle. To go from a rest state to a state of motion, we use the boost L_p (see Eq. (5.12)):

$$P|p, \sigma> = PU(L_p)P^{-1}P|\tilde{p}, \sigma> = \eta PU(L_p))P^{-1}|\tilde{p}, \sigma> . \tag{6.12}$$

From Eq. (6.3) one can easily derive

$$PU(L_p))P^{-1} = U(I_s L_p I_s) = U(L_{\underline{p}}) , \tag{6.13}$$

so that Eq. (6.12) becomes

$$P|p, \sigma> = \eta|\underline{p}, \sigma> . \tag{6.14}$$

Care must be taken if one uses the helicity representation $|p, \lambda>$. Since $\mathbf{P}$ and $\mathbf{J}$ transform as polar and axial vectors, respectively, Eq. (4.41) shows that the helicity changes sign under parity.

It is customary to introduce the operator of reflection in the (x^1, x^3) plane

$$Y = e^{-i\pi J_2} P , \tag{6.15}$$

which clearly commutes with a Lorentz transformation along the x^3-axis

$$YU(L_3(p)) = U(L_3(p))Y . \tag{6.16}$$

Let us now consider the case of a massive particle. The helicity state $|p, \lambda> \equiv |m, s; \mathbf{p}, \lambda>$, where p is chosen along the x^3-axis without loss of generality, can be obtained from a rest state $|\tilde{p}, \lambda>$ by a boost $L_3(p)$. From the above relation we get

$$Y|p, \lambda> = YU(L_3(p))|\tilde{p}, \lambda> = U(L_3(p))Y|\tilde{p}, \lambda> = \eta U(L_3(p))e^{-i\pi J_2}|\tilde{p}, \lambda> , \tag{6.17}$$

Using the properties of the rotation matrices, Eqs. (A.1), (A.5), one finds

$$D_{\lambda'\lambda}^{(s)}(0, \pi, 0) = d_{\lambda'\lambda}^{(s)}(\pi) = (-1)^{s-\lambda}\delta_{\lambda' -\lambda} , \tag{6.18}$$

so that, applying the above rotation to the rest frame, Eq. (6.17) becomes

$$Y|p, \lambda> = \eta(-1)^{s-\lambda}|p, -\lambda> . \tag{6.19}$$

Using again Eq. (6.15), one finally obtains

$$P|p, \lambda> = \eta(-1)^{s-\lambda}e^{i\pi J_2}|p, -\lambda> . \tag{6.20}$$

Usually, if $|p, \lambda>$ is a helicity state of a particle with momentum $\mathbf{p}$ along x^3, the state $|\underline{p}, \lambda>$, corresponding to momentum $-\mathbf{p}$, can be defined by

$$|\underline{p}, \lambda> = (-1)^{s-\lambda}e^{-i\pi J_2}|p, \lambda> . \tag{6.21}$$

In fact, the rotation $e^{-i\pi J_2}$ leaves the helicity unchanged, and transforms p into $\underline{p}$. However, the relative phase of $|p, \lambda>$, $|\underline{p}, \lambda>$ is not uniquely defined; one can choose it in such a way that both states $|p, \lambda>$ and $|\underline{p}, \lambda>$ reduce to the same state-vector in the rest frame.

Following the same arguments leading to Eq. (6.19), one finds

$$Y|\underline{p}, \lambda> = \eta(-1)^{s+\lambda}|\underline{p}, -\lambda> , \tag{6.22}$$

so that

$$P|\underline{p}, \lambda> = \eta(-1)^{s+\lambda}e^{i\pi J_2}|\underline{p}, -\lambda> . \tag{6.23}$$

We observe that the sign of λ in the exponent is reversed, since the boost $L_3(\underline{p})$ reverses the sign of the helicity. In fact it is

$$|\underline{p}, \lambda> = U(L_3(\underline{p}))|\tilde{p}, -\lambda> , \tag{6.24}$$

consistently with the definition (6.21). Eq. (6.24) can be easily derived from Eq. (6.21) if one takes into account that

$$U(L_3(p)) = PU(L_3(\underline{p}))P^{-1} = e^{i\pi J_2}U(L_3(\underline{p}))e^{-i\pi J_2} , \tag{6.25}$$

where use has been made of Eqs. (6.4), (6.15) and (6.16).

Eqs. (6.20), (6.23) hold also for a massless particle. However, in this case, the situation is quite different, since rest states do not exist, so that the definition (6.11) for the intrinsic parity becomes meaningless. We can regard Eqs. (6.20), (6.23) as a definition of intrinsic parity for a massless particle. The commutation relation (6.16) assures that η is independent of p.

We see that Eq. (6.21) relates, when both exist, the two states of opposite helicity. For instance, both eigenstates corresponding to $\lambda = \pm 1$ exist for the photon: they describe, as it will be shown in Section 7.3, the two states of right- and left-handed circular polarization.

In the frame of the Standard Model of electroweak interactions, which is considered in detail in Sections 9.7 and 9.7.3, neutrinos are strictly massless. It is usual to identify the neutrino ν as the particle associated with the positron in the beta-plus decay: a proton in a nucleus decays into a neutron, a positron and a neutrino. The anti-neutrino $\bar{\nu}$ is then the particle associated with the electron in the beta-minus process: a neutron in a nucleus decays into a proton, an electron and an anti-neutrino. Experimentally, it has been observed[3] that there is only one eigenstate of helicity for the neutrino ν, which is the opposite of that of the antineutrino $\bar{\nu}$, so that ν and $\bar{\nu}$ denote the eigenstates corresponding to $\lambda = -\frac{1}{2}$ and $\lambda = +\frac{1}{2}$, respectively.

By assuming *electron lepton number* $+1$ for electron and neutrino, and -1 for their antiparticles, within the Standard Model the processes involving ν and $\bar{\nu}$ are regulated by the conservation of this quantum number. Similar properties are assigned to the muon and to the tau neutrinos in association with their corresponding charged leptons, with the introduction of two other distinct quantum numbers: *muon and tau lepton numbers*. In general, the three specific lepton numbers are called *lepton flavors*; they are conserved in the frame of the Standard Model.

Actually, however, there is clear experimental evidence of *neutrino flavor oscillation* (neutrinos change their flavor during their temporal evolution), which implies that neutrinos are indeed massive particles, even though we still ignore their absolute masses. To account for the experimental evidence which indicates that neutrinos have a mass different from zero, it is necessary to go beyond the Standard Model.

From the above analysis we see that, among the one-particle states, only those describing a (massive) particle *at rest* are eigenstates of parity. The situation is more interesting for a system of two or more particles in relative motion, for which one can build parity eigenstates.

We limit ourselves to the states of two massive particles in the ℓ-s coupling representation given by Eq. (5.49). Let us start with the state (5.47) which describes the two particles in their c.m. system

$$|E, \mathbf{q}; s_1, \sigma_1; s_2, \sigma_2 > , \tag{5.47}$$

where E is the total energy and $\mathbf{q} = \frac{1}{2}(\mathbf{p}_1 - \mathbf{p}_2)$ the relative momentum. Taking into account Eqs. (5.39) and (6.14), we get immediately

$$P|E, \mathbf{q}; s_1, \sigma_1; s_2, \sigma_2 > = \eta_1 \eta_2 |E, -\mathbf{q}; s_1, \sigma_1; s_2, \sigma_2 >; , \tag{6.26}$$

where η_1 and η_2 are the intrinsic parities of the two particles.

---

[3] This property was established long ago by a famous experiment: M. Goldhaber, L. Grodzins and A.W. Sunyar, Phys. Rev. 109, 1015 (1958).

For a state of definite angular momentum, Eq. (5.48), we then obtain

$$P|E,\ell,\mu;s_1,\sigma_1;s_2,\sigma_2> = \eta_1\eta_2 \int d\hat{\mathbf{q}} Y_\ell^\mu(\hat{\mathbf{q}})|E,-\mathbf{q};s_1,\sigma_1;s_2,\sigma_2> =$$
$$= \eta_1\eta_2 \int d\hat{\mathbf{q}} Y_\ell^\mu(-\hat{\mathbf{q}})|E,\mathbf{q};s_1,\sigma_1;s_2,\sigma_2> ,$$
(6.27)

which, making use of the well-known relation

$$Y_\ell^\mu(-\hat{\mathbf{q}}) = (-1)^\ell Y_\ell^\mu(\hat{\mathbf{q}}) ,$$
(6.28)

becomes

$$P|E,\ell,\mu;s_1,\sigma_1;s_2,\sigma_2> = \eta_1\eta_2(-1)^\ell P|E,\ell,\mu;s_1,\sigma_1;s_2,\sigma_2> .$$
(6.29)

This shows that the angular momentum eigenstates are also eigenstates of parity corresponding to the eigenvalue $\eta_1\eta_2(-1)^\ell$, which is the product of the intrinsic parities times the parity of the state of relative orbital momentum. The same result holds for the eigenstates of total angular momentum J; in fact, from (6.29), (5.49) we get

$$P|E,J,j;l,s;s_1,s_2> = \eta_1\eta_2(-1)^\ell|E,J,j;l,s;s_1,s_2> .$$
(6.30)

6.2 Parity invariance

We have seen above how the states of one and two particles are transformed under parity. In particular, the one-particle states at rest are eigenstates of parity corresponding to the eigenvalues $+1$ and -1 which define the intrinsic parity (*even* and *odd*, respectively) of the particles. There is, however, some arbitrariness in the assignment of intrinsic parities.

It is known that parity is conserved to a high degree of accuracy in the strong and electromagnetic interactions. The meaning of this statement is that it is possible to assign intrinsic parities to all the particles which have strong and electromagnetic interactions in such a way that parity is always conserved in these interactions.

The intrinsic parity can be determined uniquely only for those particles whose internal quantum numbers, such as baryon number and strangeness, which are conserved in strong and electromagnetic interactions, are *zero*.

For instance, in any reaction involving nucleons (proton and neutron), one has always to deal at least with a pair of them, so that the parity assignment for the nucleons is only a matter of convention. Because of the two "superselection" rules of charge and baryon number conservation, even the relative proton-neutron parity is fixed by convention; since proton and neutron belong to the same iso-spin doublet, it is convenient to assign the same intrinsic parity to them. The standard convention is to take both proton and neutron to have *even intrinsic parity*. Then, the intrinsic parities of all the other non-strange particles can be deduced from experiment.

Let us consider, as an example, the classical argument for the determination of the intrinsic parity of the π^- meson from the reaction

$$\pi^- + d \rightarrow n + n . \tag{6.31}$$

The deuteron d is a $^3S_1$ state, so that, from (6.30) and the above assignment for the nucleon parity, we conclude that it has even parity. Parity conservation in the above reaction then implies:

$$\eta_\pi \cdot (-1)^\ell = (-1)^{\ell'} , \tag{6.32}$$

where ℓ and ℓ' are the relative π-d and n-n orbital angular momenta. It was shown that the π^- is absorbed in an S-wave, i.e. $\ell = 0$. Since the spin of d is 1, and π^- has spin zero, angular momentum conservation implies that the only possible final states in Eq. (6.31) are $^3S_1$, $^3P_1$, $^1P_1$, $^3D_1$. Moreover, the Fermi-Dirac statistics requires the two neutrons n to be in an antisymmetric state, so that only $^3P_1$ is possible, i.e. $\ell' = 1$. The above relation fixes the parity of the π^- to be *odd*.

The intrinsic parity of the photon is also determined to be odd. This is related to the fact that the photon is a quantum of the electromagnetic field and its polarization vector behaves as the (polar) vector potential **A**.

Going to strange particles, one needs to make a further convention. In fact, now one has to deal always with a pair of two strange particles, and only their *relative parity* can be determined. For instance, from the process

$$\pi^- + p \rightarrow K^0 + \Lambda^0 . \tag{6.33}$$

one could determine the relative K^0-Λ^0 parity; with the convention that Λ^0 has *even* parity as the nucleon, the parity of the K^0-meson is then fixed, and it turns out to be *odd*. as in the case of the π.

It can be shown that the relative parity of a *particle-antiparticle* pair is *odd* in the case of fermions. and *even* in the case of bosons. In general, if a particle-antiparticle system has relative orbital momentum ℓ, its parity is given by $-(-1)^\ell$ for fermions, and $(-1)^\ell$ for bosons. This follows from the transformation properties under the so-called particle-antiparticle conjugation or *charge conjugation*, denoted by C and conserved by strong and electromagnetic interactions[4].

It is well known that parity invariance (as well as charge conjugation) no longer holds in the case of weak interactions. This means that weak interactions can cause transitions among different parity eigenstates; consequently, particle states are not strictly parity eigenstates. An example of parity violation is given by the decay $K^0 \rightarrow \pi^+\pi^-$; since the spin of K^0 is zero, the $\pi^+\pi^-$ system is in a S-state, and then its parity is *even*; on the other hand, the parity of K^0 was determined in strong interaction processes to be *odd* (the assignment of *even* parity for K^0 would also lead to parity violation, since the

---

[4] See e.g. J..J. Sakurai, *Advanced Quantum Mechanics*, Addison-Wesley (1967).

parity of the final state in the decay $K^0 \to \pi^+\pi^-\pi^0$ can be determined to be odd).

A systematic investigation of the validity of parity conservation was undertaken by Lee and Yang[5], who first suggested its possible violation. The first conclusive evidence that parity is not conserved in weak decays was provided by the famous experiment of Wu *et al.*[6].

A direct test of parity invariance consists in investigating if particle interactions distinguish between right and left. Parity invariance implies that for any process there exists the specular one which occurs with the same probability. If pseudoscalar quantities, such as $\mathbf{J} \cdot \mathbf{P}$, have non-zero expectation values, then parity invariance is violated. The experiment of Wu *et al.* consists in the detection of a term of the form $\mathbf{J} \cdot \mathbf{P}$ in the β-decay of polarized ^{60}Co. In fact, they observed an angular distribution for the emitted electron of the form $1 + \alpha P \cos\theta$, where P is the ^{60}Co polarization, and θ the angle between the momentum $\mathbf{p}$ of the electron and the direction of the polarization $\mathbf{P}$. Clearly this term describes an up-down asymmetry in the electron angular distribution; a measure of it is given by the asymmetry parameter α which is proportional to the velocity of the electron.

A detailed analysis of parity non-conservation in weak interactions is out of the purposes of this book. We want only to mention here one of the most striking facts of the violation of parity invariance, i.e. the existence of a left-handed neutrino and the lack of evidence of a right-handed one.

6.3 Time reversal

In the following, we shall consider the transformation properties of a state under *time reversal*, i.e. under the time inversion operation I_t, Eq. (3.19), which is contained in the proper groups $\mathcal{L}_+$ and $\mathcal{P}_+$.

The transformation properties under the space-time inversion I_{st}, which is contained only in the full group $\mathcal{L}$ and $\mathcal{P}$, will be trivially given in terms of those under I_s and I_t.

In the four-dimensional IR of $\mathcal{L}$, I_t is given by

$$I_t = \begin{pmatrix} -1 & 0 & 0 & 0 \\ 0 & 1 & 0 & 0 \\ 0 & 0 & 1 & 0 \\ 0 & 0 & 0 & 1 \end{pmatrix} = -g \ . \tag{6.34}$$

The commutation relations of I_t with the other generators of the Poincaré group easily follow[7]:

---

[5] T.D. Lee and C.N. Yang, Phys. Rev. **104**, 254 (1956).

[6] C.S. Wu *et al.*, Phys. Rev. **105**, 1413 (1957).

[7] The commutation relations with P_0, P_i are obtained using the 5-dimensional IR of $\mathcal{P}$, where I_t is given by

$$[I_t, J_i] = 0 ,$$
$$\{I_t, K_i\} = 0 ,$$
$$\{I_t, P_0\} = 0 ,$$
$$[I_t, P_i] = 0 .$$

(6.35)

Let us denote by T the operator representing I_t in the Hilbert space: $T = U(I_t)$. Should T be unitary, Eqs. (6.35) would mean that the energy changes sign under time reversal, while the linear and angular momenta remain unchanged. This is contrary to the physical interpretation of time reversal.

Applying Eq. (6.3) with $I_x = I_t$ and $\Lambda = I$ to the one-particle state $|p, \sigma>$, we get

$$U(a, I)T|p, \sigma> = T\, e^{ip\underline{a}}|p, \sigma> \qquad (6.36)$$

since the four-vector a is transformed into $-\underline{a} = (-a^0, \mathbf{a})$. Taking into account that $p\underline{a} = \underline{p}a$, one gets

$$U(a, I)T|p, \sigma> = e^{i\underline{p}a}T|p, \sigma> \qquad (6.37)$$

if T is a unitary operator, and

$$U(a, I)T|p, \sigma> = e^{-i\underline{p}a}T|p, \sigma> \qquad (6.38)$$

if T is antiunitary. The above equations show that, if time reversal is a symmetry operation, in the sense that $T|p, \sigma>$ is a physical state corresponding to a positive energy eigenvalue, T has to be an *antiunitary* operator.

We consider now a massive particle of spin s at rest; its states $|\tilde{p}, \sigma>$ form the basis of the $(2s + 1)$-dimensional IR of the rotation group. If one applies T to one of these states, one obtains again a state of zero linear momentum, and then a state in the same Hilbert subspace. We can write

$$T|\tilde{p}, \sigma> = \sum_{\tau} C_{\tau\sigma}|\tilde{p}, \tau> , \qquad (6.39)$$

introducing a unitary matrix C, which is determined as follows. Applying a rotation R on the same state, one gets:

$$U(R)|\tilde{p}, \sigma> = \sum_{\rho} D^{(s)}_{\rho\sigma}(R)|\tilde{p}, \rho> , \qquad (6.40)$$

where $D^{(s)}(R)$ is the $(2s + 1)$-dimensional IR of $SO(3)$.

Combining Eqs.(6.39) and (6.40), since T is antiunitary, we get:

$$TU(R)|\tilde{p}, \sigma> = \sum_{\rho} D^{(s)*}_{\rho\sigma}(R)T|\tilde{p}, \rho> = \sum_{\rho\tau} D^{(s)*}_{\rho\sigma}(R)C_{\tau\rho}|\tilde{p}, \tau> , \qquad (6.41)$$

$$I_t = \begin{pmatrix} -g & 0 \\ 0 & 1 \end{pmatrix} .$$

and

$$U(R)T|\tilde{p},\sigma> = \sum_{\rho} U(R)C_{\rho\sigma}|\tilde{p},\rho> = \sum_{\rho\tau} C_{\rho\sigma}D^{(s)}_{\tau\rho}(R)|\tilde{p},\tau> \ . \qquad (6.42)$$

Taking into account the commutation properties of T with J_i and remembering Eq. (3.52), one gets

$$TU(R) = U(R)T \ , \qquad (6.43)$$

and a comparison of Eqs. (6.41), (6.42) gives

$$D^{(s)}(R)C = CD^{*(s)}(R) \ . \qquad (6.44)$$

We know that the two IR's $D^{(s)}$ and $D^{*(s)}$ are equivalent (see Section 2.3), so that Eq. (6.44) corresponds to a similarity transformation. In the case $s = \frac{1}{2}$, C can be identified with the inverse of the matrix S given in Eq. (2.42)[8]

$$C = \begin{pmatrix} 0 & -1 \\ 1 & 0 \end{pmatrix} = -i\sigma_2 = e^{-i\pi\frac{1}{2}\sigma_2} \ . \qquad (6.45)$$

The above result can be generalized to an arbitrary spin J as follows

$$C = e^{-i\pi J_2} \ , \qquad (6.46)$$

so that, making use of (6.18), Eq. (6.39) becomes

$$T|\tilde{p},\sigma> = (-1)^{s-\sigma}|\tilde{p},-\sigma> \ . \qquad (6.47)$$

We can extend this equation to states with arbitrary linear momentum, taking into account Eq. (6.4):

$$T|p,\sigma> = TU(L_p)|\tilde{p},\sigma> = U(L_p)T|\tilde{p},\sigma> = (-1)^{s-\sigma}|\underline{p},-\sigma> \ , \qquad (6.48)$$

with the usual symbol $\underline{p} = (p^0, -\mathbf{p})$.

It should be noted the change $\sigma \to -\sigma$ in the above equations; in fact, σ is the third component of the spin, and it must change sign under time reversal. Since also the components of the linear momentum change their sign, the helicity remains unchanged under time reversal.

The helicity representation allows us to find out how the states of a masless particle are modified under time reversal.

Starting from the helicity state $|p,\lambda>$ of a massive particle, obtained by a boost $L_3(p)$ from a rest state, and making use of Eqs. (6.4) and (6.24), we get

---

[8] The matrix C is determined aside from a phase factor, which is not fixed by Eq. (6.44); following the usual conventions we take it equal to 1.

$$T|p, \lambda> = U(L_3(\underline{p}))T|\tilde{p}, \lambda> = (-1)^{s-\lambda}U(L_3(\underline{p}))|\tilde{p}, -\lambda>$$
$$= (-1)^{s-\lambda}|\underline{p}, \lambda> \ . \tag{6.49}$$

On the other hand, when applied to a state $|\underline{p}, \lambda>$

$$T|\underline{p}, \lambda> = TU(L_3(\underline{p}))|\tilde{p}, -\lambda> = U(L_3(p))T|\tilde{p}, -\lambda> =$$
$$= (-1)^{s+\lambda}U(L_3(\underline{p}))|\tilde{p}, \lambda> = (-1)^{s+\lambda}|p, \lambda> \ . \tag{6.50}$$

In this form Eqs. (6.49), (6.50) hold also for the states of a massless particle.

The above results indicate a profound difference between the time reversal and the parity operators. We saw that in the case of P one *can* choose the phase in such a way that $P^2 = I$. Also T^2 must be proportional to the identity (repeating twice the time reversal operation, one has to recover the original situation), but one *cannot* modify the value of T^2 by changing the phase of T. In fact, suppose that it is replaced by $T' = e^{i\alpha}T$. Since T is anti-unitary, we get:

$$T'^2 = e^{i\alpha}Te^{i\alpha}T = e^{i\alpha}e^{-i\alpha}T^2 = T^2 \ . \tag{6.51}$$

From Eq. (6.48), we see that

$$T^2|p, \sigma> = (-1)^{s-\sigma}T|\underline{p}, -\sigma> = (-1)^{s-\sigma}(-1)^{s+\sigma}|p, \sigma> = (-1)^{2\sigma}|p, \sigma> \ . \tag{6.52}$$

The eigenvalue of T^2 is $+1$ in the case of integer spin, and -1 for half-integer spin. This result correspond to a *superselection rule*: the states with $T^2 = +1$ and $T^2 = -1$ belong to two Hilbert subspaces which are orthogonal to each other.

Before closing this Chapter, we want to discuss briefly the physical meaning of time-reversal invariance. Obviously, since T is antilinear, it cannot correspond to any observable. Unlike parity, it cannot give rise to an additional quantum number. However, time reversal invariance imposes restrictions to physical processes: if a process occurs, also the reversed process has to occur, with a rate completely defined in terms of that of the direct one (one has to take into account the kinematical changes in going from final to initial states).

It was believed that time reversal was an exact symmetry, since the known interactions define no sense of time's direction. There is experimental evidence that strong and electromagnetic interactions are invariant under time reversal, while weak interactions show a tiny violation. In fact, in 1964 the $\pi^+\pi^-$ decay of the long-lived K_2^0-meson was observed[9]; this decay violates CP invariance, and this indicates also violation of time-reversal invariance, since it is believed that all interactions are invariant under the combined operation CPT[10].

---

[9] J.H. Christenson, J.W. Cronin, V.L. Fitch, R. Turlay, Phys. Rev. Lett. 13, 138 (1964).

[10] For details on C, CP and CPT see e.g.: J.J. Sakurai, *Advanced Quantum Mechanics*, Addison-Wesley (1967); M.D. Scadron, *Advanced Quantum Theory*, Springer Verlag (1979).

Tests of time-reversal invariance are based on the following fact: if a quantity, which is odd under T, has non-zero expectation value, then it indicates the violation of the invariance. A test investigated since a long time is based on the beta-decay of polarized neutrons:

$$n \rightarrow p + e^- + \bar{\nu} . \tag{6.53}$$

Clearly, the quantity

$$\boldsymbol{\sigma}_n \cdot (\mathbf{p}_e \times \mathbf{p}_\nu) = -\mathbf{p}_p \cdot (\boldsymbol{\sigma}_n \times \mathbf{p}_e) , \tag{6.54}$$

where $\mathbf{p}_e$, $\mathbf{p}_n$ and $\mathbf{p}_p = -\mathbf{p}_\nu - \mathbf{p}_e$ are the c.m. momenta and $\boldsymbol{\sigma}_n$ represents the direction of the neutron spin, changes its sign under time inversion, while it is invariant under parity. A detection of asymmetry in the distribution of recoil protons above and below the $\boldsymbol{\sigma}_n \times \mathbf{p}_e$ plane would indicate breakdown of time-reversal invariance. However, up to now no experiment has been able to detect such asymmetry. The search is in progress and a new apparatus has been built for this purpose[11].

At present, the only direct detection of a departure from time-reversal invariance comes from the analysis of the $K^0\overline{K}^0$ meson system. The first detection was obtained in 1998[12]. It is a measurement of the difference between the rate of a process and its inverse, specifically in the comparison of the probabilities of the $\overline{K}^0$ transforming into K^0 and of K^0 into $\overline{K}^0$.

Problems

6.1. Show how the intrinsic parity of the π^0 meson can be determined from the decay $\pi^0 \rightarrow \gamma\gamma$ by measurements of the photon polarization. By similar measurements one can check that the electron-positron pair has odd relative parity, since the annihilation process $e^+e^- \rightarrow 2\gamma$ can occur in the $^1S_0$ positronium state.

6.2. Deduce from the observed process $K^- + He^4 \rightarrow {}_\Lambda He^4 + \pi^-$, that the parity of the K^- meson is odd, with the assignment $\eta_N = \eta_\Lambda = +1$ (${}_\Lambda He^4$ is a "hyperfragment", and it is equivalent to a helium atom He^4 in which a neutron is replaced by a Λ^0; both He^4 and ${}_\Lambda He^4$ have spin zero and even parity).

6.3. Show that the parity of a three spinless particle state of total angular momentum $\mathbf{J} = \boldsymbol{\ell} + \mathbf{L}$ in its c.m. system is given by $\eta_1 \cdot \eta_2 \cdot \eta_3 \cdot (-1)^{\ell+L}$, where ℓ is the relative angular momentum of particles 1 and 2, and L the angular momentum of particle 3 relative to the c.m. system of 1 and 2.

---

[11] H.P. Mumm et al., Review of Scientific Instruments 75, 5343 (2004).
[12] A. Angelopoulos et al. (CPLEAR Collaboration), Phys. Lett. B 444, 43 (1998).

6.4. The spin and parity of the ρ-meson have been determined to be 1^-. Show that the decays $\rho^0 \to \pi^0\pi^0$ and $\rho^0 \to 2\gamma$ are forbidden.

6.5. Show that parity invariance implies that the static electric dipole moment of a particle is zero.

6.6. Show that the decays $K^+ \to \pi^+\pi^0$, $K^+ \to \pi^+\pi^+\pi^-$, even if the intrinsic parity of K^+ were not known, would indicate violation of parity invariance.

6.7. Discuss how one could detect parity violation in the hyperon decay $\Lambda^0 \to p + \pi^-$ (Λ^0 has spin $\frac{1}{2}$ and its parity is assumed to be even).

6.8. Derive from the Maxwell's equations the transformation properties under time-reversal of the electric and magnetic fields $\mathbf{E}$ and $\mathbf{B}$.

6.9. Show that a non-zero electric dipole static moment of a particle would indicate, besides parity non-conservation, also violation of time reversal invariance.

7

Relativistic equations

In this Chapter, we examine the relativistic wave-functions which are obtained
starting from the one-particle states considered in Chapter 5 and making use
of the IR's of the orthochronous Lorentz group $\mathcal{L}^\uparrow$. We examine the lower
spin cases (0, 1 and $\frac{1}{2}$) and the corresponding relativistic equations both for
massive and massless particles. Higher integer and half-integer spin cases are
considered for massive particles.

7.1 The Klein-Gordon equation

In Chapter 5 it was shown how the one-particle states can be classified
according to unitary IR's of the Poincaré group. Let us consider the case
$p^2 > 0$, $p^0 > 0$, i.e. a massive particle with positive energy. Its states
$|p, \sigma> \equiv |m, s; \mathbf{p}, \sigma>$ form a complete set, so that a generic one particle
state vector $|\Phi>$ in the Hilbert space H can be expanded as

$$|\Phi> = \sum_\sigma \int \frac{d^3 p}{2p^0} \Phi_\sigma(p)|p, \sigma> , \qquad (7.1)$$

where $p^0 = +\sqrt{\mathbf{p}^2 + m^2}$ and

$$\Phi_\sigma(p) = <p, \sigma|\Phi> . \qquad (7.2)$$

If $|\Phi>$ is normalized to 1, one gets from (5.10)

$$\sum_\sigma \int \frac{d^3 p}{2p^0} |\Phi_\sigma(p)|^2 = 1 . \qquad (7.3)$$

The function $\Phi_\sigma(p)$ has the meaning of probability amplitude and is identified, in quantum mechanics, with a *wave function* in momentum space. Since
it corresponds to the eigenvalue $p^2 = m^2$, it satisfies the condition (*mass-shell
condition*)

G. Costa and G. Fogli, *Symmetries and Group Theory in Particle Physics*,
Lecture Notes in Physics 823, DOI: 10.1007/978-3-642-15482-9_7,
© Springer-Verlag Berlin Heidelberg 2012

$$(p_\mu p^\mu - m^2)\, \Phi_\sigma(p) = 0 \,, \tag{7.4}$$

which is nothing else that the *Klein-Gordon equation* in momentum space.

The Fourier transform

$$\Phi_\sigma(x) = \frac{1}{(2\pi)^{\frac{3}{2}}} \int \frac{d^3p}{2p^0}\, \Phi_\sigma(p) e^{-ipx} \,, \tag{7.5}$$

which defines the wave function in the configuration space, satisfies the Klein-Gordon equation in the coordinate representation

$$(\Box + m^2)\, \Phi_\sigma(x) = 0 \,, \tag{7.6}$$

with the usual notation $\Box = \partial_\mu \partial^\mu$.

It is interesting to examine the transformation properties of the wave function $\Phi_\sigma(p)$ under $\mathcal{P}_+^\uparrow$. Since an element of $\mathcal{P}_+^\uparrow$ transforms the state $|\Phi>$ into $U(a, \Lambda)|\Phi>$, the transformed wave function can be defined by

$$\Phi'_\sigma(p) = <p, \sigma\, |U(a, \Lambda)|\, \Phi> \,. \tag{7.7}$$

From Eqs. (7.1) and (5.14) we obtain

$$U(a, \Lambda)|\Phi> = \sum_{\sigma\sigma'} \int \frac{d^3p}{2p^0}\, \Phi_\sigma(p) e^{-ip'a} D_{\sigma'\sigma}^{(s)}(L_{p'}^{-1}\Lambda L_p)|p', \sigma'> \,, \tag{7.8}$$

where $p' = \Lambda p$. According to the definition (7.7), one gets

$$\Phi'_\sigma(p) = e^{-ipa} \sum_{\sigma'} D_{\sigma\sigma'}^{(s)}(L_p^{-1}\Lambda L_{p''})\Phi_{\sigma'}(\Lambda^{-1}p) \,, \tag{7.9}$$

where $p'' = \Lambda^{-1}p$.

As for the states of massive particle, the transformation properties of the wave function $\Phi_\sigma(p)$ under $\mathcal{P}_+^\uparrow$ are determined by the IR's of the rotation group. For a given spin s, the set $\Phi_\sigma(p)$, with $\sigma = -s, ..., +s$, corresponding to the $(2s + 1)$-component wave function, describes completely the physical system and its transformation properties. This description, however, is not relativistic covariant. Moreover, while $\Phi_\sigma(p)$ transforms in a very simple way under pure rotations, the general transformation (7.9) is formally very simple, but the explicit dependence on p, p'' is rather complicated.

In order to get a covariant formalism, we should replace the bases of the IR's of the rotation group with those of the finite-dimensional IR's of the homogeneous Lorentz group. In this way, the number of components is increased and, according to the discussion in Section 3.4, *supplementary conditions* are needed in order that the basis of an IR $D^{(j,j')}$ of $\mathcal{L}_+^\uparrow$ describes a unique value of the spin. Since it is convenient to deal with wave functions having definite transformation properties under parity, we shall use the bases of the IR's of the orthochronous group $\mathcal{L}^\uparrow$, considered in Section 3.5.

We shall consider here the wave functions for the simplest case of spin 0 and 1. The wave function for a spin zero particle is clearly a scalar (or pseudoscalar, according to parity) under $\mathcal{L}^\uparrow$. It is then described by a function $\Phi(p)$, satisfying the Klein-Gordon equation (7.4), which transforms under an element (a, Λ) of $\mathcal{P}_+^\uparrow$, according to Eq. (7.9), as

$$\Phi'(p) = e^{-ipa}\Phi(\Lambda^{-1}p) . \tag{7.10}$$

The corresponding transformation in the configuration space is obtained by Fourier transform

$$\Phi'(x) = \Phi\left(\Lambda^{-1}(x+a)\right) . \tag{7.11}$$

A particle of spin and parity $J^P = 1^\mp$ can be described in terms of the IR $\mathcal{D}^{(\frac{1}{2},\frac{1}{2},\mp)}$ of $\mathcal{L}^\uparrow$; the relative wave function transforms as a polar or axial four-vector $\Phi^\mu(p)$. According to Eq. (3.99), we have

$$\mathcal{D}^{(\frac{1}{2},\frac{1}{2},\mp)} = \mathcal{D}^{(1,\pm)} \oplus \mathcal{D}^{(0,\mp)} , \tag{7.12}$$

so that, with respect to the subgroup $O(3)$, $\Phi^\mu(p)$ contains also a scalar (or pseudoscalar) component. One needs, in this case, a supplementary condition which leaves only three independent components, corresponding to the three independent states of polarization. It is natural to impose, besides the Klein-Gordon equation

$$\left(p^2 - m^2\right)\Phi^\mu(p) = 0 , \tag{7.13}$$

the invariant condition

$$p_\mu\Phi^\mu(p) = 0 . \tag{7.14}$$

We notice that $p_\mu\Phi^\mu$ is a scalar (or pseudoscalar) quantity. In the rest frame of the particle, Eq. (7.14) becomes

$$p^0\Phi^0 = m\Phi^0 = 0 , \tag{7.15}$$

showing that the time-like component vanishes. The transformation properties under an element (a, Λ) of $\mathcal{P}_+^\uparrow$ are given by

$$\Phi'^\mu(p) = e^{-ipa}\Lambda^\mu{}_\nu\Phi^\nu(\Lambda^{-1}p) . \tag{7.16}$$

In the coordinate representation, the above equations become

$$\left(\Box + m^2\right)\Phi^\mu(x) = 0 , \tag{7.17}$$

$$\partial_\mu\Phi^\mu(x) = 0 , \tag{7.18}$$

$$\Phi'^\mu(x) = \Lambda^\mu{}_\nu\Phi^\nu\left(\Lambda^{-1}(x+a)\right) . \tag{7.19}$$

The general wave function $\Phi^\mu(x)$ is usually expanded as follows

$$\Phi^\mu(x) = \frac{1}{(2\pi)^{\frac{3}{2}}} \sum_{\lambda=1}^{3} \int \frac{d^3p}{2p^0} \, \epsilon^\mu(p,\lambda) \left\{ a(p,\lambda)e^{-ipx} + b^*(p,\lambda)e^{+ipx} \right\} , \tag{7.20}$$

in terms of the plane wave (positive and negative energy) solutions $\sim e^{\mp ipx}$. Since $\Phi^\mu(x)$ has only three independent components, we have introduced in (7.20) three independent four-vectors, $\epsilon^\mu(p, \lambda)$ ($\lambda = 1, 2, 3$), usually called *polarization vectors*, since they specify the state of polarization. It is possible to satisfy Eq. (7.14) for each polarization, independently, by choosing

$$\epsilon(p, 1) = (0, 1, 0, 0) ,$$

$$\epsilon(p, 2) = (0, 0, 1, 0) , \qquad (7.21)$$

$$\epsilon(p, 3) = \left(\frac{p^3}{m}, 0, 0, \frac{p^0}{m} \right) ,$$

where, for the sake of simplicity, we have taken the momentum along the x^3-axis. Thus we have

$$p_\mu \epsilon^\mu(p, \lambda) = 0 , \qquad (7.22)$$

and the vectors are normalized in such a way that

$$\epsilon_\mu(p, \lambda) \epsilon^\mu(p, \nu) = g_{\lambda\nu} . \qquad (7.23)$$

We see that $\epsilon(p, \lambda)$ correspond to transverse polarization for $\lambda = 1, 2$ and to longitudinal polarization for $\lambda = 3$. In the rest frame also the time-component of $\epsilon(p, 3)$ vanishes and the polarization vectors reduce to unit vectors along the axes x^1, x^2, x^3. It is instructive to consider a generic unit vector $\mathbf{n} = (\sin\theta \cos\phi, \sin\theta \sin\phi, \cos\theta)$ corresponding to polarization along a direction specified by the polar angles θ, ϕ. One can immediately write the following proportionality relations with the spherical harmonics $Y_1(\theta, \phi)$:

$$-\sqrt{\tfrac{1}{2}}(n^1 + in^2) \sim Y_1^1(\theta, \phi) ,$$

$$\sqrt{\tfrac{1}{2}}(n^1 - in^2) \sim Y_1^{-1}(\theta, \phi) , \qquad (7.24)$$

$$n^3 \sim Y_1^0(\theta, \phi) .$$

It is clear that the two combinations $(n^1 \pm in^2)$ of transverse polarizations correspond to spin component ± 1 along x^3, and n^3 to spin component 0 along x^3. In a frame of reference in motion along x^3, analogous combinations can be written in terms of (7.21), corresponding to the helicity eigenvalues ± 1 and 0.

7.2 Extension to higher integer spins

The previous considerations can be generalized to the cases of higher integer spins. The wave function of a massive particle of spin s can be obtained from the irreducible tensor of rank s

$$\Phi^{\mu_1 \mu_2 \cdots \mu_s}(p) , \qquad (7.25)$$

which is *completely symmetrical and traceless*

$$g_{\mu\nu}\Phi^{\mu\nu\cdots\mu_s}(p) = 0 \ . \qquad (7.26)$$

Moreover, it satisfies the Klein-Gordon equation

$$\left(p^2 - m^2\right)\Phi^{\mu_1\mu_2\cdots\mu_s}(p) = 0 \ . \qquad (7.27)$$

Such tensor can be taken as the basis of the IR $D^{(\frac{s}{2},\frac{s}{2})}$ of $\mathcal{L}_+^\uparrow$ (we disregard here parity considerations; parity can be included going to $\mathcal{L}^\uparrow$, as indicated in the previous Section). It reduces with respect to the subgroup $SO(3)$ as follows (see Eq. (3.67))

$$D^{(\frac{s}{2},\frac{s}{2})} = D^{(s)} \oplus D^{(s-1)} \oplus \ldots \oplus D^{(0)} \ . \qquad (7.28)$$

The s^2 components corresponding to $D^{(0)}$, ..., $D^{(s-1)}$ are eliminated by the supplementary condition

$$p_\mu \Phi^{\mu\mu_2\cdots\mu_s}(p) = 0 \ . \qquad (7.29)$$

In the rest frame of the particle, the above condition shows that all components of the type $\Phi^{0\mu_2\cdots\mu_s}$, with at least one index $\mu_i = 0$, vanish. The above equations can be trivially written in the coordinate representation.

As an example, let us consider the case of spin 2. The wave function is the symmetrical traceless tensor

$$\Phi^{\mu\nu}(p) = \Phi^{\nu\mu}(p) \ , \qquad g_{\mu\nu}\Phi^{\mu\nu}(p) = 0 \ , \qquad (7.30)$$

which has 9 independent components, corresponding to spin values 0, 1 and 2. The condition

$$p_\mu \Phi^{\mu\nu}(p) = 0 \qquad (7.31)$$

eliminates the 4 components corresponding to spin 0 and 1.

7.3 The Maxwell equations

The above considerations can be applied, with the appropriate modifications, to the case of massless particles. We saw in Section 5.3 that, in this case, there exist at most *two distinct states of helicity* (i.e. polarization) for a spin different from zero.

We shall examine only the case of a spin 1 massless particle, specifically the photon. The wave function, denoted by $A^\mu(x)$ in the configuration space, is still described by a four-vector which satisfies the equations

$$\Box A^\mu(x) = 0 \ , \qquad (7.32)$$

$$\partial_\mu A^\mu(x) = 0 \ . \qquad (7.33)$$

In terms of the antisymmetric tensor $F^{\mu\nu}$ defined by

$$F^{\mu\nu} = \partial^\mu A^\nu(x) - \partial^\nu A^\mu(x) , \tag{7.34}$$

they become

$$\partial_\mu F^{\mu\nu} = 0 , \tag{7.35}$$

i.e. the Maxwell equations in covariant form (the tensor $F^{\mu\nu}$ describes, in fact, the electromagnetic fields). We note that Eqs. (7.32) and (7.35) are equivalent only if the supplementary condition (7.33), called *Lorentz condition*, is satisfied.

In the present case, however, we have to reduce to *two* the number of independent components of $A^\mu(x)$. One makes use of the fact that the theory of a massless vector field is invariant under the transformation

$$A_\mu(x) \to A'_\mu(x) = A_\mu(x) + \partial_\mu \chi(x) , \tag{7.36}$$

which is called "gauge transformation of the second kind". The above transformation, where $\chi(x)$ is an arbitrary scalar function, leaves the tensor $F^{\mu\nu}(x)$ invariant; Eqs. (7.32), (7.33) are also unchanged if $\chi(x)$ satisfies the equation

$$\Box \chi(x) = 0 . \tag{7.37}$$

This condition leaves still a great deal of arbitrariness in the possible choices of $\chi(x)$. By a convenient choice, corresponding to the so-called *radiation* or *Coulomb gauge*, one can impose, besides Eq. (7.33),

$$\mathbf{\nabla} \cdot \mathbf{A}(x) = 0 . \tag{7.38}$$

The physical meaning of Eq. (7.38) becomes clear going to the Fourier transform $a^\mu(k)$

$$A^\mu(x) = \frac{1}{(2\pi)^{\frac{3}{2}}} \int \frac{d^3 p}{2p^0} a^\mu(k) e^{-ikx} , \tag{7.39}$$

which gives

$$\mathbf{k} \cdot \mathbf{a}(k) = 0 . \tag{7.40}$$

The two independent components of $a^\mu(k)$ are those perpendicular to the direction of motion $\mathbf{k}$, and they correspond to the two independent transverse polarization states of the photon.

As done for the massive vector field, it is convenient to expand $A^\mu(x)$ in terms of the plane wave solutions of the Klein-Gordon equation (we note that $A^\mu(x)$ is a real field)

$$A^\mu(x) = \frac{1}{(2\pi)^{\frac{3}{2}}} \int \frac{d^3 k}{2k^0} \sum_{\lambda=0}^{3} \epsilon^\mu(k, \lambda) \left\{ a(k, \lambda) e^{-ikx} + a^*(k, \lambda) e^{+ikx} \right\} . \tag{7.41}$$

In the present case of massless particle, it is convenient to introduce four polarization vectors $\epsilon(k, \lambda)$, $(\lambda = 0, 1, 2, 3)$. We choose them in the following way (taking $\mathbf{k}$ along x^3):

$$\epsilon(k,0) = (1,0,0,0) \,,$$

$$\epsilon(k,1) = (0,1,0,0) \,,$$

$$\epsilon(k,2) = (0,0,1,0) \,,$$

$$\epsilon(k,3) = \left(0,0,0,\frac{k^3}{k^0}\right) \,.$$

(7.42)

Since $m = 0$, it is no longer possible to satisfy the analogue of Eq. (7.22) for all λ; in fact, from (7.42) one gets

$$k_\mu \epsilon^\mu(k,\lambda) = 0 \qquad (\lambda = 1,2) \,,$$

$$k_\mu \epsilon^\mu(k,0) = +k^0 \,,$$

$$k_\mu \epsilon^\mu(k,3) = -k^0 \,.$$

(7.43)

The Lorentz condition (7.33) then implies a simple relation between the longitudinal and time-like components

$$a(k,3) - a(k,0) = 0 \,.$$ (7.44)

Using the gauge invariance, we can replace, according to (7.36), Eq. (7.41) by

$$A'_\mu(x) = \frac{1}{(2\pi)^{\frac{3}{2}}} \int \frac{d^3k}{2k^0} \left\{ \left[\sum_{\lambda=0}^{3} \epsilon_\mu(k,\lambda)a(k,\lambda) + ik_\mu \tilde{\chi}(k) \right] e^{-ikx} + c.c. \right\} \,,$$

(7.45)

where $\tilde{\chi}(k)$ is defined by

$$\chi(x) = \frac{1}{(2\pi)^{\frac{3}{2}}} \int \frac{d^3k}{2k^0} \left\{ \tilde{\chi}(k)e^{-ikx} + \tilde{\chi}^*(k)e^{+ikx} \right\} \,.$$ (7.46)

Equation (7.45) defines new components $a'(k,\lambda)$, which are related to $a(k,\lambda)$ by

$$\sum_{\lambda=0}^{3} \epsilon^\mu(k,\lambda) \left\{ a'(k,\lambda) - a(k,\lambda) \right\} = ik^\mu \tilde{\chi}(k) \,.$$ (7.47)

This equation is satisfied by

$$a'(k,\lambda) = a(k,\lambda) \qquad (\lambda = 1,2) \,,$$

$$a'(k,0) = a(k,0) + ik^0 \tilde{\chi}(k) \,,$$

$$a'(k,3) = a(k,3) + ik^0 \tilde{\chi}(k) \,,$$

(7.48)

and one can choose $\tilde{\chi}(k)$ in such a way to eliminate both $a'(k,0)$ and $a'(k,3)$, which are equal according to (7.44).

Then one is left with two independent components, relative to the polarization vectors $\epsilon(k,1)$ and $\epsilon(k,2)$. They represent linear polarizations, along to the x^1 and x^2 axes, respectively. Defining, in analogy to (7.24)

$$\epsilon(k,+) = \{\epsilon(k,1) + i\epsilon(k,2)\} \ ,$$
$$\epsilon(k,-) = \{\epsilon(k,1) - i\epsilon(k,2)\} \ ,$$
(7.49)

we obtain a description in terms of *right-handed and left-handed circular polarization*; they correspond to helicity $+1$ and -1, respectively.

7.4 The Dirac equation

In the following, we shall derive in detail the relativistic equations for half-integer spin particles according to the scheme outlined in Section 7.1.

We shall examine here the case of a massive spin $\frac{1}{2}$ particle. The appropriate IR of the group $\mathcal{L}^\uparrow$ is in this case $\mathcal{D}^{(\frac{1}{2},0)}$, which contains, according to Eq. (3.103), both lowest spinor IR's of $\mathcal{L}^\uparrow_+$

$$\mathcal{D}^{(\frac{1}{2},0)} = D^{(\frac{1}{2},0)} \oplus D^{(0,\frac{1}{2})} \ .$$
(3.103)

As seen in Section 3.4, the bases of these two IR's of $\mathcal{L}^\uparrow_+$ consist in the spinors ξ and $\eta^\dagger$, and the following correspondence holds

$$D^{(\frac{1}{2},0)} \to A \ ,$$
$$D^{(0,\frac{1}{2})} \to (A^\dagger)^{-1} \ .$$
(7.50)

In Section 3.5 it was noticed that the two IR's are interchanged into one another by space inversion.

The above considerations lead naturally to the introduction of a four-component (Dirac) spinor defined by

$$\psi = \begin{pmatrix} \xi \\ \eta^\dagger \end{pmatrix} \ .$$
(7.51)

We write explicitly its transformation properties under $\mathcal{L}^\uparrow_+$:

$$\psi \to \psi' = S(\Lambda)\,\psi \ ,$$
(7.52)

where

$$S(\Lambda) = \begin{pmatrix} A & 0 \\ 0 & (A^\dagger)^{-1} \end{pmatrix} \ ,$$
(7.53)

and under space inversion (see Eq. (3.93))

$$\psi \rightarrow \psi' = \begin{pmatrix} \xi' \\ \eta^{\dagger'} \end{pmatrix} = i \begin{pmatrix} 0 & I \\ I & 0 \end{pmatrix} \begin{pmatrix} \xi \\ \eta^{\dagger} \end{pmatrix} . \tag{7.54}$$

We shall now investigate the connection with the one particle states $|p, \sigma>$, where in this case $\sigma = \pm \frac{1}{2}$.

According to Eq. (7.1), we can define a two-component (spinor) wave function $\Phi_\sigma(p)$, satisfying the mass-shell condition (7.4). Its transformation properties under $\mathcal{P}_+^\uparrow$ are given by (7.9), which we rewrite here, neglecting translations for the sake of simplicity,

$$\Phi'_\sigma(p) = \sum_{\sigma'} D_{\sigma\sigma'}^{(\frac{1}{2})}(L_p^{-1} \Lambda L_{p''}) \, \Phi_{\sigma'}(\Lambda^{-1}p) , \tag{7.55}$$

where $p'' = \Lambda^{-1}p$.

It is more convenient to introduce the following spinor wave-function[1]

$$\phi_\tau(p) = \sum_\sigma D_{\tau\sigma}(L_p) \, \Phi_\sigma(p) . \tag{7.56}$$

From this definition and Eq. (7.55), one can derive the transformation properties of $\phi_\tau(p)$ under $\mathcal{L}_+^\uparrow$:

$$\phi'_\tau(p) = \sum_\rho D_{\tau\rho}(\Lambda) \, \phi_\rho(\Lambda^{-1}p) . \tag{7.57}$$

Note that $\phi_\tau(p)$ transforms in a much simpler way than $\Phi_\sigma(p)$: in fact Eq. (7.57) is the analogue of (7.16) for the two-dimensional IR $D^{(\frac{1}{2},0)}(\Lambda)$ of $\mathcal{L}_+^\uparrow$.

Together with $\phi_\tau(p)$, we define another spinor wave function

$$\chi_\tau(p) = \sum_\sigma D_{\tau\sigma}(L_p^{\dagger^{-1}}) \, \Phi_\sigma(p) . \tag{7.58}$$

Its transformation properties under $\mathcal{L}_+^\uparrow$ are obtained from (7.16), taking into account the unitarity of $D^{(\frac{1}{2})}(L_p^{-1}\Lambda L_{p''})$

$$\chi'_\tau(p) = \sum_\rho D_{\tau\rho}(\Lambda^{\dagger^{-1}}) \, \chi_\rho(\Lambda^{-1}p) . \tag{7.59}$$

From the correspondence (7.50) we see that the two-component wave functions $\phi_\tau(p)$ and $\chi_\tau(p)$ transform as the spinors ξ and $\eta^\dagger$, respectively. A four-component (Dirac) wave function is then defined by

---

[1] Here and in the following we denote simply by $D(\Lambda)$ an element of the IR $D^{(\frac{1}{2},0)}(\Lambda)$ of $\mathcal{L}_+^\uparrow$. $D(\Lambda)$, in general, is not unitary and $D(\Lambda^{\dagger^{-1}})$ is an element of the IR $D^{(0,\frac{1}{2})}(\Lambda)$. For a pure rotation R, $D(R)$ reduce to the unitary IR $D^{(\frac{1}{2})}(R)$ of $SO(3)$.

$$\psi(p) = \begin{pmatrix} \phi(p) \\ \chi(p) \end{pmatrix} , \tag{7.60}$$

which transforms similarly to (7.52) under $\mathcal{L}_+^\uparrow$, i.e.

$$\psi'(p) = S(A)\,\psi(\Lambda^{-1}p) . \tag{7.61}$$

The two spinor wave functions $\phi(p)$ and $\chi(p)$ are not independent, but they are related to one another. In terms of $\psi(p)$, this relation becomes the Dirac equation. In fact, let us write Eqs. (7.56), (7.58) in matrix form

$$\phi(p) = D(L_p)\Phi(p) ,$$
$$\chi(p) = D(L_p^{\dagger-1})\Phi(p) . \tag{7.62}$$

Eliminating $\Phi(p)$ one gets

$$\phi(p) = D(L_p)D(L_p^\dagger)\,\chi(p) , \tag{7.63}$$

or

$$\chi(p) = D(L_p^{\dagger-1})D(L_p^{-1})\,\phi(p) . \tag{7.64}$$

From the correspondence (7.50) and the explicit expression (3.70), i.e.

$$D(L_p) = e^{-\frac{1}{2}\psi\boldsymbol{\sigma}\cdot\mathbf{n}} , \tag{7.65}$$

where $\mathbf{n} = \mathbf{p}/|\mathbf{p}|$, $\cosh\psi = \gamma = p^0/m$, Eqs. (7.63), (7.64) become

$$\phi(p) = \frac{1}{m}(p^0 - \boldsymbol{\sigma}\cdot\mathbf{p})\,\chi(p) , \tag{7.66}$$

$$\chi(p) = \frac{1}{m}(p^0 + \boldsymbol{\sigma}\cdot\mathbf{p})\,\phi(p) . \tag{7.67}$$

The above equations can be written in compact form, in terms of the Dirac spinor (7.60). Introducing the four matrices

$$\gamma^0 = \begin{pmatrix} 0 & I \\ I & 0 \end{pmatrix} , \qquad \gamma^i = \begin{pmatrix} 0 & \sigma_i \\ -\sigma_i & 0 \end{pmatrix} , \tag{7.68}$$

and the notation

$$\not{p} = \gamma^\mu p_\mu , \tag{7.69}$$

one gets

$$(\not{p} - m)\psi(p) = 0 , \tag{7.70}$$

which is the Dirac equation.

The four γ-matrices satisfy the anticommutation relations

$$\gamma^\mu\gamma^\nu + \gamma^\nu\gamma^\mu = 2g^{\mu\nu} , \tag{7.71}$$

and the hermiticity (and anti-hermiticity) properties

$$\gamma^{0\dagger} = \gamma^0 , \qquad \gamma^{i\dagger} = -\gamma^i . \tag{7.72}$$

These properties are abstracted from the explicit expression (7.68), which represents only a particular choice. Another useful choice is that in which γ^0 is diagonal

$$\gamma^{0\prime} = \begin{pmatrix} I & 0 \\ 0 & -I \end{pmatrix} , \qquad \gamma^{i\prime} = \gamma^i = \begin{pmatrix} 0 & \sigma_i \\ -\sigma_i & 0 \end{pmatrix} . \tag{7.73}$$

The representations (7.68), (7.73) are related by

$$\gamma^{\mu\prime} = V\gamma^\mu V^{-1} , \tag{7.74}$$

where

$$V = \frac{1}{\sqrt{2}} \begin{pmatrix} I & I \\ -I & I \end{pmatrix} . \tag{7.75}$$

In general, the γ-matrices are defined by (7.71) and it is not necessary, for many purposes, to use a specific representation.

One can check that the Dirac equation (7.70) is invariant under a Lorentz (in general Poincaré) transformation, provided the γ-matrices satisfy the condition

$$S^{-1}(\Lambda)\gamma^\mu S(\Lambda) = \Lambda^\mu{}_\nu \gamma^\nu , \tag{7.76}$$

where the matrix $S(\Lambda)$ is defined by Eqs. (7.52) and (7.53).

The *adjoint spinor*, defined by

$$\overline{\psi}(p) = \psi^\dagger(p)\gamma^0 , \tag{7.77}$$

satisfies the equation

$$\overline{\psi}(p)(\not{p} - m) = 0 , \tag{7.78}$$

The corresponding Dirac equations for the coordinate space wave functions $\psi(x)$ and $\overline{\psi}(x)$ are given by

$$(i\gamma^\mu \partial_\mu - m)\,\psi(x) = 0 , \tag{7.79}$$

$$\overline{\psi}(x)(i\gamma^\mu \overleftarrow{\partial}_\mu - m) \equiv i\partial_\mu\overline{\psi}(x)\gamma^\mu - m\overline{\psi}(x) = 0 , \tag{7.80}$$

as one can easily check making use of the Fourier transform (7.5).

Finally, we list the transformation properties of the Dirac spinor $\psi(x)$ under the discrete operations. They are obtained from (7.60) and the transformation properties of ξ and $\eta^\dagger$ given in Section 3.5. We first define a fifth matrix[2]

---

[2] Definitions which differ by a factor (-1) or $(\pm i)$ can be found in the literature. The choice (7.81) corresponds to a hermitian matrix.

$$\gamma_5 = \gamma^5 = i\gamma^0\gamma^1\gamma^2\gamma^3 \,, \tag{7.81}$$

which satisfies

$$\gamma^\mu\gamma^5 + \gamma^5\gamma^\mu = 0 \tag{7.82}$$

and, in the representation (7.68), is given explicitly by

$$\gamma_5 = \begin{pmatrix} I & 0 \\ 0 & -I \end{pmatrix} . \tag{7.83}$$

Then we get (with $\underline{p} = (p^0, -\mathbf{p})$)

$$\psi(p) \xrightarrow{I_s} \psi'(p) = i\gamma^0\psi(\underline{p}) \,, \tag{7.84}$$

$$\psi(p) \xrightarrow{I_t} \psi'(p) = -\gamma^0\gamma_5\psi(-\underline{p}) \,, \tag{7.85}$$

$$\psi(p) \xrightarrow{I_{st}} \psi'(p) = i\gamma_5\psi(-p) \,. \tag{7.86}$$

In terms of the spinor $\psi(p)$, one can build bilinear quantities with a definite transformation properties under $\mathcal{L}^\uparrow$, as follows

scalar	$\overline{\psi}(p)\psi(p) \,,$
pseudoscalar	$\overline{\psi}(p)\gamma_5\psi(p) \,,$
four–vector	$\overline{\psi}(p)\gamma^\mu\psi(p) \,,$
pseudo–vector	$\overline{\psi}(p)\gamma_5\gamma^\mu\psi(p) \,,$
antisymmetric tensor	$\overline{\psi}(p)[\gamma^\mu, \gamma^\nu]\psi(p) \,.$

$$\tag{7.87}$$

7.5 The Dirac equation for massless particles

In the case of a masless particle, the Dirac equation (7.70) becomes

$$\not{p}\psi(p) = 0 \,. \tag{7.88}$$

Since $\psi(p)$ satisfies the Klein-Gordon Eq. (7.4) with $m = 0$, we have

$$\left(p^{0^2} - \mathbf{p}^2\right)\psi(p) = 0 \,, \tag{7.89}$$

i.e. $p^0 = \pm|\mathbf{p}|$.

In terms of the two-component spinors $\phi(p)$ and $\chi(p)$, Eq. (7.88) can be splitted into two equations

$$(p^0 + \boldsymbol{\sigma} \cdot \mathbf{p})\,\phi(p) = 0 \,, \tag{7.90}$$

$$(p^0 - \boldsymbol{\sigma} \cdot \mathbf{p})\,\chi(p) = 0 \,, \tag{7.91}$$

The two-component spinors are eigenfunctions of helicity $\lambda = \boldsymbol{\sigma} \cdot \mathbf{p}/|\mathbf{p}|$. For positive energy $p^0 = +|\mathbf{p}|$, one gets

$$\frac{\boldsymbol{\sigma} \cdot \mathbf{p}}{|\mathbf{p}|}\phi(p) = -\phi(p) \, , \tag{7.92}$$

$$\frac{\boldsymbol{\sigma} \cdot \mathbf{p}}{|\mathbf{p}|}\chi(p) = +\chi(p) \, , \tag{7.93}$$

i.e. $\phi(p)$ corresponds to the eigenvalue $\lambda = -1$ and $\chi(p)$ to $\lambda = +1$. The two eigenvalues are interchanged for the negative energy solutions $p^0 = -|\mathbf{p}|$.

As mentioned already in Section 6.1, there is clear experimental evidence that the *neutrino* has negative helicity, i.e. spin antiparallel to the momentum $\mathbf{p}$. If we choose, by convention, the neutrino to be the particle, we have to describe it with the wave-function $\phi(p)$ satisfying Eq. (7.90) with $p^0 = +|\mathbf{p}|$. Its antiparticle, the *antineutrino*, will be described by the solution of Eq. (7.90) corresponding to $p^0 = -|\mathbf{p}|$, and *positive helicity*. We assume that neutrinos are massless; this is a good approximation for the present purposes, since their masses are much smaller than the electron mass.

We see that a neutrino-antineutrino pair can be described in terms of the two-components spinor Eq. (7.90), which is called *Weyl equation*. Since $\phi(p)$ corresponds to the IR $D^{(\frac{1}{2},0)}$ of $\mathcal{L}_+^\uparrow$, it is clear that the two-component theory is not invariant under parity. In fact, a state with $p^0 = +|\mathbf{p}|$ and $\lambda = -1$ is transformed under parity ($\mathbf{p} \to -\mathbf{p}$, $\boldsymbol{\sigma} \to \boldsymbol{\sigma}$) into a state with $p^0 = +|\mathbf{p}|$ and $\lambda = +1$, which does not exist for the two-component spinor $\phi(p)$. This description is consistent with the experimental evidence: neutrinos and antineutrinos appear only in the helicity eigenstates -1 and $+1$, respectively.

The Weyl two-component theory is equivalent to a Dirac description, in which the four-component spinors are required to satisfy a specific condition. One can easily verify that the linear combinations

$$\psi_R(p) = \tfrac{1}{2}(I + \gamma_5)\,\psi(p) = \begin{pmatrix} \phi(p) \\ 0 \end{pmatrix} \tag{7.94}$$

and

$$\psi_L(p) = \tfrac{1}{2}(I - \gamma_5)\,\psi(p) = \begin{pmatrix} 0 \\ \chi(p) \end{pmatrix} \tag{7.95}$$

which are solutions of the Dirac equation (7.88) have the required properties. They are eigenfunctions of γ_5 corresponding to the two eigenvalues $+1$ and -1, which are denoted as positive and negative *chirality*, respectively. The above relation is easily obtained using the explicit expression (7.83) for γ_5. Dirac spinors of the type (7.94) and (7.95) are selected by imposing the condition

$$\tfrac{1}{2}(I \mp \gamma_5)\,\psi(p) = 0 \, , \tag{7.96}$$

which is clearly invariant under the restricted Lorentz group $\mathcal{L}_+^\uparrow$.

For massless particles, helicity coincides with chirality and is relativistic invariant, i.e. it is not affected by a Lorentz boost. However, for massive particles, helicity and chirality must be distinguished. In particular, mass eigenstates have no definite chirality but they can be simultaneously helicity eigenstates.

In the case of neutrinos, the difference between helicity and chirality eigenvalues is of the order of $(m_\nu/|p|)^2$ and, in practice, it can be neglected since the neutrino masses are very small in comparison with the usual values of the momentum.

7.6 Extension to higher half-integer spins

The previous analysis of spin $\frac{1}{2}$ particles can be extended to massive fermions with higher spin following two main approaches. In the first approach, due to Fierz and Pauli[3], the wave function is a higher rank spinor built from the spinors ξ and $\eta^\dagger$, which are the bases of the IR's $D^{(\frac{1}{2},0)}$ and $D^{(0,\frac{1}{2})}$, while in the second, due to Rarita and Schwinger[4], the wave function has mixed transformation properties, i.e. it transforms as a Dirac spinor and a completely symmetric tensor; supplementary conditions are applied in both approaches.

We shall follow the Rarita-Schwinger formalism and consider explicitly the case of spin $\frac{3}{2}$. In this case, the wave function is described by a set of four Dirac spinors

$$\psi^\mu = \begin{pmatrix} \psi_1^\mu \\ \psi_2^\mu \\ \psi_3^\mu \\ \psi_4^\mu \end{pmatrix} \qquad (\mu = 0, 1, 2, 3) , \qquad (7.97)$$

which, in turn, can be considered the components of a four-vector. They satisfy the Dirac equation

$$(\not{p} - m)\psi^\mu(p) = 0 , \qquad (7.98)$$

and one has to impose

$$\gamma_\mu \psi^\mu(p) = 0 . \qquad (7.99)$$

The spin-tensor $\psi^\mu(p)$ corresponds to the basis of the direct product representation

$$D^{(\frac{1}{2},\frac{1}{2})} \otimes \left(D^{(\frac{1}{2},0)} \oplus D^{(0,\frac{1}{2})} \right) = D^{(1,\frac{1}{2})} \oplus D^{(\frac{1}{2},1)} \oplus D^{(\frac{1}{2},0)} \oplus D^{(0,\frac{1}{2})} , \qquad (7.100)$$

[3] H. Fierz and W. Pauli, Proc. Roy. Soc. **A173**, 211 (1939).
[4] W. Rarita and J. Schwinger, Phys. Rev. **60**, 61 (1941).

so that it contains, besides the spin $\frac{3}{2}$ part, also a spin $\frac{1}{2}$ part (see Eq. (3.103)). The latter part is eliminated by the condition (7.99). In fact, the quantity

$$\gamma_5 \gamma_\mu \psi^\mu(p) \tag{7.101}$$

behaves as a Dirac spinor and one can easily show that it satisfies the Dirac equation; it is clearly eliminated by the condition (7.99). From (7.98) and (7.99) the condition

$$p_\mu \psi^\mu(p) = 0 \tag{7.102}$$

follows. It can be obtained by multiplying Eq. (7.98) on the left by γ_μ and making use of the anticommutation properties (7.71).

One can check that the independent components are four, which is the right number needed for the description of the spin $\frac{3}{2}$ states. We know that, in the rest frame of the particle, a Dirac spinor reduces to a *two-component* spinor, e.g. to $\phi(p)$ if one choose $p^0 > 0$. Making use of the representation (7.73) for the γ-matrices, Eq. (7.99) gives

$$\begin{pmatrix} \phi^0 - \boldsymbol{\sigma} \cdot \boldsymbol{\chi} \\ \chi^0 - \boldsymbol{\sigma} \cdot \boldsymbol{\phi} \end{pmatrix} = 0 . \tag{7.103}$$

Since at rest $\chi = 0$, for $p^0 > 0$ we get

$$\begin{aligned} \phi^0 &= 0 , \\ \sigma_i \phi^i &= 0 , \end{aligned} \tag{7.104}$$

so that we are left with only 2×2 independent components.

Following a similar approach, one can obtain the wave function for higher half-integer spin cases. We introduce the spin-tensor $\psi^{\mu_1 \cdots \mu_s}$, completely symmetrical in the tensor indeces $\mu_1 ... \mu_s$. It satisfies the Dirac equation

$$(\not{p} - m)\psi^{\mu_1 \cdots \mu_s}(p) = 0 \tag{7.105}$$

and the supplementary condition

$$\gamma_\mu \psi^{\mu \mu_2 \cdots \mu_s}(p) = 0 . \tag{7.106}$$

From the above equations one can easily derive the conditions

$$p_\mu \psi^{\mu \mu_2 \cdots \mu_s}(p) = 0 \tag{7.107}$$

and

$$g_{\mu\nu} \psi^{\mu\nu \cdots \mu_s}(p) = 0 . \tag{7.108}$$

The first condition is obtained by the same procedure indicated for (7.99), and the traceless condition is simply obtained by multiplying Eq. (7.106) on the left by γ_ν and using the symmetry of $\psi^{\mu_1 \cdots \mu_s}$ and the anticommutation relations of the γ-matrices. With respect to the tensor components we see that the above conditions (7.107), (7.108) are those needed to describe the integer spin s (see Section 7.2). Then, the spin-tensor $\psi^{\mu_1 \cdots \mu_s}$ corresponds to the combination of spin s and spin $\frac{1}{2}$: the total spin can be $s + \frac{1}{2}$ and $s - \frac{1}{2}$, but the lower value is eliminated by the condition (7.107).

Problems

7.1. Discuss the alternative possibility of describing the wave function of a spin 1 particle in terms of the IR $D^{(1,0)}$ of $\mathcal{L}_+^\uparrow$.

7.2. Show that the wave equation (7.17) and the supplementary condition (7.18) are equivalent to the so-called Proca equations

$$\partial_\mu f^{\mu\nu} + m^2 \Phi^\nu = 0 \qquad \text{where} \qquad f^{\mu\nu} = \partial^\mu \Phi^\nu - \partial^\nu \Phi^\mu \ .$$

7.3. Check the Lorentz covariance of the Dirac equation making use of Eq. (7.76).

7.4. Derive explicitly $S(\Lambda)$ of Eq. (7.76) in the form

$$S(\Lambda) = e^{-\frac{1}{4} i \sigma_{\mu\nu} \omega^{\mu\nu}} \qquad \text{with} \qquad \sigma_{\mu\nu} = \tfrac{1}{2} i [\gamma_\mu, \gamma_\nu] \ .$$

7.5. Show that the bilinear quantities (7.87) have the stated transformation properties under $\mathcal{L}^\uparrow$.

7.6. Determine the behaviour under time reversal of the bilinear quantities (7.87).

8

Unitary symmetries

This Chapter is devoted to the analysis of some applications of group theory to particle physics and, in particular, to the wide use of unitary groups and their representations. Their role has been extremely important in the investigation of the particle interactions, since they led to the discovery of hidden symmetries and new invariance principles which have, in general, no analogues in classical physics and which provide the basic frame for all theoretical schemes.

8.1 Introduction

The investigation of the structure of matter shows a hierarchy of levels, each of which is, to some extent, independent of the others. At the first level, ordinary matter is described in terms of atoms composed of nuclei and electrons; at a smaller scale, atomic nuclei reveal their structure, which is described in terms of nucleons (protons and neutrons); at a much smaller scale, even nucleons appear to be composed of more fundamental objects: the quarks. To go from the level of nucleons to that of quarks took a few decades of the last century and required a lot of theoretical and experimental endeavor and investigations.

Group theory and symmetry principles were applied to the description of the phenomena in all levels of matter; in particular, they were extensively applied to the study of atomic nuclei and elementary particles. We think that, for understanding the importance of group theory for physics, it is very useful to give a brief historical review of the different steps through which the physics of elementary particles passed and developed.

We shall concentrate in the following on several aspects at the level of elementary particles, from nucleons to quark, where the use of unitary symmetry has been extremely useful.

With the study of cosmic ray interactions and the advent of high energy accelerators, lot of new particles were discovered: some seemed to be excited states of nucleons and pions; others showed new strange properties, and they were called strange particles. A new quantum number, called *strangeness* and denoted by S, was assigned to these particles to distinguish them from nucleons and pions characterized by $S = 0$. All these particles were participating in strong interactions and they were denoted as *hadrons*. A striking feature of hadrons is that they occur in small families of particles with approximately equal masses. The members of each family are very few (no more than four); they have the same spin, parity, strangeness, etc. and they differ only in their electric charges. This fact suggests to classify the different families according to the lower representations of $SU(2)$. In analogy with the ordinary spin, the concept of *isotopic spin* (or *isospin*), first introduced by Heisenberg, was employed to describe the new internal symmetry, interpreted as *charge independence* of the strong forces. This symmetry would be exact in the absence of electromagnetic interactions, but it can be applied with a very good approximation to nuclear and particle reactions, since the electric charge effects are much weaker than those of strong interactions.

The analysis of the hadronic spectra of hadrons revealed a higher symmetry which provided a deeper insight in the classification and properties of hadrons. Different isospin multiplets with different values of S could be grouped in higher multiplets of dimensionalities 8 and 10. Each component of the multiplet was identified by two quantum numbers, i.e. the *electric charge* and the *strangeness*, and therefore one was looking for a group of rank 2, containing the isospin $SU(2)$ group as a subgroup. The right choice was the $SU(3)$ group, introduced independently by Gell-Mann and Ne'eman, and whose irreducible representations could accomodate all known families of hadrons. The $SU(3)$ symmetry appeared as an approximate and broken symmetry, but it was very useful because also its breaking could be described in terms of group representations.

First, one arrived, from the analysis of the structure of the hadron multiplets, to the introduction of three *quarks* as the fundamental constituents of matter. These quarks, named *up*, *down* and *strange*, explained the origin of the approximate mass degeneracy of the members of each multiplet, since these quarks are much lighter than the typical hadron scale. Each quark was identified by a specific quantum number, called *flavor*, and consequently the symmetry group was named *flavor* $SU(3)$. Then other peculiar features of the hadron multiplets required the introduction of an additional property of quarks, called *color*, and of a more fundamental symmetry, i.e. the *color* $SU(3)$, which up to now seems to be an exact symmetry.

In the following, after a general discussion of the different kinds of symmetries employed in the theory of elementary particles, we shall examine in detail the unitary symmetries, based on the unitary groups. First of all, we shall consider the group $U(1)$ and its role in describing the conservation of additive quantum numbers. Then we shall consider the isospin $SU(2)$ symmetry, which was applied successfully both to nuclear and particle physics, and the flavor $SU(3)$ symmetry with which it was possible to classify all ordinary and strange hadrons. As already pointed out, they are approximate symmetries, emerging from the approximate mass degeneracy of the hadronic states.

Besides the three quarks, which are the basic ingredient of the flavor $SU(3)$ symmetry, later experimental discoveries indicated the existence of new heavier quarks, first of all the *charm* quark. This led to the introduction of higher approximate symmetry groups, such as $SU(4)$; however, they were of limited utility, due to the large mass differences involved and, consequently, to a big symmetry breaking. A different extension of $SU(3)$ was the combination of the flavor symmetry with the ordinary spin $SU(2)$ symmetry in the hybrid group $SU(6)$. Even if nowadays these extensions have lost most of their interest, they had an important role in the development of the particle theory and, since they represent useful applications of the unitary groups, we shall discuss them briefly.

On the other hand, it was the approximate flavor $SU(3)$ symmetry which lead indirectly to the introduction of the exact color $SU(3)$ symmetry. We shall examine this symmetry in a separate Section, deferring the discussion of its implications to the next Chapter.

8.2 Generalities on symmetries of elementary particles

It is well known that the symmetry properties of a mechanical system, expressed as invariance under the transformations of a group, lead to physical implications for the quantum mechanical states of the system.

As discussed in detail in Chapter 5, the states of one or more particles in the Hilbert space are described in terms of the IR's of the Poincaré group $\mathcal{P}$. This follows from the assumption that the physical systems are invariant under any Poincaré transformations, i.e. space-time translations, rotations and Lorentz transformations, which we can define as *geometrical* transformations. In particular, the states of a massive particle are denoted by $|m, s; \mathbf{p}, \sigma>$ or simply $|p, \sigma>$.

Let us suppose for a moment that we are dealing with a larger group $\mathcal{P} \otimes \mathcal{G}$, where $\mathcal{P}$ is the Poincaré group and $\mathcal{G}$ a compact group whose transformations do not involve space-time coordinates or spin, and whose generators Q_α $(\alpha = 1, 2, \ldots, n)$ commute with those of the Poincaré group

$$[Q_\alpha, P_\mu] = 0 , \tag{8.1}$$

$$[Q_\alpha, M_{\mu\nu}] = 0 . \tag{8.2}$$

In particular, defining $M^2 = P^2$, one has

$$[Q_\alpha, M^2] = 0 . \tag{8.3}$$

In the Hilbert space, the transformations of the group $\mathcal{G}$ will be described by *unitary* operators, and the generators Q_α by *hermitian* operators which satisfy a Lie algebra

$$[Q_\alpha, Q_\beta] = c_{\alpha\beta\gamma} Q_\gamma . \tag{8.4}$$

The states in the Hilbert space can be labelled by the eigenvalues of the Casimir operator and of other additional commuting operators. We denote by λ the set of these eigenvalues and by $|\lambda>$ the corresponding eigenstates.

In the direct product Hilbert space, the quantum states will be described by

$$|m, s; \mathbf{p}, \sigma> \otimes |\lambda> , \qquad (8.5)$$

where now $|\lambda>$ is the basis of a IR of $\mathcal{G}$ of dimension n_λ. It follows immediately, from the assumption (8.3), that the states represented by (8.5) are a set of n_λ states with the *same mass* and the *same spin*, i.e. they are a set of n_λ *degenerate* states.

We call the above type of symmetry an *exact internal symmetry*; it implies *degeneracy* among the states of the system.

On the other hand, symmetries imply conservation laws which, in turn, imply that transitions between different states are forbidden if the quantum numbers are not conserved: then one has *selection rules*. The situation is summarized s follows:

Symmetry group $\rightarrow$ degeneracy of quantum states $\rightarrow$ multiplets
    $\downarrow\uparrow$
Conservation laws $\rightarrow$ vanishing matrix elements $\rightarrow$ selection rules
    $\downarrow\uparrow$
Tensor operators $\rightarrow$ Wigner-Eckart theorem $\rightarrow$ intensity rules

If the dynamics of a mechanical system is known and, in particular, if one knows its Lagrangian, one knows also its invariance properties under the transformations of a symmetry group $\mathcal{G}$. Then one can construct conserved quantities, making use of the well-known Noether's theorem [1] and, following the arrows, deduce all the implications of the symmetry.

However, since a dynamical theory of elementary particles is not available, we have no symmetry principle to start from. On the other hand, we can observe that particles occur in multiplets, that certain reactions are forbidden, and that there are indications of the existence of intensity rules. One can infer that there are *degeneracies, conservation laws*, and therefore an underlying *invariance principle*. In other words, the arrows are followed in the opposite direction; one then discovers new quantum numbers which are conserved, and a group of transformations under which the interactions should be invariant. All this will help to build a model possessing this kind of symmetry, and eventually to formulate, step by step, the Lagrangian of the theory.

In the above, we have assumed to deal with *exact* symmetries. However, in most cases, it occurs that the symmetry is only approximate; therefore, the commutator (8.3) is no longer zero:

$$[Q_\alpha, M^2] \neq 0 . \qquad (8.6)$$

---

[1] See Section 10.2 and references therein.

One says that the symmetry is *broken*, but often the breaking is small and, moreover, it follows a specific pattern, i.e. the operator M^2, no longer a scalar under $\mathcal{G}$, contains additional terms with specific tensor transformation properties. Correspondingly, mass splittings appear among the members of the same multiplet, so that the states, in general, are no longer degenerate. If the splittings are not too large, symmetry is still a useful concept, even if it is only an *approximate symmetry*.

Other kinds of symmetries, which it would not be appropriate to call *internal*, were used in particle physics, e.g. those for which the generators Q_α do not commute with all the $M_{\mu\nu}$:

$$[Q_\alpha, M_{\mu\nu}] \neq 0 . \tag{8.7}$$

Embedding in a group $\mathcal{H}$ an internal group $\mathcal{G}$ and a subgroup of $\mathcal{P}$, one can have the following situation: the IR's of $\mathcal{H}$ contain now different values of spin. We have then a sort of *hybrid* symmetry where internal quantum numbers and space-time variables are mixed. We shall not examine this kind of symmetry in detail, but limit ourselves to mention the model based on the group $SU(6)$, which combines the internal symmetry $SU(3)$ with the spin symmetry $SU(2)$. It was considered with much interest in the middle of the sixties of last century, but it was only partially successful since it applied only to static, non-relativistic situations.

All these symmetries correspond to invariance properties under *global* transformations. Except for a few exact symmetries based on the group $U(1)$, those based on higher groups are, in general, approximate, but they are useful for the classification of hadrons, for obtaining mass formulae and other symmetry relations.

A different kind of symmetries, based on a different use of unitary groups, was introduced at the same time in field theory. They are called *gauge* symmetries, because they imply invariance under *local* transformations. They played a fundamental role in building the field theory of elementary particles. These symmetries will be discussed in Chapter 9.

8.3 $U(1)$ invariance and Additive Quantum Numbers

We consider here the symmetries corresponding to the group $U(1)$ and related to additive quantum numbers which are conserved in the reactions among elementary particles, and some of which were discovered from the appearance of selection rules.

Let us consider, to be specific, the *electric charge*. Charge is described by a Hermitian operator Q in the Hilbert space, and its eigenvalues by q. Let us consider the eigenvalue equation

$$Q|\Phi> = q|\Phi> . \tag{8.8}$$

In particular, the state of a particle with charge q, according to Eq. (8.5), is denoted by $|m, s; \mathbf{p}, \sigma> \otimes |q>$, or in short notation simply by $|p, \sigma, q>$.

We know that electric charge is strictly conserved, so that states with different charges do not mix. This conservation property corresponds to the set of transformations

$$|\Phi'> = e^{-i\lambda q}|\Phi> \ . \tag{8.9}$$

In fact, since not even the relative phase can be measured, we can introduce a *unitary operator*

$$U = e^{-i\lambda Q} \tag{8.10}$$

and the unitary transformations

$$|\Phi'> = U|\Phi> \ , \tag{8.11}$$

which give rise to equivalent states.

Suppose that $|\Phi>$ is a momentum eigenstate

$$P^\mu|\Phi> = p^\mu|\Phi> \ , \tag{8.12}$$

so that, in particular, denoting by H the Hamiltonian, one has

$$H|\Phi> = p^0|\Phi> \ . \tag{8.13}$$

Invariance under the transformation (8.11) implies:

$$UHU^{-1}|\Phi'> = p^0|\Phi'> \ , \tag{8.14}$$

and if $|\Phi>$ is eigenstate also of Q, $|\Phi'>$ will differ only for a phase from $|\Phi>$, so that

$$UHU^{-1} = H \ . \tag{8.15}$$

Using an infinitesimal transformation,

$$U = 1 - i\lambda Q \ , \tag{8.16}$$

we get the commutation relation

$$[H, Q] = 0 \ , \tag{8.17}$$

which is the quantum mechanical formulation of the conservation principle of electric charge.

The transformations (8.10) form a one-parameter (Abelian) group. It is the group of one-dimensional unitary transformations, denoted by $U(1)$. Its irreducible representations are all one-dimensional and consist in the phase $e^{-i\lambda q}$.

As in the case of electric charge, the conservation of other additive quantum numbers can be expressed in terms of phase transformations, corresponding to $U(1)$ invariance. We know that there are in nature other kinds of "charges" which are believed to be conserved, such as the baryon number. Other charges, such as the hypercharge, are approximately conserved. We list in the following some of the known additive quantum numbers.

- *Electric charge.* The validity of the conservation principle of the electric charge is based on the experimental value of the lifetime of the electron, for which there is the lower limit of 4.6×10^{26} years (at 90% C.L.).

- *Baryon number.* The introduction of this conserved quantum number is based on the stability of the proton: the experimental lower limit for its mean life is of the order of 10^{32} years[2]. Baryon number B is a quantum number associated with strongly interacting fermions (baryons); it is taken equal to +1 for baryons and to −1 for antibaryons. However, it is generally believed that the baryon number is not strictly conserved, but that it should be violated at extremely high energies. As a consequence, the proton would not be stable, as predicted by the Grand Unified Theories. The violation of baryon number is required also by the Big Bang model, in order to explain how the present matter-antimatter asymmetry can be originated from a symmetrical initial state.

- *Lepton number.* The analysis of the reactions among leptons gives evidence for three *lepton numbers*, which up to now appear to be conserved in the charged lepton decays, but violated in neutrino oscillations. They are denoted by L_e, L_μ, L_τ and are attributed, respectively, to the lepton pairs (ν_e, e^-), (ν_μ, μ^-), (ν_τ, τ^-). Their eigenvalues are +1 for particles and −1 for antiparticles.
  The quantity $L = L_e + L_\mu + L_\tau$ is a good quantum number which, however, is violated in Grand Unified Theories, in analogy with the baryon number. Its violation is also required to provide a reasonable mechanism for generating the neutrino masses, but the question is still open.

- *Hypercharge.* This quantum number was introduced in connection with the meta-stability of some kind of hadrons, such as hyperons and K-mesons. It is conserved in strong interactions, but it is violated by weak interactions. It is defined by $Y = S + B$, where B is the baryon number and S a new additive quantum number called *strangeness*. The hypercharge was a key ingredient in the classification of hadrons, but now we know that S corresponds to only one of the six different *flavors* related to the six kinds of quarks.

In general, strictly conserved quantum numbers give rise to absolute *selection rules*. Suppose that there is no matrix element connecting states with different charges. As pointed out already, it is not possible to determine the common phase of a set of state vectors, but only the relative phase can be measured. However, if we have invariance under a phase transformation, it means that not even the relative phase can be measured. A physical state cannot be a superposition of states with different charges; in other words, all physical states must be always eigenstates of charge. We have then different Hilbert subspaces, and different sets of state vectors belonging to different

---

[2] The Review of Particle Physics; C. Amsler *et al.*, Phys. Lett. B 667, 1 (2008).

subspaces have non-measurable relative phase. We say that there is a *super-selection rule* that forbids to compare phases of various state vectors: this is the case of electric charge.

8.4 Isospin invariance

Strong interactions of elementary particles exhibit important symmetries, the discovery of which is a step forward in the understanding of their properties. In this Section we examine the so-called *isospin invariance*, which is described in terms of the group $SU(2)$.

One of the most striking features of *hadrons* is that they occur in multiplets, each element of a multiplet having the same spin and parity, and roughly the same mass. This occurrence of sets of approximately degenerate states suggests a symmetry principle. One is led to introduce a group of invariance and to describe the multiplets in terms of the irreducible representations (IR's) of this group. The group transformations interchange the elements of a multiplet among themselves, thus implying invariance laws. The success of this scheme lies on the fact that such invariance laws have been found to hold even if only approximately. Since one can characterize the various classes of fundamental interactions by coupling constants that differ in order of magnitude, one can assume that the symmetry *holds exactly* within the domain of some of the interactions, while the others are responsible for the *symmetry breaking*.

All hadrons can then be classified in charge multiplets, and the elements of a given multiplet differ only for the electromagnetic properties (they have same spin, parity, flavor, etc., but *different electric charge*). The masses of the particles in a multiplet are very close: their differences (of the order of a few MeV) are attributed to the electromagnetic interactions. Some example, relative to the lightest mesons and baryons, are given in Table 8.1.

If we imagine to *switch off the electromagnetic interactions*, any element of a given multiplet becomes equivalent to any other element. In other words, each multiplet contains a set of *degenerate states*, and it corresponds to an IR of the symmetry group.

The symmetry group is $SU(2)$: in fact, one is looking for a compact group of rank 1 with finite and unitary IR's, since there is only a quantum observable, the electric charge, to specify the members of each set of degenerate states. and no more. There are three groups satisfying these requirements: $SU(2)$, $SO(3)$ and $Sp(2)$. $SU(2)$ and $Sp(2)$ are isomorphic to each other, and homomorphic to $SO(3)$. Since the latter is not simply connected and it has also double-valued representations, it is more convenient to choose $SU(2)$. The above arguments are of mathematical character; of course one can check a posteriori if $SU(2)$ is a symmetry of nature, by comparing its prediction with the experimental observations.

Table 8.1. Some multiplets of hadrons.

	kind of multiplet	particles	J^P	$Y = B + S$
	singlet	Λ^0	$\frac{1}{2}^+$	0
	doublet	$n\ p$	$\frac{1}{2}^+$	+1
		$\Xi^-\ \Xi^0$	$\frac{1}{2}^+$	−1
Baryons $(B = 1)$	triplet	$\Sigma^+ \Sigma^0 \Sigma^-$	$\frac{1}{2}^+$	0
	quadruplet	$\Delta^{++}\Delta^+\Delta^0\Delta^-$	$\frac{3}{2}^+$	+1
	singlet	η	0^-	0
Mesons $(B = 0)$	doublet	$K^+ K^0$	0^-	+1
		$K^- \overline{K}^0$	0^-	−1
	triplet	$\pi^+ \pi^0 \pi^-$	0^-	0

8.4.1 Preliminary considerations

The group $SU(2)$ and its IR's have been examined in Chapter 2, in connection with space rotations. For the sake of convenience, we rewrite in the following some of the formulae given already in Sections 2.2 and 2.3. We denote here by $I_i (i = 1, 2, 3)$ the generators of the Lie algebra which, in the self-representation, are given by

$$I_i = \tfrac{1}{2}\sigma_i , \tag{8.18}$$

in terms of the three Pauli matrices (2.22). They satisfy the commutation relations

$$[I_i, I_j] = i\epsilon_{ijk}I_k , \tag{8.19}$$

and

$$[I^2, I_i] = 0 \qquad \text{with} \qquad I^2 = \sum_{i=1}^{3} I_i^2 . \tag{8.20}$$

Each element of $SU(2)$ is a 2×2 unitary unimodular matrix and can be written as (the α_i's are three real parameters)

$$U = e^{-i\sum_{i=1}^{3}\alpha_i\sigma_i} . \tag{8.21}$$

The I_i's are linear operators with the same formal properties of the angular momentum operators; they are called *isospin operators*. In analogy with ordinary spin, one diagonalizes I^2 and the third component I_3.

Each IR is characterized by the eigenvalue $I(I + 1)$ of I^2 (the Casimir operator of the group) and is denoted by $D^{(I)}$. The dimension of $D^{(I)}$ is $(2I + 1)$ and corresponds to the $(2I + 1)$ eigenvalues of I_3. The basis of $D^{(I)}$ is a vector of component $|I; I_3> \equiv \xi_I^{I_3}$ in a $(2I + 1)$-dimensional space.

We denote by ξ and ξ^* the two-component bases of the self-representation U and of the conjugate representation U^*. They are given by

$$\xi \equiv \begin{pmatrix} \xi^1 \\ \xi^2 \end{pmatrix}, \qquad \xi^* \equiv \begin{pmatrix} \xi_1 \\ \xi_2 \end{pmatrix}, \tag{8.22}$$

and they transform according to

$$\xi \to \xi' = U\xi \qquad \text{i.e.} \qquad \xi'^a = \sum_{b=1}^2 U_{ab}\xi^b, \tag{8.23}$$

and

$$\xi^* \to \xi'^* = U^*\xi^* \qquad \text{i.e.} \qquad \xi'_a = \sum_{b=1}^2 U^*_{ab}\xi_b. \tag{8.24}$$

We recall (see Section 2.3) that the two representations U and U^* are equivalent, since they are related by a similarity transformation

$$U^* = SUS^{-1}, \tag{8.25}$$

with

$$S = i\sigma_2 = \begin{pmatrix} 0 & 1 \\ -1 & 0 \end{pmatrix}. \tag{8.26}$$

The matrix $S = i\sigma_2$ plays the role of metric tensor and it is identified with the antisymmetric tensor $\epsilon^{ab} = \epsilon_{ab}$. It follows that the vector

$$\bar{\xi} = S^{-1}\xi^* = \begin{pmatrix} -\xi_2 \\ \xi_1 \end{pmatrix} \tag{8.27}$$

defined by Eq. (2.43) has the same transformation properties of ξ, i.e.

$$\bar{\xi} \to \bar{\xi}' = U\bar{\xi} \tag{8.28}$$

and its components are given by

$$\bar{\xi}^a = -\epsilon^{ab}\xi_b. \tag{8.29}$$

The bases of the higher IR's, $\zeta_I^{I_3}$, are constructed starting from ξ^a and ξ_a; equivalently one can make use of only one kind of vectors, say ξ^a. The general method is described in Appendix C: one takes the direct product of the bases of two (or more) IR's and decomposes it into the direct sum of irreducible tensors.

In this connection, let us recall the direct product expansion

$$D^{(I)} \otimes D^{(I')} = D^{(I+I')} \oplus D^{(I+I'-1)} \oplus \cdots \oplus D^{(|I-I'|)} . \tag{8.30}$$

The basic element for the representation $D^{(I)} \otimes D^{(I')}$ is given by the tensor

$$\zeta_I^{I_3} \cdot \zeta_{I'}^{I'_3} . \tag{8.31}$$

This tensor describes a set of eigenstates of I^2, I'^2, I_3, I'_3. It is more useful to obtain a set of eigenstates of $\mathcal{I}^2$, $\mathcal{I}_3$, and I^2, I'^2, where $\mathcal{I}$ is given by the usual addition law of angular momenta:

$$\mathcal{I} = \mathbf{I} + \mathbf{I}' . \tag{8.32}$$

The corresponding eigenstates form the bases of the IR's on the r.h.s. of Eq. (8.30) and are given by

$$\zeta_{\mathcal{I}}^{\mathcal{I}_3}(I, I') = \sum_{I_3, I'_3} <I_3, I'_3|\mathcal{I}, \mathcal{I}_3> \zeta_I^{I_3} \zeta_{I'}^{I'_3} , \tag{8.33}$$

where the quantities

$$<I_3, I'_3|\mathcal{I}, \mathcal{I}_3> \equiv <I, I', \mathcal{I}; I_3, I'_3, \mathcal{I}_3> \tag{8.34}$$

are the *Clebsch-Gordan coefficients*. They are the same coefficients used in the addition of angular momenta and they are tabulated, for a few cases, in Appendix A.

As a relevant case, let us consider the direct product decomposition

$$D^{(\frac{1}{2})} \otimes D^{(\frac{1}{2})} = D^{(1)} \oplus D^{(0)} . \tag{8.35}$$

Let us consider the tensor $\zeta^{ab} = \xi^a \xi^b$: one can identify an antisymmetric

$$\zeta^{[ab]} = \frac{1}{\sqrt{2}}(\xi^1 \xi^2 - \xi^2 \xi^1) \tag{8.36}$$

and a symmetric part

$$\zeta^{\{ab\}} = \begin{cases} \xi^1 \xi^1 \\ \frac{1}{\sqrt{2}}(\xi^1 \xi^2 + \xi^2 \xi^1) \\ \xi^2 \xi^2 \end{cases} , \tag{8.37}$$

the factor $\frac{1}{\sqrt{2}}$ being introduced for normalization. Since the symmetry properties of the tensors are invariant under the unitary group, then one can identify

$\zeta^{[ab]}$ as the basis of the $D^{(0)}$ ($I = 0, I_3 = 0$) and $\zeta^{\{ab\}}$ as the basis of the $D^{(1)}$ ($I = 1, I_3 = +1, 0, -1$). In this way the decomposition (8.35) corresponds, in terms of basic tensors, to

$$\zeta^{ab} = \xi^a \xi^b = \zeta^{[ab]} + \zeta^{\{ab\}} . \tag{8.38}$$

Let us now consider the mixed tensor $\zeta^a{}_b$; it can be decomposed as follows

$$\zeta^a{}_b = \xi^a \xi_b = \hat{\zeta}^a{}_b + \tfrac{1}{2}\delta^a{}_b \mathrm{Tr}(\zeta^c{}_{c'}) , \tag{8.39}$$

where the trace $\tfrac{1}{2}\mathrm{Tr}(\zeta)$ is invariant and the traceless tensor

$$\hat{\zeta}^a{}_b = \xi^a \xi_b - \tfrac{1}{2}\delta^a{}_b \xi^c \xi_c , \tag{8.40}$$

is irreducible. They are the bases of the same IR's $D^{(0)}$ and $D^{(1)}$, respectively. In fact, the tensors ζ^{ab} and $\zeta^a{}_b$ are equivalent, since

$$\zeta^a{}_b = \xi^a \xi_b = \epsilon_{bc} \xi^a \xi^c = \epsilon_{bc} \zeta^{ac} , \tag{8.41}$$

and in particular

$$\tfrac{1}{2}\delta^a{}_b \mathrm{Tr}(\zeta^c{}_{c'}) = \epsilon_{bc} \zeta^{[ac]} , \qquad \hat{\zeta}^a{}_b = \epsilon_{bc} \zeta^{\{ac\}} . \tag{8.42}$$

8.4.2 Isospin classification of hadrons

The classification of the elementary particles into *isospin multiplets* is now easily carried out by identifying each multiplet as the basis of a IR of $SU(2)$. If one wants to take into account also the baryon number B, one has to go from $SU(2)$ to $U(2)$ which is locally isomorphic to $SU(2) \otimes U(1)$: the phase transformations of $U(1)$ are related to an additive quantum number, identified in this case with B. We note that in going from a IR of $U(2)$ to the conjugate IR, B changes sign.

Let us consider the *nucleon* $\mathcal{N}$ ($B = 1$), which consists of the two states p, n: it can be identified with the basis of the self-representation of $SU(2)$

$$\mathcal{N} = \begin{pmatrix} p \\ n \end{pmatrix} . \tag{8.43}$$

The antinucleon $\overline{\mathcal{N}}$ ($B = -1$) can be associated to the conjugate representation (8.27)

$$\overline{\mathcal{N}} = \begin{pmatrix} -\overline{n} \\ \overline{p} \end{pmatrix} , \tag{8.44}$$

i.e. $\overline{\mathcal{N}}$ has the same transformation properties of $\mathcal{N}$.

The charge of each element of the multiplet is related to the third component I_3 through the well-known Gell-Mann-Nishijima formula [3].

---

[3] T. Nakano and N. Nishijima, Progr. Theor. Phys. 10, 581 (1955); M. Gell-Mann, Nuovo Cimento 4, 848 (1956).

A phenomenological Lagrangian for strong interactions, expressed in terms of non-strange and strange hadrons, must be invariant under two different phase transformations: one for the conservation of the electric charge Q, and the other for the conservation of the baryonic number B. Assuming isospin invariance, the former is equivalent to the invariance under rotations around the I_3 axis in the isospin space plus the invariance under a further phase transformation, which can be expressed as the conservation of the *hypercharge* $Y = B + S$, satisfying Eq. (8.45).

$$Q = I_3 + \tfrac{1}{2}(B + S) = I_3 + \tfrac{1}{2}Y \ , \tag{8.45}$$

where S is the *strangeness* and Y the *hypercharge*.

The situation is similar for the two doublets of K and $\overline{K}$ *mesons*

$$K = \begin{pmatrix} K^+ \\ K^0 \end{pmatrix} \ , \qquad \overline{K} = \begin{pmatrix} -\overline{K}^0 \\ K^- \end{pmatrix} \ , \tag{8.46}$$

which have $Y = +1$ and $Y = -1$, respectively.

Let us consider now the *two nucleon* $\mathcal{N}\mathcal{N}$ system. It corresponds to the tensor $\mathcal{N}^a\mathcal{N}^b$, so that, according to Eqs. (8.36) and (8.37), we identify the (normalized) states

$$
\begin{array}{lll}
I = 0 & I_3 = \quad 0 & \frac{1}{\sqrt{2}}(pn - np) \ , \\[4pt]
& \left\{ \begin{array}{l} I_3 = +1 \qquad pp \\[4pt] I_3 = \quad 0 \qquad \frac{1}{\sqrt{2}}(pn + np) \\[4pt] I_3 = -1 \qquad nn \end{array} \right. & \\
I = 1 & &
\end{array}
\tag{8.47}
$$

The *nucleon-antinucleon* $\mathcal{N}\overline{\mathcal{N}}$ system can be described in terms of the decomposition of the mixed tensor $\mathcal{N}^a\mathcal{N}_b$ (Eqs. (8.39) and (8.44)):

$$
\begin{array}{lll}
I = 0 & I_3 = \quad 0 & \frac{1}{\sqrt{2}}(p\overline{p} + n\overline{n}) \ , \\[4pt]
& \left\{ \begin{array}{l} I_3 = +1 \qquad -p\overline{n} \\[4pt] I_3 = \quad 0 \qquad \frac{1}{\sqrt{2}}(p\overline{p} - n\overline{n}) \\[4pt] I_3 = -1 \qquad n\overline{p} \end{array} \right. & \\
I = 1 & &
\end{array}
\tag{8.48}
$$

The η *meson* is an isosinglet $(I = 0)$. Since it has the same quantum numbers of the $(I = 0)$ $\mathcal{N}\overline{\mathcal{N}}$, taking into account the decomposition in Eq. (8.39), one can make the identification

$$\eta = \frac{1}{\sqrt{2}} \mathrm{Tr} \left(\zeta^a{}_b \right) \ . \tag{8.49}$$

The π *meson* is an isotriplet $(I = 1)$. It has the same quantum numbers of the $(I = 1)$ $\mathcal{N}\overline{\mathcal{N}}$ system, so that it can be described in terms of the tensor

$$\pi^a{}_b = \hat{\zeta}^a{}_b = \xi^a \xi_b - \tfrac{1}{2}\delta^a{}_b \xi^c \xi_c = \tfrac{1}{\sqrt{2}}(\sigma_i)^a{}_b \pi^i \,, \tag{8.50}$$

where in the last term we have introduced the isovector $\boldsymbol{\pi}$ of components

$$\pi_i = \tfrac{1}{\sqrt{2}}\mathrm{Tr}(\pi\sigma_i) \,, \tag{8.51}$$

normalized so that

$$\pi^a{}_b \pi^b{}_a = \pi^i \pi_i \,. \tag{8.52}$$

In matrix form, taking into account the correspondence between the quantum numbers of $\pi^\pm$, π^0 and those of the $\mathcal{N}\mathcal{N}$ states, one gets

$$\pi = \begin{pmatrix} \tfrac{1}{2}(\xi^1 \xi_1 - \xi^2 \xi_2) & \xi^1 \xi_2 \\ \xi^2 \xi_1 & \tfrac{1}{2}(\xi^2 \xi_2 - \xi^1 \xi_1) \end{pmatrix} = \begin{pmatrix} \tfrac{1}{\sqrt{2}}\pi^0 & \pi^+ \\ \pi^- & -\tfrac{1}{\sqrt{2}}\pi^0 \end{pmatrix} \,, \tag{8.53}$$

and, by comparison with Eq. (8.51),

$$\begin{aligned} \pi^\pm &= \tfrac{1}{\sqrt{2}}(\pi_1 \mp i\pi_2) \,, \\ \pi^0 &= \pi_3 \,. \end{aligned} \tag{8.54}$$

Finally, the $\pi\mathcal{N}$ *system* can be described on the basis of the decomposition

$$D^{(1)} \otimes D^{(\frac{1}{2})} = D^{(\frac{1}{2})} \oplus D^{(\frac{3}{2})} \,. \tag{8.55}$$

and the states can be classified in a doublet $(I = \tfrac{1}{2}, I_3 = \pm\tfrac{1}{2})$ and a quadruplet $(I = \tfrac{3}{2}, I_3 = \pm\tfrac{3}{2}, \pm\tfrac{1}{2})$

$$I = \tfrac{1}{2} \begin{cases} I_3 = \tfrac{1}{2} & \sqrt{\tfrac{1}{3}}\, p\pi^0 - \sqrt{\tfrac{2}{3}}\, n\pi^+ \\[2mm] I_3 = -\tfrac{1}{2} & \sqrt{\tfrac{2}{3}}\, p\pi^- - \sqrt{\tfrac{1}{3}}\, n\pi^0 \end{cases} \,,$$

$$I = \tfrac{3}{2} \begin{cases} I_3 = \tfrac{3}{2} & p\pi^+ \\[2mm] I_3 = \tfrac{1}{2} & \sqrt{\tfrac{2}{3}}\, p\pi^0 + \sqrt{\tfrac{1}{3}}\, n\pi^+ \\[2mm] I_3 = -\tfrac{1}{2} & \sqrt{\tfrac{1}{3}}\, p\pi^- + \sqrt{\tfrac{2}{3}}\, n\pi^0 \\[2mm] I_3 = -\tfrac{3}{2} & n\pi^- \end{cases} \,, \tag{8.56}$$

where use has been made of Eq. (8.33) and of the Tables of the Clebsch-Gordan coefficients of Appendix A.

Following similar procedure, one can classify all hadrons into isospin multiplets with any flavour quantum numbers. At first sight, all this appears as a formal game. Physics enters when one assumes that strong interactions depend on the total isospin I of the system and not on the third component I_3; in other words, that strong interactions are *charge independent*. This invariance property is equivalent to the *conservation of the total isotopic spin*: the third component is also conserved, since it is related to the charge by Eq. (8.45).

Charge independence means, for instance, that for a system of two nucleons, instead of four possible amplitudes corresponding to the four charge states, there are only two amplitudes, corresponding to the isospin singlet and the isospin triplet. Similarly, in a system of a nucleon and a pion, strong interactions distinguish only between the $I = \frac{1}{2}$ and the $I = \frac{3}{2}$ states.

In general, charge independence means that, given a set of particles, what matters for strong interactions are only the symmetry properties of the states, and not their difference in the electric charge.

Let us finally remark that in the framework of exact $SU(2)$ symmetry, particles which belong to the same multiplet are to be considered as identical: then a generalized Pauli principle applies in this case.

As a specific example of isospin invariance, we consider the pion-nucleon scattering. In general, in terms of the scattering S-matrix, isospin invariance means that the S operator commutes with isospin

$$[S, I_i] = 0 \,, \tag{8.57}$$

i.e. S connects states with same I and I_3.

Moreover, introducing the shift (raising and lowering) operators

$$I_\pm = I_1 \pm iI_2 \,, \tag{8.58}$$

one gets

$$[S, I_\pm] = 0 \,, \tag{8.59}$$

and one can show that

$$I_\pm|I, I_3> = C_\pm(I, I_3)|I, I_3 \pm 1> \,, \tag{8.60}$$

where $C_\pm$ is a coefficient which is not necessary to specify here.

From Eq. (8.59) it follows

$$<I_3|S|I_3> = <I_3 + 1|S|I_3 + 1> \,, \tag{8.61}$$

i.e. the S-matrix elements do not depend on the third component of the I-spin.

As a physical consequence, the reaction between two particles of isospin I and I' is completely described in terms of a number of independet S-matrix elements or *amplitudes* equal to the number of IR's into which the direct product $D^{(I)} \otimes D^{(I')}$ is decomposed. This number is, in general, much smaller than the number of charge states.

For instance, in the case of the $\pi - \mathcal{N}$ system, one needs only two amplitudes, A_1, A_3, corresponding to the transition between states of isospin $I = \frac{1}{2}$, $I = \frac{3}{2}$, respectively. Each given state of charge can be expressed in terms of the total isospin states, according to (see Table A.3 in Appendix A)

$$p\pi^+ = \xi_{3/2}^{3/2} \, ,$$

$$p\pi^- = \sqrt{\tfrac{1}{3}}\, \xi_{3/2}^{-1/2} - \sqrt{\tfrac{2}{3}}\, \xi_{1/2}^{-1/2} \, , \qquad (8.62)$$

$$n\pi^0 = \sqrt{\tfrac{2}{3}}\, \xi_{3/2}^{-1/2} + \sqrt{\tfrac{1}{3}}\, \xi_{1/2}^{-1/2} \, .$$

We shall now write the amplitudes and cross sections for some specific reactions in Table 8.2. They are simply obtained by expressing the initial and final states in terms of the isospin eigenstates and taking into account that the latter are orthogonal.

Table 8.2. Analysis of some reactions in terms of I-spin amplitudes.

reaction	amplitude	cross section
$\pi^+ p \to \pi^+ p$	A_3	$\sim \lvert A_3 \rvert^2$
$\pi^- p \to \pi^- p$	$\frac{1}{3}A_3 + \frac{2}{3}A_1$	$\sim \frac{1}{9}\lvert A_3 + 2A_1 \rvert^2$
$\pi^- p \to \pi^0 n$	$\frac{\sqrt{2}}{3}A_3 - \frac{\sqrt{2}}{3}A_1$	$\sim \frac{2}{9}\lvert A_3 - A_1 \rvert^2$

8.5 $SU(3)$ invariance

In this Section we examine the so-called $SU(3)$ symmetry which was proposed independently by Murray Gell-Mann and Yuval Ne'eman[4] in 1964 to account for the regularity appearing in the spectrum of hadrons.

In order to understand the role of this symmetry in the development of particle physics, it is useful to make some historical remarks. In fact, even if nowadays this symmetry has lost part of its interest, it was very important in indicating the right way for the discovery of the hidden properties of the subnuclear world.

We know that strong interactions are responsible for the nuclear forces; in fact, they bind nucleons (protons and neutrons) in atomic nuclei. In 1935 Hideki Yukawa formulated the hypothesis that strong interactions are mediated by scalar (spin zero) bosons[5]. The pion ($\pi^\pm$, π^0), discovered in 1947[6], appeared to be this hypothetical particle. Its mass was found to be $m_\pi \approx 140$ MeV, in agreement with the range of the interaction $\sim 1.4 \times 10^{-13}$ cm.

---

[4] M. Gell-Mann and Y. Ne'eman, *The Eightfold Way*, Benjamin, New York (1964).

[5] H. Yukawa, Proc. Phys. Math. Soc. Japan 17, 48 (1935).

[6] C. M. G. Lattes, H. Muirhead, G. P. S. Occhialini, and C. F. Powell, Nature 159, 694 (1947).

For a few years, the world of strongly interacting particles was limited to nucleons and pions. But, with the advent of high energy accelerators, a large variety of others particles was discovered during the decades 1950 and 1960. They are named *hadrons* (for strongly interacting particles), and they are separated into *baryons* and *mesons*, in correspondence with half-integer and integer spin, respectively. Among the baryons, some look like excited states of the nucleons (which we shall call *ordinary* particles), while others require the introduction of a new quantum number, called *strangeness* and denoted by S. The situation is analogous in the case of mesons: some appears to be of the same type of the pions, while others, called strange mesons, possess a value of S different from zero. As mentioned in Subsection 8.3, the strangeness S was often replaced by the *hypercharge* $Y = S + B$, where B is the baryon number.

At the time of the introduction of the $SU(3)$ symmetry, only these two kinds of hadrons were known. Later on, other types of hadrons were discovered, which required the introduction of extra additive quantum numbers, as it will be briefly discussed in the following.

The spectrum of ordinary and strange baryons and mesons consists in hundreds of states with increasing values of mass and spin ($s = \frac{1}{2}, \frac{3}{2}, \frac{5}{2}, \ldots$ for baryons, and $s = 0, 1, 2, \ldots$ for mesons). A peculiar regularity was soon manifested in the spectrum of these states: both baryons and mesons can be grouped into multiplets, the components of each multiplet being close in mass, and having the same spin and parity. Baryons appear in singlets, octets and decuplets; mesons only in singlets and octets. It was this property which led to the introduction of the group $SU(3)$ and to the hadron classification according to the so-called *eightfold way*.

In this Section, we shall discuss the properties of the group $SU(3)$ and its physical applications.

8.5.1 From $SU(2)$ to $SU(3)$

Looking at a table of hadrons[7], one notices immediately peculiar regularities. One can distribute hadrons in *supermultiplets*, made up of *isospin* multiplets, characterized by the same *baryon number, spin, parity*. One needs now a higher degree of abstraction to claim for a symmetry principle, since particle belonging to the same supermultiplet have rather large mass differences (even of the order of few hundreds MeV). The symmetry one is looking for is badly broken, and there is no known interaction which causes the breaking as the electromagnetic interaction in the case of isospin. To extend the analogy, one should invent two subclasses of interactions: *superstrong* and *medium strong*: if only the former were present (by switching off the latter) one would obtain completely degenerate multiplets of states and the symmetry would be exact. However, such a separation is artificial, and it is by no means necessary for the use of a broken symmetry.

---

[7] The Review of Particle Physics; C. Amsler *et al.*, Phys. Lett. B 667, 1 (2008).

If one limits oneself to consider only ordinary and strange hadrons, the different isospin multiplets in a supermultiplet are characterized, in general, by different values of the hypercharge Y. The introduction of this additive quantum number was made in order to explain the so-called *associate production*, and to take into account the paradox between the strong interactions of pairs of strange hadrons and the very slow hadron decays, i.e. the very long lifetimes. The strangeness S, or equivalently the hypercharge $Y = S + B$, is assumed to be conserved in strong interactions and violated in weak interactions, which are responsible for the decays. A value for Y can be assigned to each hadron, as can be seen in the few examples given in Table 8.1.

Then each element of a supermultiplet is characterized by two internal quantum numbers: I_3, the third component of the isospin, and Y, the hypercharge, which are related to the electric charge by the Gell-Mann Nishijima formula, already given in Eq. (8.45)

If one assumes that there is an underlying symmetry group $\mathcal{G}$ such that the observed supermultiplets can be classified according to its irreducible representations, one has to look for a group of *rank* 2 (it must have two commuting generators corresponding to I_3 and Y).

The group $\mathcal{G}$ must contain $SU(2)$ as a subgroup. Keeping to a compact, simple group, we could have, in principle, the following choices (see Table 1.2): $SU(3)$, $Sp(4)$, $O(5)$ ($Sp(4)$ and $O(5)$ have the same algebra, i.e. they are homomorphic) and G_2. One then looks for the lowest IR's of these groups: one needs to know, besides the dimension of each IR, its content in terms of the IR's of $SU(2)$. The situation for the lowest IR's is described in Table 8.3.

Table 8.3. $SU(2)$ content of the groups $SU(3)$, $O(5)$, $Sp(4)$, G_2.

group	dimension of the lowest IR's (isospin content)[(*)]
$SU(3)$	$1(0)$, $3\left(0, \frac{1}{2}\right)$, $6\left(0, \frac{1}{2}, 1\right)$, $8\left(0, \frac{1}{2}, \frac{1}{2}, 1\right)$, $10\left(0, \frac{1}{2}, 1, \frac{3}{2}\right)$
$O(5)$	$1(0)$, $4\left(\frac{1}{2}, \frac{1}{2}\right)$, $5(0, 0, 1)$, $10(0, 1, 1, 1)$, $14(0, 0, 0, 1, 1, 2)$
$Sp(4)$	$1(0)$, $4\left(0, 0, \frac{1}{2}\right)$, $5\left(0, \frac{1}{2}, \frac{1}{2}\right)$, $10\left(0, 0, 0, \frac{1}{2}, \frac{1}{2}, 1\right)$, $14\left(0, \frac{1}{2}, \frac{1}{2}, 1, 1, 1\right)$
G_2	$1(0)$, $7\left(\frac{1}{2}, \frac{1}{2}, 1\right)$, $14\left(0, 0, 0, 1, \frac{3}{2}, \frac{3}{2}\right)$, $27\left(0, \frac{1}{2}, \frac{1}{2}, 1, 1, 1, \frac{3}{2}, \frac{3}{2}, 2\right)$

[(*)] We notice that the dimensions of the IR's of $O(5)$ and $Sp(4)$ are the same; they differ, however, according to the usual conventions, with respect to their $SU(2)$ content.

The data show that hadrons occur in multiplets of dimension 1, 8 and 10; so that the only candidate for $\mathcal{G}$ is $SU(3)$. Moreover, the isospin content exhibited by the hadron multiplets agrees with what required by $SU(3)$.

Of course, these features are not sufficient to prove that $SU(3)$ has something to do with Nature. In fact, one has to examine its physical consequences, but, before going to physics, we shall make a short mathematical digression.

8.5.2 Irreducible representations of $SU(3)$

For a discussion of the $SU(3)$ classification of hadrons, we need to know the irreducible representations of this group.

A very useful tool in determining the $SU(3)$ IR's and their $SU(2)$ content is provided by the use of the *Young tableaux*. A short account of this technique is given in the Appendices B and C, to which we shall refer for details.

$SU(3)$ is the group of unitary unimodular 3×3 matrices. Denoting by U a generic element, it must satisfy the relations

$$U^\dagger U = I \qquad \text{i.e.} \qquad (U^\dagger)^\alpha{}_\beta U^\beta{}_\gamma = \delta^\alpha{}_\gamma \,, \tag{8.63}$$

$$\det U = 1 \qquad \text{i.e.} \qquad \epsilon_{\alpha'\beta'\gamma'} U^{\alpha'}{}_\alpha U^{\beta'}{}_\beta U^{\gamma'}{}_\gamma = \epsilon_{\alpha\beta\gamma} \,, \tag{8.64}$$

where $\epsilon_{\alpha\beta\gamma}$ is the completely antisymmetric tensor ($\epsilon_{123} = 1$) and sum over identical (lower and upper) indeces is implied.

From the above conditions it follows that the generic element U depends on 8 real independent parameters. The basic element of the self-representation is the *controvariant* vector

$$\xi = \begin{pmatrix} \xi^1 \\ \xi^2 \\ \xi^3 \end{pmatrix} \,, \tag{8.65}$$

transforming as

$$\xi' = U\xi \qquad \text{i.e.} \qquad \xi'^\alpha = U^\alpha{}_\beta \xi^\beta \,. \tag{8.66}$$

A *covariant* vector is defined by the quantity

$$\eta = (\eta_1 \ \ \eta_2 \ \ \eta_3) \,, \tag{8.67}$$

which transforms according to

$$\eta' = \eta U^\dagger \qquad \text{i.e.} \qquad \eta'_\alpha = \eta_\beta U^{\dagger\beta}{}_\alpha \,. \tag{8.68}$$

One can show immediately that the quantity

$$\eta\xi = \eta_\alpha \xi^\alpha \tag{8.69}$$

is *invariant* under the group transformations.

We notice that the quantity

$$\xi^\dagger = \left(\xi^{1*} \ \ \xi^{2*} \ \ \xi^{3*} \right) \,, \tag{8.70}$$

transforms as a covariant vector and its components can be denoted by ξ_α.

A higher tensor $\zeta^{\alpha\beta\gamma}$ is defined by the transformation properties

$$\zeta'^{\,\alpha\beta\gamma\cdots} = U^\alpha_{\;\rho}U^\beta_{\;\sigma}U^\gamma_{\;\tau}\cdots\zeta^{\rho\sigma\tau\cdots}\,, \tag{8.71}$$

and it can be built in terms of the ξ^α components

$$\zeta'^{\,\alpha\beta\gamma\cdots} = \xi^\alpha\xi^\beta\xi^\gamma\cdots\,. \tag{8.72}$$

It can be taken as element of the basis of the direct product representation $U\otimes U\otimes U\ldots$, which is *reducible*. The decomposition of this representation into a sum of irreducible representations can be carried out by decomposing the reducible tensor $\zeta'^{\,\alpha\beta\gamma\cdots}$ into irreducible ones. The reduction of a tensor can be easily performed making use of the following circumstance (see Appendix C): a tensor having the symmetry properties of a Young tableau is *irreducible*. For instance, a tensor of third rank can be decomposed as follows

$$\zeta^{\alpha\beta\gamma} \;=\; \zeta^{\{\alpha\beta\gamma\}} \;+\; \zeta^{\{\alpha\beta\}\gamma} \;+\; \zeta^{\{\alpha\gamma\}\beta} \;+\; \zeta^{[\alpha\beta\gamma]}\,. \tag{8.73}$$

Under each term we have drawn the corresponding Young tableau: $\{\alpha\beta\ldots\}$ and $[\alpha\beta\ldots]$ indicate respectively complete symmetry and antisymmetry among the indeces α, β, $\ldots$. Each tensor on the r.h.s. of Eq. (8.73) identifies a IR of the group $SU(3)$.

Eq. (8.71) does not define the most general tensor, which can have mixed (both covariant and controvariant) indeces $\zeta^{\alpha\beta\cdots}_{\kappa\lambda\cdots}$ and transforms as

$$\zeta'^{\,\mu\nu\cdots}_{\;\;\sigma\tau\cdots} = U^\mu_{\;\alpha}U^\nu_{\;\beta}\zeta^{\alpha\beta\cdots}_{\kappa\lambda\cdots}U^{\dagger\,\kappa}_{\;\;\sigma}U^{\dagger\,\lambda}_{\;\;\tau}\,. \tag{8.74}$$

We note that the tensors $\delta^\alpha_{\;\beta}$ and $\epsilon_{\alpha\beta\gamma} = \epsilon^{\alpha\beta\gamma}$ are *invariant* (see Eqs. (8.63) and (8.64)).

We remark that one can obtain all the IR's of $SU(3)$ starting from only one kind of tensors, say controvariant, without needing mixed tensors. We limit ourselves here to a simple example. Let us consider the second rank mixed tensor $\zeta^\alpha_{\;\beta} = \xi^\alpha\xi_\beta$. Clearly its trace $\zeta^\alpha_{\;\alpha}$ is invariant, so that it can be decomposed as follows

$$\zeta^\alpha_{\;\beta} = \tfrac{1}{3}\delta^\alpha_{\;\beta}\zeta^\gamma_{\;\gamma} + \left(\zeta^\alpha_{\;\beta} - \tfrac{1}{3}\delta^\alpha_{\;\beta}\zeta^\gamma_{\;\gamma}\right)\,, \tag{8.75}$$

where the traceless tensor

$$\hat{\zeta}^\alpha_{\;\beta} = \zeta^\alpha_{\;\beta} - \tfrac{1}{3}\delta^\alpha_{\;\beta}\zeta^\gamma_{\;\gamma} \tag{8.76}$$

is irreducible. We see that it is equivalent to the tensor $\zeta^{\{\mu\nu\}\sigma}$ appearing in Eq. (8.73), since $\epsilon_{\beta\gamma\delta}$ is invariant, and one can write:

$$\zeta^\alpha_\beta = \epsilon_{\beta\gamma\delta}\zeta^{\{\alpha\gamma\}\delta} . \tag{8.77}$$

We note that the two tensors $\hat{\zeta}^\alpha_\beta$ and $\zeta^{\{\alpha\beta\}\gamma}$ are no longer equivalent in going from $SU(3)$ to the group $U(3)$ which is locally isomorphic to $SU(3) \otimes U(1)$; in fact the tensor $\epsilon^{\alpha\beta\gamma}$ is no longer invariant. $U(3)$ is employed when one wants to take into account also the *baryon number* B, which can be related to $U(1)$; the two IR's of $U(3)$ corresponding to $\hat{\zeta}^\alpha_\beta$ and $\zeta^{\{\alpha\beta\}\gamma}$ are distinguished by the eigenvalue of B which can be taken to be 0 and 1.

In general, each irreducible tensor, and therefore each IR of $SU(3)$, is characterized by a Young tableau. A general Young tableau for $SU(3)$ has at most three rows consisting of $\lambda_1, \lambda_2, \lambda_3$ boxes

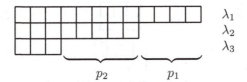

but the two numbers

$$\begin{aligned} p_1 &= \lambda_1 - \lambda_2 \\ p_2 &= \lambda_2 - \lambda_3 \end{aligned} \tag{8.78}$$

are sufficient to characterize the IR (in practice, the 3-box columns can be erased in all tableaux).

We recall Eq. (C.29) from Appendix C, which gives the dimension of the IR (i.e. the number of independent components of the basic tensor):

$$N = \tfrac{1}{2}(p_1 + 1)(p_2 + 1)(p_1 + p_2 + 2) . \tag{8.79}$$

Usually a IR is denoted by $D(p_1, p_2)$. In general, for $p_1 \neq p_2$ (suppose $p_1 > p_2$) there are two different IR's with the same dimension: $D(p_1, p_2)$ and $D(p_2, p_1)$. They can be simply denoted by N and $\overline{N}$, respectively The $\overline{N}$ representation is called *conjugate* to N; if $p_1 = p_2$, $\overline{N} = N$ and the IR is said to be *self-conjugate*.

In Table 8.4 we list the lowest IR's of $SU(3)$ which are most important for the applications to hadron physics.

In general, one can distinguish three classes of IR's of $SU(3)$, in correspondence with the numbers $k = 0, 1, 2$ (modulo 3) of boxes of the related Young tableaux. Alternatively, for each $D(p_1, p_2)$ one can define *triality* the number

$$\tau = p_1 - p_2 . \tag{8.80}$$

Table 8.4. Lowest IR's of $SU(3)$ of physical interest.

IR	N or $\overline{N}$	Young tableau	basic tensor
$D(0,0)$	1	•	$\epsilon_{\alpha\beta\gamma}$
$D(1,0)$	3	□	ξ^{α}
$D(0,1)$	$\overline{3}$		ξ_{α}
$D(1,1)$	8		$\zeta^{\{\alpha\beta\}\gamma}$ or $\hat{\zeta}^{\alpha}{}_{\beta}$
$D(3,0)$	10		$\zeta^{\{\alpha\beta\gamma\}}$
$D(0,3)$	$\overline{10}$		$\zeta_{\{\alpha\beta\gamma\}}$
$D(2,2)$	27		$\zeta^{\{\alpha\beta\gamma\delta\}\{\mu\nu\}}$

We give also a list of direct product decompositions into IR's, which are useful for applications; they can be easily obtained following the rules given in Appendix C.

$$
\begin{aligned}
3 \otimes \overline{3} &= 1 \oplus 8 \,, \\
3 \otimes 3 &= \overline{3} \oplus 6 \,, \\
3 \otimes 3 \otimes 3 &= 1 \oplus 8 \oplus 8 \oplus 10 \,, \\
8 \otimes 8 &= 1 \oplus 8 \oplus 8 \oplus 10 \oplus \overline{10} \oplus 27 \,, \\
8 \otimes 10 &= 8 \oplus 10 \oplus 27 \oplus 35 \,, \\
10 \otimes \overline{10} &= 1 \oplus 8 \oplus 27 \oplus 64 \,.
\end{aligned}
\tag{8.81}
$$

The numbers p_1, p_2 are sufficient to characterize the IR's of $SU(3)$ and the basic irreducible tensors; however, we need to specify also the components of a given tensor. This can be done specifying the $SU(2)$ content of the IR's. It is clear that $SU(3)$ contains the subgroup $SU(2)$. In fact one can single out three different $SU(2)$ subgroups, as it will be discussed in Subsection 8.5.5. We associate a subgroup $SU(2)$ to isospin and $U(1)$ to hypercharge, and we make use of the subgroup $SU(2)_I \otimes U(1)_Y$.

A IR of $SU(3)$ is no longer irreducible when considered as a representation of the subgroup $SU(2)_I \otimes U(1)_Y$, but it can be decomposed into a direct sum of IR's of this subgroup. We give here the decompositions of the IR's of $SU(3)$ which are more important in particle physics:

$$
\begin{aligned}
8 &= (2,1) \oplus (1,0) \oplus (3,0) \oplus (2,-1) \,, \\
10 &= (4,1) \oplus (3,0) \oplus (2,-1) \oplus (1,-2) \,, \\
\overline{10} &= (4,-1) \oplus (3,0) \oplus (2,1) \oplus (1,2) \,, \\
27 &= (3,2) \oplus (4,1) \oplus (2,1) \oplus (5,0) \oplus (3,0) \oplus \\
&\quad \oplus (1,0) \oplus (4,-1) \oplus (2,-1) \oplus (3,-2) \,.
\end{aligned}
\tag{8.82}
$$

The first number in parenthesis gives the dimension of the IR of $SU(2)_I$, and therefore specifies completely the IR, i.e. the isospin; the second number gives directly, by convenient choice of the scale, the hypercharge of each isospin multiplet. With the same convention, one gets for the fundamental representations:

$$
\begin{aligned}
3 &= \left(2, \tfrac{1}{3}\right) \oplus \left(1, -\tfrac{2}{3}\right) \,, \\
\overline{3} &= \left(2, -\tfrac{1}{3}\right) \oplus \left(1, \tfrac{2}{3}\right) \,.
\end{aligned}
\tag{8.83}
$$

8.5.3 Lie algebra of $SU(3)$

Before going to the $SU(3)$ classification of hadrons, it is convenient to consider the general properties of its Lie algebra. In analogy with $SU(2)$, one can write the elements of the $SU(3)$ group in the form

$$
U = \exp\left\{ i \sum_{k=1}^{8} a_k \lambda_k \right\} \,,
\tag{8.84}
$$

in terms of eight 3×3 traceless hermitian matrices λ_k and eight real parameters a_k.

One can easily convince oneself that there exist 8 independent matrices of this type, for which it is convenient to choose the following form introduced by Gell-Mann:

$$
\lambda_1 = \begin{pmatrix} 0 & 1 & 0 \\ 1 & 0 & 0 \\ 0 & 0 & 0 \end{pmatrix}, \quad
\lambda_2 = \begin{pmatrix} 0 & -i & 0 \\ i & 0 & 0 \\ 0 & 0 & 0 \end{pmatrix}, \quad
\lambda_3 = \begin{pmatrix} 1 & 0 & 0 \\ 0 & -1 & 0 \\ 0 & 0 & 0 \end{pmatrix},
$$

$$
\lambda_4 = \begin{pmatrix} 0 & 0 & 1 \\ 0 & 0 & 0 \\ 1 & 0 & 0 \end{pmatrix}, \quad
\lambda_5 = \begin{pmatrix} 0 & 0 & -i \\ 0 & 0 & 0 \\ i & 0 & 0 \end{pmatrix}, \quad
\lambda_6 = \begin{pmatrix} 0 & 0 & 0 \\ 0 & 0 & 1 \\ 0 & 1 & 0 \end{pmatrix},
\tag{8.85}
$$

$$
\lambda_7 = \begin{pmatrix} 0 & 0 & 0 \\ 0 & 0 & -i \\ 0 & i & 0 \end{pmatrix}, \quad
\lambda_8 = \sqrt{\tfrac{1}{3}} \begin{pmatrix} 1 & 0 & 0 \\ 0 & 1 & 0 \\ 0 & 0 & -2 \end{pmatrix} \,.
$$

where the matrices λ_1, λ_2, λ_3 are nothing else that the three Pauli matrices bordered with a row and a column of zeros.

The λ matrices satisfy the following commutation and anticommutation relations:

$$[\lambda_i, \lambda_j] = 2i f_{ijk} \lambda_k ,\tag{8.86}$$

$$\{\lambda_i, \lambda_j\} = 2d_{ijk}\lambda_k + \tfrac{4}{3}\delta_{ij}1 ,\tag{8.87}$$

the completely antisymmetric coefficients f_{ijk} (structure constants of the Lie algebra) and the completely symmetric ones d_{ijk} being given in Table 8.5.

Table 8.5. Coefficients f_{ijk} and d_{ijk}.

(ijk)	f_{ijk}	(ijk)	d_{ijk}	(ijk)	d_{ijk}
123	1	118	$\frac{1}{\sqrt{3}}$	355	$\frac{1}{2}$
147	$\frac{1}{2}$	146	$\frac{1}{2}$	366	$-\frac{1}{2}$
156	$-\frac{1}{2}$	157	$\frac{1}{2}$	377	$-\frac{1}{2}$
246	$\frac{1}{2}$	228	$\frac{1}{\sqrt{3}}$	448	$-\frac{1}{2\sqrt{3}}$
257	$\frac{1}{2}$	247	$-\frac{1}{2}$	558	$-\frac{1}{2\sqrt{3}}$
345	$\frac{1}{2}$	256	$\frac{1}{2}$	668	$-\frac{1}{2\sqrt{3}}$
367	$-\frac{1}{2}$	338	$\frac{1}{\sqrt{3}}$	778	$-\frac{1}{2\sqrt{3}}$
458	$\frac{\sqrt{3}}{2}$	344	$\frac{1}{2}$	888	$-\frac{1}{\sqrt{3}}$
678	$\frac{\sqrt{3}}{2}$				

The following relations are also very useful:

$$\text{Tr}\,(\lambda_\alpha \lambda_\beta) = \delta_{\alpha\beta} ,\tag{8.88}$$

$$\text{Tr}\,([\lambda_\alpha, \lambda_\beta]\lambda_\gamma) = 4i f_{\alpha\beta\gamma} ,\tag{8.89}$$

$$\text{Tr}\,(\{\lambda_\alpha, \lambda_\beta\}\lambda_\gamma) = 4d_{\alpha\beta\gamma} .\tag{8.90}$$

Similarly to the case of $SU(2)$, the generators of $SU(3)$, in the self-representation, are defined by

$$F_i = \tfrac{1}{2}\lambda_i .\tag{8.91}$$

They satisfy the commutation relations

$$[F_i, F_j] = i f_{ijk} F_k .\tag{8.92}$$

Out of the 8 generators F_i, only 2 are diagonalized, F_3, F_8: this corresponds to the fact that the algebra of $SU(3)$ has rank 2. We know that the IR's of $SU(3)$ are characterized by two parameters: given an IR, it is convenient to use the eigenvalues of F_3 and F_8 (which correspond to the third component I_3 of the isospin and to the hypercharge Y) to label the basic elements of the IR. The eigenvalues of I_3 and Y are given, in the 3-dimensional representation $D(1,0)$, by the following identification:

$$I_3 = F_3 \,, \tag{8.93}$$

$$Y = \frac{2}{\sqrt{3}} F_8 \,. \tag{8.94}$$

From these, one can obtain the eigenvalues of I_3 and Y for the higher IR's[8]; for the cases of interest they can be obtained from Eq. (8.82).

It is convenient to make use of a graphical representation of the IR's of $SU(3)$, namely of the *weight diagrams* considered in Subsection 1.3.2. To each element of an IR we associate a two-dimensional weight, whose components are the eigenvalues of F_3 and F_8. If the dimensionality of the IR is n, we have then n weights (some of which may be zero), which can be drawn in a plane. In this way, one obtains for each IR a regular pattern, which characterizes the representation. For convenience, we give on the axes the eigenvalues of I_3 and Y: however, in order to preserve the symmetry of the weight diagram, the scale units of I_3 and Y axes are taken in the ratio $1 : \frac{\sqrt{3}}{2}$.

The diagrams corresponding to the lower representations are shown in Figs. 8.1 and 8.2. The point inside a small circle in Fig. 8.2 stands for two degenerate components.

8.5.4 $SU(3)$ classification of hadrons

As anticipated in Subsection 8.5.1, the analysis of the hadron spectrum shows that $SU(3)$ is a group of approximate symmetry for strong interactions. It appears that all mesons (with spin-parity $J^P = 0^-, 1^-, 1^+, 2^+, \ldots$) can be arranged in *octets* and *singlets*, and all baryons ($J^P = \frac{1}{2}^+, \frac{3}{2}^+, \frac{1}{2}^-, \frac{3}{2}^-, \ldots$) into *singlets*, *octets* and *decuplets*.

In Figs. 8.3 and 8.4 we represent the lowest meson ($J^P = 0^-$) and baryon ($J^P = \frac{1}{2}^+$) states. They are stable under strong interactions and decay via weak and electromagnetic interactions, except for the proton which is absolutely stable.

Besides the weight diagrams, we represent also the split levels of the octets: the numbers on the right are the experimental values of the masses (in MeV).

---

[8] Given the three-dimensional IR for the generators F_i, it is easy to obtain the matrix F_i in any other IR by the following procedure. For a given IR we consider the irreducible tensor written in terms of products of the type $\xi^\alpha \xi^\beta \xi^\gamma \ldots$ Then the matrix elements of F_i can be obtained making use of the relation

$$F_i \xi^\alpha \xi^\beta \xi^\gamma \ldots = \left(F_i \xi^\alpha \right) \xi^\beta \xi^\gamma \ldots + \xi^\alpha \left(F_i \xi^\beta \right) \xi^\gamma \ldots + \xi^\alpha \xi^\beta \left(F_i \xi^\gamma \right) \ldots + \ldots$$

$$= \left(F_i \right)^\alpha_{\ \alpha'} \xi^{\alpha'} \xi^\beta \xi^\gamma \ldots + \xi^\alpha \left(F_i \right)^\beta_{\ \beta'} \xi^{\beta'} \xi^\gamma \ldots + \xi^\alpha \xi^\beta \left(F_i \right)^\gamma_{\ \gamma'} \xi^{\gamma'} \ldots + \ldots$$

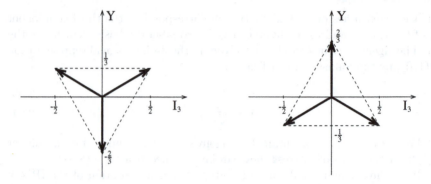

Fig. 8.1. Weight diagrams for the IR's 3 and $\bar{3}$.

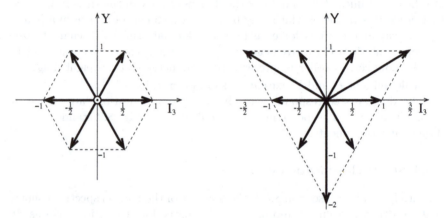

Fig. 8.2. Weight diagrams for the IR's 8 and 10.

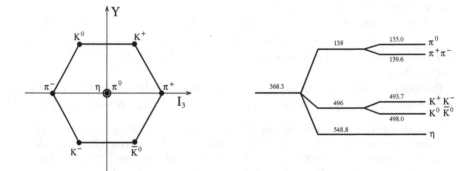

Fig. 8.3. Weight diagram and mass levels of the meson 0^- octet.

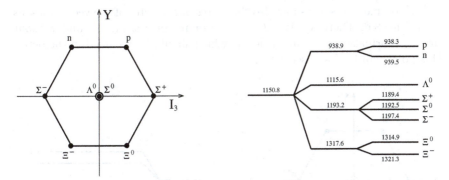

Fig. 8.4. Weight diagram and mass levels of the baryon $\frac{1}{2}^{+}$ octet.

In the meson octet particle-antiparticle pairs appear, and have, therefore, the same masses. An antibaryon octet corresponds to the baryon octet, with the same masses, and opposite Y, B quantum numbers.

As discussed in Subsection 8.4.2, one assumes that, in the absence of electromagnetic interactions, there is a degeneracy among the masses of each isospin multiplet: the numbers above the intermediate lines in the graphs correspond to the average values. If $SU(3)$ were an exact symmetry, all the levels in an octet would be degenerate, i.e. all the members of the octet would have the same mass (a rough estimate of the average value is indicated in Figs. 8.3 and 8.4).

In Fig. 8.5 we show another important multiplet, namely the decuplet of the $\frac{3}{2}^{+}$ baryons. For the sake of simplicity, the actual splitting of the levels due to the electromagnetic interactions is not reported in the figure.

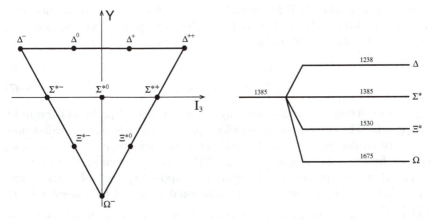

Fig. 8.5. Weight diagram and mass levels of the baryon $\frac{3}{2}^{+}$ decuplet.

For completeness, we show also the states of the octet of the vector mesons 1^- in Fig. 8.6. There are two $I = 0$ vector mesons: ω and ϕ, and without further information one cannot decide which should be included in the octet. This point will be discussed later.

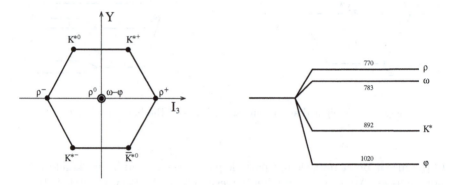

Fig. 8.6. Weight diagram and mass levels of the vector meson 1^- octet.

This type of classification was proposed by Gell-Mann and Ne'eman and it is known as the *eightfold way*[9].

We note that the multiplets 1, 8 and 10 correspond to IR's of *zero triality*. The direct products of IR's of this class give rise to other IR's of zero triality. One would expect that hadronic states, which correspond to baryon-meson or mesonic resonances, should belong also to multiplets higher than octets and decuplets. In fact, according to Eqs. (8.81), one has

$$8 \otimes 8 = 1 \oplus 8 \oplus 8 \oplus 10 \oplus \overline{10} \oplus 27 . \tag{8.95}$$

It was proposed by Gell-Mann and Zweig[10] that one can associate fictitious particles - called *quarks* - to the IR 3, and the corresponding antiparticles, i.e. *antiquarks*, to the $\overline{3}$. The product decompositions

$$3 \otimes \overline{3} = 1 \oplus 8 , \tag{8.96}$$

$$3 \otimes 3 \otimes 3 = 1 \oplus 8 \oplus 8 \oplus 10 \tag{8.97}$$

suggest that mesons are constituted by a *quark-antiquark* pair and baryons by *three quarks*. The three quarks are called *up*, *down* and *strange*; we shall denote the quark triplet by $q = (u, d, s)$, and the antiquark triplet by $\bar{q} = (\bar{u}, \bar{d}, \bar{s})$; they are fitted in the weight diagrams of Fig. 8.1, as shown in Fig. 8.7.

The above interpretation of hadrons as composite systems of quarks would imply that quarks are *particles* with fractional values of baryon number B,

---

[9] M. Gell-Mann and Y. Ne'eman, *The Eightfold Way*, Benjamin, New York (1964).

[10] M. Gell-Mann, Phys. Lett. 8 (1964) 214; G. Zweig, CERN Reports TH 401, TH 402 (1964), unpublished.

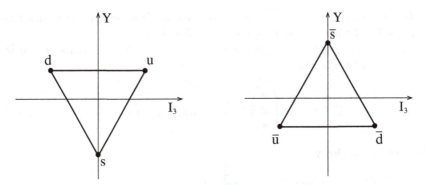

Fig. 8.7. Triplet of quarks and antiquarks.

hypercharge Y and electric charge Q. The internal quantum numbers for the two q and $\bar{q}$ triplets are summarized in Table 8.6 (see Eq. (8.83)). Moreover, they are assumed to have spin $\frac{1}{2}$.

Table 8.6. Quantum numbers of quarks and antiquarks.

	B	S	Y	I	I_3	Q
u	$\frac{1}{3}$	0	$\frac{1}{3}$	$\frac{1}{2}$	$\frac{1}{2}$	$\frac{2}{3}$
d	$\frac{1}{3}$	0	$\frac{1}{3}$	$\frac{1}{2}$	$-\frac{1}{2}$	$-\frac{1}{3}$
s	$\frac{1}{3}$	-1	$-\frac{2}{3}$	0	0	$-\frac{1}{3}$
$\bar{u}$	$-\frac{1}{3}$	0	$-\frac{1}{3}$	$\frac{1}{2}$	$-\frac{1}{2}$	$-\frac{2}{3}$
$\bar{d}$	$-\frac{1}{3}$	0	$-\frac{1}{3}$	$\frac{1}{2}$	$\frac{1}{2}$	$\frac{1}{3}$
$\bar{s}$	$-\frac{1}{3}$	$+1$	$\frac{2}{3}$	0	0	$\frac{1}{3}$

For the present purposes, it is sufficient to consider the introduction of quarks as a useful mathematical device. However, several independent experiments provided indirect evidence of their existence inside hadrons[11], so that quarks are considered as real particles, even if they never appear in free states, but always bound inside hadrons.

Multiplets not fitting in the above scheme (e.g. decuplets of mesons, or baryons belonging to the IR 27) are called *exotic*; they would require the addition of extra $q\bar{q}$ pairs to their constituents. The evidence for such states is still controversial.

---

[11] G. Sterman *et al.*, Rev. Mod. Phys. **67**, 157 (1995).

Let us now consider the octets in more details. It is very useful to describe them as 3×3 matrices, making use of the λ-matrices.

First we express the traceless tensor $\hat{\zeta}^\alpha{}_\beta$ (Eq. (8.76)) in terms of q, $\bar{q}$ by making the identification

$$(\xi^\alpha) \to q = \begin{pmatrix} u \\ d \\ s \end{pmatrix} \quad , \quad (\xi_\alpha) \to \bar{q} = (\bar{u} \ \bar{d} \ \bar{s}) \ . \tag{8.98}$$

One obtains in this way

$$\hat{\zeta} = \begin{pmatrix} \frac{1}{3}(2u\bar{u} - d\bar{d} - s\bar{s}) & u\bar{d} & u\bar{s} \\ d\bar{u} & \frac{1}{3}(-u\bar{u} + 2d\bar{d} - s\bar{s}) & d\bar{s} \\ s\bar{u} & s\bar{d} & \frac{1}{3}(-u\bar{u} - d\bar{d} + 2s\bar{s}) \end{pmatrix} . \tag{8.99}$$

Let us now consider the case of the octet of pseudoscalar $J^P = 0^-$ mesons. Since the octet corresponds to the traceless tensor $\hat{\zeta}^\alpha{}_\beta$, we can write the 0^- states in the matrix form

$$\mathcal{P} = \begin{pmatrix} \frac{1}{\sqrt{2}}\pi^0 + \frac{1}{\sqrt{6}}\eta & \pi^+ & K^+ \\ \pi^- & -\frac{1}{\sqrt{2}}\pi^0 + \frac{1}{\sqrt{6}}\eta & K^0 \\ K^- & \overline{K}^0 & -\frac{2}{\sqrt{6}}\eta \end{pmatrix} . \tag{8.100}$$

In fact strange mesons are easily recognized by their quantum numbers ($Y = \pm 1$), pions by their isospin classification (see Eq. (8.53)) and the remaining piece, taking into account the normalization, as the η meson.

This procedure gives immediately the quark content of the different mesons; in particular, one has

$$\pi^0 = \frac{1}{\sqrt{2}} \left(u\bar{u} - d\bar{d} \right) \ , \tag{8.101}$$

$$\eta = \frac{1}{\sqrt{6}} \left(u\bar{u} + d\bar{d} - 2s\bar{s} \right) \ . \tag{8.102}$$

For the other mesons, the quark content is immediately read off by comparison of (8.100) with (8.99); for convenience, the explicit expressions are given in Table 8.7.

It is known that a ninth pseudoscalar meson exists: it is the η' (958) with $I = 0$, $Y = 0$. It can be considered an $SU(3)$ 0^- *singlet*; then its quark content is given by

$$\eta' = \frac{1}{\sqrt{3}} \left(u\bar{u} + d\bar{d} + s\bar{s} \right) \ . \tag{8.103}$$

We notice that there is no difference between the quantum numbers of η and η', but η has a lower mass than η'. We shall see in Subsection 8.5.7 that the physical states contain a small mixing between η and η'.

Another octet is represented by the 1^- *vector mesons* (see Fig. 8.6)

$$\mathcal{V} = \begin{pmatrix} \frac{1}{\sqrt{2}}\rho^0 + \frac{1}{\sqrt{6}}\omega_8 & \rho^+ & K^{*+} \\ \rho^- & -\frac{1}{\sqrt{2}}\rho^0 + \frac{1}{\sqrt{6}}\omega_8 & K^{*0} \\ K^{*-} & \overline{K}^{*0} & -\frac{2}{\sqrt{6}}\omega_8 \end{pmatrix} . \tag{8.104}$$

Experimentally, nine 1^- vector mesons are known, two of which, ω and ϕ, with the same quantum numbers, $I = Y = 0$. At this stage, in the absence of a criterion for discriminate between them, we denote with ω_8 the isoscalar member of the octet and introduce an isoscalar (1^-) $SU(3)$ *singlet*

$$\omega_1 = \frac{1}{\sqrt{3}}\left(u\bar{u} + d\bar{d} + s\bar{s}\right) , \tag{8.105}$$

leaving open the problem of the correspondence between ω_1, ω_8 and the physical states ω, ϕ. The quark content of the vector mesons is the same of the scalar ones, as indicated in Table 8.7.

Table 8.7. Quark content of the lowest meson states.

0^- octet	1^- octet	$q\bar{q}$ structure
K^+	K^{*+}	$u\bar{s}$
K^0	K^{*0}	$d\bar{s}$
π^+	ρ^+	$u\bar{d}$
π^0	ρ^0	$\frac{1}{\sqrt{2}}\left(u\bar{u} - d\bar{d}\right)$
η	ω_8	$\frac{1}{\sqrt{6}}\left(u\bar{u} + d\bar{d} - 2s\bar{s}\right)$
π^-	ρ^-	$d\bar{u}$
$\overline{K}^0$	$\overline{K}^{*0}$	$s\bar{d}$
K^-	K^{*-}	$s\bar{u}$

Once we have built an octet as a 3×3 matrix, we can forget about the quark content exhibited in (8.99), and use directly (8.100) and the graphical representations of Fig. 8.3 and Fig. 8.4. We can then write for the $\frac{1}{2}^+$ baryon octet

$$\mathcal{B} = \begin{pmatrix} \frac{1}{\sqrt{2}}\Sigma^0 + \frac{1}{\sqrt{6}}\Lambda^0 & \Sigma^+ & p \\ \Sigma^- & -\frac{1}{\sqrt{2}}\Sigma^0 + \frac{1}{\sqrt{6}}\Lambda^0 & n \\ \Xi^- & \Xi^0 & -\frac{2}{\sqrt{6}}\Lambda^0 \end{pmatrix} . \tag{8.106}$$

An alternative way of representing the meson and baryon octets is in terms of eight-component vectors P^i, V^i, B^i $(i = 1, \ldots, 8)$, by identifying

$$\mathcal{P} = \tfrac{1}{\sqrt{2}}\lambda_i P^i \ , \quad \mathcal{V} = \tfrac{1}{\sqrt{2}}\lambda_i V^i \ , \quad \mathcal{B} = \tfrac{1}{\sqrt{2}}\lambda_i B^i \ , \tag{8.107}$$

where the sum over i from 1 to 8 is implied, and the normalization conditions

$$\mathcal{P}^\alpha{}_\beta \mathcal{P}^\beta{}_\alpha = P^i P_i \ , \quad \mathcal{V}^\alpha{}_\beta \mathcal{V}^\beta{}_\alpha = V^i V_i \ , \quad \mathcal{B}^\alpha{}_\beta \mathcal{B}^\beta{}_\alpha = B^i B_i \ , \tag{8.108}$$

are satisfied.

The relations (8.107) can be inverted taking into account Eq. (8.88)

$$P_i = \tfrac{1}{\sqrt{2}}\mathrm{Tr}\,(\mathcal{P}\lambda_i) \ , \quad V_i = \tfrac{1}{\sqrt{2}}\mathrm{Tr}\,(\mathcal{V}\lambda_i) \ , \quad B_i = \tfrac{1}{\sqrt{2}}\mathrm{Tr}\,(\mathcal{B}\lambda_i) \ , \tag{8.109}$$

from which[12]

$$P_1 = \tfrac{1}{\sqrt{2}}(\pi^+ + \pi^-) \ , \ P_4 = \tfrac{1}{\sqrt{2}}(K^+ + K^-) \ , \ P_6 = \tfrac{1}{\sqrt{2}}(K^0 + \overline{K}^0) \ , \ P_3 = \pi^0 \ ,$$
$$P_2 = \tfrac{1}{\sqrt{2}}(\pi^+ - \pi^-) \ , \ P_5 = \tfrac{1}{\sqrt{2}}(K^+ - K^-) \ , \ P_7 = \tfrac{1}{\sqrt{2}}(K^0 - \overline{K}^0) \ , \ P_8 = \eta \ ,$$
$$\tag{8.110}$$

and similarly for the V_i and B_i.

A complete description of the meson and baryon octets $\mathcal{P}$ and $\mathcal{B}$ should require the introduction of the baryon number B which would differentiate the two octets. Then one should go from $SU(3)$ to $U(3)$: as already noticed, the IR's tensor $\hat{\zeta}^\alpha{}_\beta$ and $\zeta^{\{\alpha\beta\}\gamma}$, which are equivalent for $SU(3)$, correspond to inequivalent IR's of $U(3)$. In the frame of $U(3)$, only $\mathcal{P}$, for which $B = 0$, can be described by $\hat{\zeta}^\alpha{}_\beta$, while $\mathcal{B}$ $(B = 1)$ is described by the tensor $\zeta^{\{\alpha\beta\}\gamma}$. From this tensor, one can obtain immediately the quark content of the baryons. Since the states have mixed symmetry, one gets two solutions relative to the possibility of symmetrizing according to the two standard Young tableaux. The two solutions are listed in Table 8.8 for all members of the octet.

For sake of completeness, it is convenient to give also the $\tfrac{1}{2}$ antibaryon $(B = -1)$ octet

$$\overline{\mathcal{B}} = \begin{pmatrix} \tfrac{1}{\sqrt{2}}\overline{\Sigma}^0 + \tfrac{1}{\sqrt{6}}\overline{\Lambda}^0 & \overline{\Sigma}^- & \overline{\Xi}^- \\ \overline{\Sigma}^+ & -\tfrac{1}{\sqrt{2}}\overline{\Sigma}^0 + \tfrac{1}{\sqrt{6}}\overline{\Lambda}^0 & \overline{\Xi}^0 \\ \overline{p} & \overline{n} & -\tfrac{2}{\sqrt{6}}\overline{\Lambda}^0 \end{pmatrix} . \tag{8.111}$$

Finally, we consider the $\tfrac{3}{2}^+$ baryon decuplet, which, according to Table 8.4, corresponds to the completely symmetric tensor $\zeta^{\{\alpha\beta\gamma\}}$. Each member of the decuplet, given the values of I_3 and Y, can be immediately identified with a component of the tensor, and each component can be expressed in terms of quarks. The explicit expressions are reported in Table 8.9.

---

[12] Eqs. (8.107) show that the set of λ-matrices, with respect to the index i, has the same transformation properties of an octet.

Table 8.8. Quark content of the $J^P = \frac{1}{2}^+$ baryon octet.

baryon	symmetric in the first pair of quarks	antisymmetric in the first pair of quarks
p	$\frac{1}{\sqrt{6}}(2uud - udu - duu)$	$\frac{1}{\sqrt{2}}(udu - duu)$
n	$\frac{1}{\sqrt{6}}(udd - dud - 2ddu)$	$\frac{1}{\sqrt{2}}(udd - dud)$
Σ^+	$\frac{1}{\sqrt{6}}(usu + suu - 2uus)$	$\frac{1}{\sqrt{2}}(suu - usu)$
Σ^0	$\frac{1}{\sqrt{12}}(usd+sud+dsu+sdu-2uds-2dus)$	$\frac{1}{2}(sud - usd + sdu - dsu)$
Λ^0	$\frac{1}{2}(usd + sud - dsu - sdu)$	$\frac{1}{\sqrt{12}}(2uds-2dus+sdu-dsu+usd-sud)$
Σ^-	$\frac{1}{\sqrt{6}}(dsd + sdd - 2dds)$	$\frac{1}{\sqrt{2}}(sdd - dsd)$
Ξ^0	$\frac{1}{\sqrt{6}}(2ssu - uss - sus)$	$\frac{1}{\sqrt{2}}(sus - uss)$
Ξ^-	$\frac{1}{\sqrt{6}}(2ssd - dss - sds)$	$\frac{1}{\sqrt{2}}(sds - dss)$

Table 8.9. Quark content of the $J^P = \frac{3}{2}^+$ baryon decuplet.

baryon	quark content
Δ^{++}	uuu
Δ^+	$\frac{1}{\sqrt{3}}(uud + udu + duu)$
Δ^0	$\frac{1}{\sqrt{3}}(udd + dud + ddu)$
Δ^-	ddd
Σ^{*+}	$\frac{1}{\sqrt{3}}(uus + usu + suu)$
Σ^{*0}	$\frac{1}{\sqrt{6}}(uds + usd + dus + dsu + sud + sdu)$
Σ^{*-}	$\frac{1}{\sqrt{3}}(dds + dsd + sdd)$
Ξ^{*0}	$\frac{1}{\sqrt{3}}(uss + sus + ssu)$
Ξ^{*-}	$\frac{1}{\sqrt{3}}(dss + sds + ssd)$
Ω^-	sss

8.5.5 *I*-spin, *U*-spin and *V*-spin

For the classification of hadrons, we made use of the $SU(2)$ isospin subgroup of $SU(3)$. We want to point out that there are alternative ways, besides the use of I and Y, of labelling the states of an $SU(3)$ multiplet.

In fact, one can identify *three* different subgroups $SU(2)$ in $SU(3)$. In the set of the 8 λ-matrices one can indeed find three subsets which generates $SU(2)$ subgroups; obviously these subsets do not commute among them.

We know already that the matrices

$$I_i = \tfrac{1}{2}\lambda_i \qquad (i = 1, 2, 3) \tag{8.112}$$

can be taken as the generators of the isospin group $SU(2)_I$. One can immediately check from Eq. (8.86) and Table 8.5 that they satisfy the commutation relations

$$[I_i, I_j] = i\epsilon_{ijk}I_k \;, \tag{8.113}$$

$$[I_i, Y] = 0 \;. \tag{8.114}$$

Let us define two other sets:

$$\begin{aligned} U_1 &= \tfrac{1}{2}\lambda_6 \;, \\ U_2 &= \tfrac{1}{2}\lambda_7 \;, \\ U_3 &= \tfrac{1}{4}\left(\sqrt{3}\lambda_8 - \lambda_3\right) \;, \end{aligned} \tag{8.115}$$

and

$$\begin{aligned} V_1 &= \tfrac{1}{2}\lambda_4 \;, \\ V_2 &= \tfrac{1}{2}\lambda_5 \;, \\ V_3 &= \tfrac{1}{4}\left(\lambda_3 + \sqrt{3}\lambda_8\right) \;. \end{aligned} \tag{8.116}$$

Using again Eq. (8.86) and Table 8.5 for the f_{ijk} coefficients, we get

$$[U_i, U_j] = i\epsilon_{ijk}U_k \;, \tag{8.117}$$

$$[V_i, V_j] = i\epsilon_{ijk}V_k \;, \tag{8.118}$$

which show that also U_i and V_i generate $SU(2)$ subgroups (denoted in the following by $SU(2)_U$ and $SU(2)_V$). For this reason, besides the isospin I or I-*spin*, one defines the *U-spin* and the *V-spin*; in analogy with the shift operators $I_\pm = I_1 \pm iI_2$ (Eq. (8.58)), we define also the operators $U_\pm = U_1 \pm iU_2$ and $V_\pm = V_1 \pm iV_2$. Their action is represented in Fig. 8.8.

What is the use of these sets of generators? Let us consider first the U-spin. Its introduction is particularly useful, since the U_i's commute with the electric charge Q defined in (8.45)

$$[U_i, Q] = 0 \;, \tag{8.119}$$

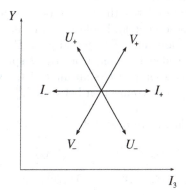

Fig. 8.8. *Action of the shifts operator $I_\pm$, $U_\pm$, $V_\pm$ in the plane (Y, I_3).*

as can be checked immediately. Expressing U_3 in terms of I_3 and Y (Eqs. (8.91), (8.93), (8.94)) one gets

$$U_3 = \tfrac{1}{2}\left(\tfrac{3}{2}Y - I_3\right) . \tag{8.120}$$

The above relation holds, in general, for any representation of $SU(3)$. We examine here the $\tfrac{1}{2}^+$ baryon octet, described by the weight diagram of Fig. 8.4. In the same weight diagram shown in Fig. 8.9 one can individuate two axes U_3 and $-Q$ obtained by rotation of the I_3, Y axes through an angle $\tfrac{5}{6}\pi$ (one has to remember that the actual I_3 and Y scales are in the ratio $1 : \tfrac{\sqrt{3}}{2}$).

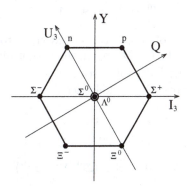

Fig. 8.9. *I-spin and U-spin for the baryon octet.*

Everything said in the case of $SU(2)_I$ can be repeated for $SU(2)_U$ by simply replacing the isospin multiplets (characterized by the hypercharge Y) with the U-spin multiplets (characterized by the electric charge Q). Since one is dealing explicitly with Q, the use of the subgroup $SU(3)_U$ is very convenient in the case of electromagnetic interactions.

Let us consider in more detail the baryon octet (the same considerations can be applied to the meson octet). Looking at Fig. 8.9, we see that the octet consists in two U-spin doublets (Σ^+, p and Ξ^-, Σ^-), one triplet and one singlet. Some care is needed in going from the I-spin multiplets (Σ^+, Σ^0, Σ^-) and Λ^0 to the corresponding U-spin multiplets, which we shall denote by (Ξ^0, Σ^0_u, n) and Λ^0_u. The connection between Σ^0_u, Λ^0_u and Σ^0, Λ^0 can be obtained as follows.

From the third of Eqs. (8.109) we get

$$B_3 \equiv \Sigma^0 = \tfrac{1}{\sqrt{2}} \mathrm{Tr}\,(\mathcal{B}\lambda_3) \ ,$$
$$B_8 \equiv \Lambda^0 = \tfrac{1}{\sqrt{2}} \mathrm{Tr}\,(\mathcal{B}\lambda_8) \ . \tag{8.121}$$

The states Σ^0_u and Λ^0_u are obtained performing the rotation described in Fig. 8.9

$$\Sigma^0_u \equiv B'_3 = \tfrac{1}{\sqrt{2}} \mathrm{Tr}\,(\mathcal{B}\lambda'_3) \ ,$$
$$\Lambda^0_u \equiv B'_8 = \tfrac{1}{\sqrt{2}} \mathrm{Tr}\,(\mathcal{B}\lambda'_8) \ , \tag{8.122}$$

where

$$\lambda'_3 = -\tfrac{1}{2}\lambda_3 + \tfrac{\sqrt{3}}{2}\lambda_8 \ ,$$
$$\lambda'_8 = \tfrac{\sqrt{3}}{2}\lambda_3 + \tfrac{1}{2}\lambda_8 \ . \tag{8.123}$$

Comparing Eqs. (8.122) and (8.123), one obtains

$$\Sigma^0_u = -\tfrac{1}{2}\Sigma^0 + \tfrac{\sqrt{3}}{2}\Lambda^0 \ ,$$
$$\Lambda^0_u = \tfrac{\sqrt{3}}{2}\Sigma^0 + \tfrac{1}{2}\Lambda^0 \ . \tag{8.124}$$

This analysis in terms of U-spin can be extended to the other multiplets of $SU(3)$, making use of the U_3, Q axes in the weight diagrams. For instance, in the case of the baryon $\tfrac{3}{2}^+$ decuplet (see Fig. 8.5), one obtains the following decomposition in terms of $SU(2)_U$ multiplets:

$$
\begin{aligned}
U=0 \,, \ Q=2 & \qquad (\Delta^{++}) \ , \\
U=\tfrac{1}{2} \,, \ Q=1 & \qquad (\Delta^+, \Sigma^{*+}) \ , \\
U=1 \,, \ Q=0 & \qquad (\Delta^0, \Sigma^{*0}, \Xi^{*0}) \ , \\
U=\tfrac{3}{2} \,, \ Q=-1 & \qquad (\Delta^-, \Sigma^{*-}, \Xi^{*-}, \Omega^-) \ .
\end{aligned}
\tag{8.125}
$$

On similar lines, one could carry out the analysis of the V-spin. In this case one uses the eigenvalues of V_3 and $Y' = I_3 - \tfrac{1}{2}Y$ to label the components of the $SU(3)$ multiplets. A graphical interpretation is shown in Fig. 8.10, which is the analogue of Fig. 8.9: V_3 and Y' correspond to the orthogonal

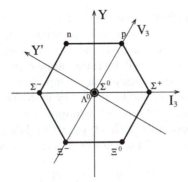

Fig. 8.10. *I*-spin and *V*-spin for the baryon octet.

axes obtained from (I_3, Y) by rotation through an angle $\frac{2}{3}\pi$. The use of the *V*-spin is limited to the weak interactions of hadrons.

Finally, we report in Fig. (8.11) the values of the matrix elements of the shift operators $I_\pm$, $U_\pm$, $V_\pm$ between two states of the baryon octet $\frac{1}{2}^+$. Their explicit calculation is proposed in Problem 8.6.

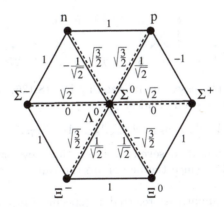

Fig. 8.11. *Values of the matrix elements of the shift operators* $I_\pm$, $U_\pm$, $V_\pm$ *between two states of the baryon octet* $\frac{1}{2}^+$*. The dotted lines join the various states to the singlet.*

8.5.6 The use of $SU(3)$ as exact symmetry

We shall illustrate here, with some simple examples, the use of $SU(3)$ in elementary particle physics.

One can have two kinds of applications, depending on whether one assumes exact $SU(3)$ invariance, or approximate invariance in which the symmetry breaking occurs according to a definite pattern. It is clear that the predictions obtained from exact symmetry will be rather approximate, since we know from the start that the mass differencs among components of a multiplet are not at all negligible. However, in the limit of very high energy, one would expect that these differences should become less important.

Let us consider an example of this type, which is a generalization of what mentioned in Subsection 8.4.2 for the isospin case, namely the *meson-baryon reactions*

$$\mathcal{P}_\alpha + \mathcal{B}_\beta = \mathcal{P}'_\gamma + \mathcal{B}'_\delta , \tag{8.126}$$

where $\mathcal{P}_\alpha$, $\mathcal{P}'_\gamma$ and $\mathcal{B}_\beta$, $\mathcal{B}'_\delta$ are members of the 0^- meson and $\frac{1}{2}^+$ baryon octet, respectively. The $\mathcal{M} + \mathcal{B}$ system belongs to the basis of the direct product representation 8×8, which is decomposed as follows (see Eqs. (8.81))

$$8 \otimes 8 = 1 \oplus 8 \oplus 8 \oplus 10 \oplus \overline{10} \oplus 27 . \tag{8.127}$$

In this decomposition two octets appear, which can be distinguished by their symmetry (8_s) and antisymmetry (8_a) with respect to the two starting octets ($\mathcal{P}$ and $\mathcal{B}$). Assuming exact invariance, only the amplitudes corresponding to the following transitions can be different from zero:

$$
\begin{array}{ccccccccccc}
1 & \oplus & 8 & \oplus & 8 & \oplus & 10 & \oplus & \overline{10} & \oplus & 27 \\
\downarrow & & \downarrow & \times & \downarrow & & \downarrow & & \downarrow & & \downarrow \\
1 & \oplus & 8 & \oplus & 8 & \oplus & 10 & \oplus & \overline{10} & \oplus & 27
\end{array}
$$

Then, all possible processes summarized by Eq. (8.126) can be described in terms of eight invariant amplitudes:

$$A_1, A_{8ss}, A_{8aa}, A_{8sa}, A_{8as}, A_{10}, A_{\overline{10}}, A_{27} . \tag{8.128}$$

This number is reduced to seven by using time reversal invariance which requires $A_{8sa} = A_{8as}$ since the octets are the same in the initial and final states.

The invariant amplitudes referred to specific isospin and hypercharge states (physical amplitudes) can be expressed in terms of the $SU(3)$ amplitudes following a procedure analogous to what done for $SU(2)$. One makes use of expansions in terms of $SU(3)$ Clebsch-Gordan coefficients, which can be written as product of $SU(2)$ Clebsch-Gordan coefficients and the so-called *isoscalar factors*.

These coefficients are not reproduced here. Comprehensive tables are found in the book by Lichtenberg[13] or in the original papers[14]. We refer to the same

---

[13] D.B. Lichtenberg, *Unitary Symmetry and Elementary Particles* , Academic Press, (1970).

[14] J.J. de Swart, Review of Modern Physics 35, 916 (1961); P. McNamee and F. Chilton, Rev. of Mod. Phys. 36, 1005 (1964).

book also for the Wigner-Eckart theorem, which can be extended without difficulties from $SU(2)$ to $SU(3)$.

By expressing, in the above example, the physical amplitudes in terms of the seven amplitudes (8.128), one gets relations (intensity rules) of the following type:

$$A\left(\pi^+ p \to K^+ \Sigma^+\right) = A\left(K^+ p \to K^+ p\right) - A\left(\pi^+ p \to \pi^+ p\right) ,$$
$$A\left(K^- p \to \pi^- \Sigma^+\right) = A\left(K^- p \to K^- p\right) - A\left(\pi^- p \to \pi^- p\right) . \tag{8.129}$$

Indeed, the above example is simply a consequence of U-invariance. Detailed analysis of two-body reactions can be found in Chapter IV of the book by Gourdin[15].

As a second example of exact symmetry, we shall briefly discuss the problem of *meson-baryon coupling* $(\overline{B}B\mathcal{P})$, B and $\mathcal{P}$ being $SU(3)$ octets. In this occasion we shall illustrate the usefulness of the matrix description of the meson and baryon octets. The interaction Lagrangian has to be invariant under $SU(3)$, i.e. it has to behave as the one-dimensional 1 IR. According to Eq. (8.127), if we first combine B and $\overline{B}$ we obtain two octets, each of which can be contracted with the $\mathcal{P}$-octet to give the 1 IR. This shows that the interaction Lagrangian contains, in general, two independent parameters. It is usual to identify the two octets with the symmetrical and antisymmetrical combinations, called F and D respectively: starting from B and $\overline{B}$ given by Eqs. (8.106), (8.111) we can distinguish

$$\left[B, \overline{B}\right]^\alpha_{\ \beta} = \left(B\overline{B}\right)^\alpha_{\ \beta} - \left(\overline{B}B\right)^\alpha_{\ \beta} , \tag{8.130}$$

$$\{B, \overline{B}\}^\alpha_{\ \beta} = \left(B\overline{B}\right)^\alpha_{\ \beta} + \left(\overline{B}B\right)^\alpha_{\ \beta} , \tag{8.131}$$

where

$$\left(B\overline{B}\right)^\alpha_{\ \beta} = B^\alpha_{\ \gamma} \overline{B}^\gamma_{\ \beta} . \tag{8.132}$$

By contraction with $\mathcal{P}$ given by Eq. (8.100), one gets, as far as only the $SU(3)$ structure is concerned, an effective interaction Lagrangian of the form

$$L_{\text{int}} = \sqrt{2}\left\{g_F \text{Tr}\left(\mathcal{P}[B\overline{B}]\right) + g_D \text{Tr}\left(\mathcal{P}\{B\overline{B}\}\right)\right\} =$$
$$= \sqrt{2}\,g\left\{(1-\alpha)\text{Tr}\left(\mathcal{P}[B\overline{B}]\right) + \alpha \text{Tr}\left(\mathcal{P}(\{B\overline{B}\})\right)\right\} , \tag{8.133}$$

where

$$g_F = g(1-\alpha) , \qquad g_D = g\alpha , \tag{8.134}$$

the so-called D/F ratio being defined by

$$D/F = \frac{g_D}{g_F} = \frac{\alpha}{1-\alpha} . \tag{8.135}$$

---

[15] M. Gourdin, *Unitary Symmetries and their applications to high energy physics* North Holland (1967).

We can rewrite Eq. (8.133) in terms of the P_i, B_i components, making use of Eqs. (8.107) and of the properties of the λ-matrices given in Subsection 8.5.3:

$$L_{\text{int}} = \sqrt{2}\, g \left\{ (1 - \alpha) \left(\overline{B} F_k B\right) P_k + \alpha \left(\overline{B} D_k B\right) P_k \right\}, \qquad (8.136)$$

where we have introduced two sets of matrices, F_k and D_k ($k = 1, \ldots, 8$), defined by:

$$(F_k)_{ij} = -i f_{ijk}, \qquad (8.137)$$

$$(D_k)_{ij} = d_{ijk}. \qquad (8.138)$$

It can be shown that the F-matrices satisfy the commutation relations (8.92)); indeed, they are nothing else that the $SU(3)$ generators expressed in the *adjoint* 8-dimensional representation.

On the other hand, the D-matrices satisfy the commutation relations

$$[F_i, D_j] = i f_{ijk} D_k; \qquad (8.139)$$

they do not provide a representation of the $SU(3)$ Lie algebra, but they transform as an octet. Moreover, one can prove the useful relation

$$D_i = \tfrac{2}{3} d_{ijk} F_j F_k. \qquad (8.140)$$

Going back to Eq. (8.136), one easily sees that, by expressing $\mathcal{P}$ and $\mathcal{B}$ in terms of meson and baryon states, one gets terms of the type:

$$L_{\text{int}} = g \left\{ \alpha \left[\overline{p}p - \overline{n}n + \tfrac{2}{\sqrt{3}} \left(\overline{\Sigma}^0 \Lambda^0 - \overline{\Lambda}^0 \Sigma^0 \right) - \overline{\Xi}^0 \Xi^0 + \overline{\Xi}^- \Xi^- \right] \pi^0 + \ldots \right.$$
$$\left. + (1 - \alpha) \left[\overline{p}p - \overline{n}n + 2\overline{\Sigma}^+ \Sigma^+ - 2\overline{\Sigma}^- \Sigma^- + \overline{\Xi}^0 \Xi^0 - \overline{\Xi}^- \Xi^- \right] \pi^0 + \ldots \right\}, \qquad (8.141)$$

so that $g_D = g\alpha$ is the coupling constant relative to $\overline{p}p\pi^0$.

The ratio D/F of Eq. (8.135) occurs in any situation in which three octets are combined to form a scalar. Thus all the $\overline{B}BP$ couplings are usually expressed as function of g and D/F.

The above procedure can be applied, *mutatis mutandis*, to all $8 \otimes 8 \otimes 8$ couplings, such as $\mathcal{VPP}$, $\mathcal{VVV}$, $\mathcal{P}$ and $\mathcal{V}$ being the pseudoscalar and vector meson octets.

Finally, we point out that all the couplings of the type $\overline{\mathcal{D}}\mathcal{BP}$, $\mathcal{D}$ being the baryon decuplet, can be described in terms of only one constant. In fact, one can write

$$L_{\text{int}} = g' \operatorname{Tr} \left(\epsilon^{\alpha\delta\lambda} \overline{\mathcal{D}}_{\alpha\beta\gamma} \mathcal{B}^{\beta}_{\ \delta} \mathcal{P}^{\gamma}_{\ \lambda} \right), \qquad (8.142)$$

and this correponds to the fact that, in the decomposition $8 \otimes 8$, the IR 10 appears only once.

8.5.7 The use of $SU(3)$ as broken symmetry

We present in the following some interesting applications of $SU(3)$ as a broken symmetry. Of course, useful results are obtained by making specific assumptions about the symmetry breaking pattern.

a) *Gell-Mann - Okubo mass formulae* [16]

We have often pointed out that the states of an $SU(3)$ multiplet present, in general, large mass differences. Such mass splittings indicate that the behaviour of the mass operator M deviates appreciably from that of a scalar under $SU(3)$. A relation which is very well satisfied for the masses of the $\frac{1}{2}^{+}$ baryon octet follows from the assumption

$$M_{(8)} = m_0 \text{Tr} \left(\overline{\mathcal{B}}\mathcal{B}\right) + \tfrac{1}{2}m_1 \text{Tr} \left(\{\overline{\mathcal{B}}, \mathcal{B}\}\lambda_8\right) - \tfrac{1}{2}m_2 \text{Tr} \left([\overline{\mathcal{B}}, \mathcal{B}]\lambda_8\right) , \qquad (8.143)$$

The first term gives obviously the same value for the eight baryons; the other two terms give rise to symmetry breaking: they transform as the eighth component ($I = 0$, $Y = 0$) of an octet. The form (8.143) is justified by the following reasons:

- One assumes exact isospin invariance (the electromagnetic mass differences are neglected), so that M behaves as a scalar ($I = 0$) under $SU(2)_I$.

- All terms in M correspond to $Y = 0$ (no mixing between different hypercharge states must occur since Y is conserved);

- A $I = 0$, $Y = 0$ component can be found in the $SU(3)$ IR's of the type $D(p,p)$ such as 1, 8, 27, etc.. One considers the lowest non-trivial case, i.e. the octet; according to Eq. (8.127) there are two independent ways of building an octet from two others. Eqs. (8.130), (8.131) identify the two combinations.

Following the same procedure used to obtain Eq. (8.136) from (8.133), Eq. (8.143) can be written as

$$M_{(8)} = m_0 \left(\overline{B}B\right) + m_1 \left(\overline{B}D_8 B\right) + m_2 \left(\overline{B}F_8 B\right) . \qquad (8.144)$$

We know from Eq. (8.94) that $F_8 = \frac{\sqrt{3}}{2}Y$ and it can be shown, by using (8.140), that

$$D_8 = \tfrac{1}{\sqrt{3}}\left[I(I+1) - \tfrac{1}{4}Y^2\right] - \tfrac{1}{\sqrt{3}} . \qquad (8.145)$$

Then Eq. (8.144) can be replaced by $\left(\overline{B}m_8 B\right)$ with

$$m_8(I,Y) = a + bY + c\left[I(I+1) - \tfrac{1}{4}Y^2\right] , \qquad (8.146)$$

where a, b and c are independent parameters.

---

[16] M. Gell-Mann and Y. Ne'eman, *The eightfold way*, Benjamin, New York (1964); S. Okubo, Progr. of Theor. Phys. 27, 949 (1962).

The $\frac{1}{2}^+$ *octet* contains four isospin multiplets (N, Σ, Λ, Ξ); the corresponding masses can be related by eliminating the three unknown parameters. One gets the famous relation

$$\tfrac{1}{2}(m_N + m_\Xi) = \tfrac{1}{4}(3m_\Lambda + m_\Sigma) , \qquad (8.147)$$

which is well satisfied by the experimental values, the discrepancy being less than 1%.

A similar mass formula can be used for the meson octets. In this case the masses of the two isospin doublets are the same, since they are antiparticles of each other. This implies that for mesons one has to put $b = 0$ in (8.146). Moreover, it is customary to use squared mass relations, rather than linear ones (this could be related to the fact that the mass term in the Lagrangian is linear for baryons and quadratic for mesons).

For the 0^- *octet* one gets

$$m_K^2 = \tfrac{1}{4}m_\pi^2 + \tfrac{3}{4}m_\eta^2 , \qquad (8.148)$$

which is satisfied experimentally within few %, but with less precision than (8.147).

In the case of the $\frac{3}{2}^+$ *decuplet* $\mathcal{D}$, in agreement to the fact that 8 appears only once in the decomposition $10 \otimes \overline{10}$ (8.81) one has to write, instead of (8.143),

$$M_{(10)} = m_0' \left(\overline{\mathcal{D}}_{\alpha\beta\gamma}\mathcal{D}^{\alpha\beta\gamma}\right) + m_1' \left(\overline{\mathcal{D}}_{\alpha\beta\gamma}\mathcal{D}^{\alpha\beta\delta}(\lambda_8)^\gamma{}_\delta\right) . \qquad (8.149)$$

In terms of the decuplet components one obtains

$$m_{(10)}(I, Y) = m_0' + a'Y , \qquad (8.150)$$

which gives for the $\frac{3}{2}^+$ decuplet the equal spacing mass formula

$$m_\Omega - m_{\Xi^*} = m_{\Xi^*} - m_{\Sigma^*} = m_{\Sigma^*} - m_\Delta , \qquad (8.151)$$

very well satisfied experimentally[17].

Finally, we note that Eq. (8.146) can be shown to hold, in general, for any multiplet[18]; it reduces to the simpler form (8.150) for the IR's with triangular weight diagram such as 10, for which I and Y are related by

---

[17] It is interesting to point out that the existence and the properties of the Ω^- particle ($J^P = \frac{3}{2}^+$, mass $\sim$ 1685 MeV, decay through weak interaction to $\Xi^0\pi^-$, $\Xi^-\pi^0$ and $\Lambda^0 K^-$ and not through strong interactions as the other members of the decuplet) were exactly predicted: see M. Gell-Mann, Proc. of the Int. Conf. on High Energy Physics, CERN (1962), p. 805. A particle with such properties was discovered two years later (see V.E. Barnes et al., Phys. Rev. Lett. 12, 204 (1964).

[18] S. Okubo, Progr. of Theor. Phys. 27, 949 (1962).

$$I = 1 + \tfrac{1}{2}Y \, . \tag{8.152}$$

b) ϕ-ω *mixing*

The mass relation (8.146) is not satisfactory when applied to the case of the 1^- *vector mesons*. The experimental situation is the following: there are two isodoublets K^* and $\overline{K}^*$, an isotriplet, ρ, and two isosinglets, ω (783) and ϕ (1020). Neither ω nor ϕ satisfies the mass formula, which would predict an isoscalar ω_8 (930), its mass being given by

$$m_8^2 = \tfrac{1}{3}\left(2m_{K^*}^2 - m_\rho^2\right) \, . \tag{8.153}$$

One can imagine that ω and ϕ, which have the same quantum numbers, do not represent pure $SU(3)$ octet (ω_8) and singlet (ω_1) states, but they are, instead, the orthogonal mixtures

$$
\begin{aligned}
\omega &= \sin\theta \, \omega_8 + \cos\theta \, \omega_1 \, , \\
\phi &= \cos\theta \, \omega_8 - \sin\theta \, \omega_1 \, .
\end{aligned}
\tag{8.154}
$$

This kind of mixing is indeed allowed by the symmetry breaking introduced in connection with the Gell-Mann − Okubo mass formula. Due to the ω_1−ω_8 mixing, the (squared) mass matrix contains off-diagonal terms:

$$
m^2 = \begin{pmatrix} m_8^2 & m_{18}^2 \\ m_{18}^2 & m_1^2 \end{pmatrix} \, . \tag{8.155}
$$

The physical states ω and ϕ correspond to the diagonalization of m^2. One gets easily the following relations

$$
\begin{aligned}
m_\phi^2 + m_\omega^2 &= m_1^2 + m_8^2 \, , \\
\left(m_{18}^2\right)^2 &= m_1^2 m_8^2 - m_\phi^2 m_\omega^2 \, .
\end{aligned}
\tag{8.156}
$$

and

$$\tan 2\theta = \frac{2m_{18}^2}{m_8^2 - m_1^2} \, . \tag{8.157}$$

They are sufficient, together with Eq. (8.153), to determine the mixing angle θ, which turns out to be $\theta = 39 \pm 1°$. We note that the value $\sin\theta = \frac{1}{\sqrt{3}}$ corresponds to the so-called *ideal mixing*, for which the quark content of the two states in Eq. (8.154) become

$$
\begin{aligned}
\omega &= \tfrac{1}{\sqrt{2}}(u\bar{u} + d\bar{d}) \\
\phi &= s\bar{s} \, .
\end{aligned}
\tag{8.158}
$$

An independent determination of this parameter, consistent with this value, can be obtained in terms of the branching ratio $(\omega \to e^+ e^-)/(\phi \to e^+ e^-)$[19].

---

[19] R.F. Dashen and D.H. Sharp, Phys. Rev. 133 B, 1585 (1964).

If one extends the same mixing mechanism to the pseudoscalar meson octet, η should not be considered a pure octet state, but taken as a mixed state with the η' (958) meson. The corresponding mixing angle is $\theta = 10 \pm 1°$.

c) *Electromagnetic mass differences*

The $SU(3)$ breaking interaction introduced in connection with Eq. (8.143) is isospin conserving, so that members of a given isospin multiplet are still degenerate. However, all isospin multiplets exhibit mass splittings, which are assumed to be of electromagnetic origin.

The form of the electromagnetic interactions is known exactly, and therefore one knows also their structure with respect to $SU(3)$.

We point out that the electric charge Q, in the IR 3, is given by (see Eq. (8.123))

$$ Q = \tfrac{1}{\sqrt{3}}\lambda_8' = \tfrac{1}{2}\lambda_3 + \tfrac{1}{2\sqrt{3}}\lambda_8 \ . \tag{8.159} $$

From this we can infer that the photon, which is coupled with Q, transforms as the $U = 0$ component of an octet. From the point of view of isospin, it contains both an isoscalar and an isovector term transforming, respectively, as the $I = 0$ and the $I = 1$, $I_3 = 0$ components of an octet. From Eq. (8.119)) it follows also that the electromagnetic interactions conserve U-spin.

Then the electromagnetic mass splitting δm can depend only on U and Q and the general mass operator can be written as

$$ m = m(I, Y) + \delta m(U, Q) \ . \tag{8.160} $$

In the case of the $\tfrac{1}{2}^+$ octet (Fig. 8.9) one gets immediately

$$ \delta m_p = \delta m_{\Sigma^+} \ , \quad \delta m_n = \delta m_{\Xi^0} \ , \quad \delta m_{\Sigma^-} = \delta m_{\Xi^-} \ , \tag{8.161} $$

and (8.160) gives

$$ \begin{aligned} \delta m_n - \delta m_p &= m_n - m_p \\ \delta m_{\Xi^-} - \delta m_{\Xi^0} &= m_{\Xi^-} - m_{\Xi^0} \\ \delta m_{\Sigma^+} - \delta m_{\Sigma^-} &= m_{\Sigma^+} - m_{\Sigma^-} \end{aligned} \tag{8.162} $$

Combining (8.161) and (8.162), we obtain the Coleman-Glashow relation

$$ m_n - m_p + m_{\Xi^-} - m_{\Xi^0} = m_{\Sigma^-} - m_{\Sigma^+} \ , \tag{8.163} $$

which is in excellent agreement with the observed values. This agreement indicates that the interference between strong and electromagnetic interactions can be neglected, at least in connection with the above equations.

Similar conclusions can be obtained in the case of the $\tfrac{3}{2}^+$ decuplet. In the case of the meson octets, the analogue of Eq. (8.163) is merely a consequence of CPT and contains no information on U-spin invariance.

d) *Magnetic moments*

Interesting relations can be obtained also among the magnetic moments of the baryons of the $\tfrac{1}{2}^+$ octet. We assume, in general, that the magnetic moment transforms as the electric charge $Q \approx \lambda'_8$, i.e. as the $U = 0$ component of an octet. The situation is analogous to that of Eq. (8.143). For the baryon $\frac{1}{2}^+$ octet we can write

$$\mu = \mu_1 \text{Tr} \left(\{\overline{\mathcal{B}}, \mathcal{B}\} \lambda'_8 \right) + \mu_2 \text{Tr} \left([\overline{\mathcal{B}}, \mathcal{B}] \lambda'_8 \right) . \tag{8.164}$$

However, without making use of the above formula, one can get several relations in a simpler way. First of all, one obtains the following relation among the members of the Σ isotriplet:

$$\mu_{\Sigma^+} + \mu_{\Sigma^-} = 2\mu_{\Sigma^0} , \tag{8.165}$$

making use only of the $SU(2)_I$ transformation properties.

Moreover, recalling that Q commutes with the U-spin generators (see Eq. (8.119)), one can conclude that the baryons belonging to the same U-spin multiplet have the same magnetic moment. Looking at Fig. 8.9 and expressing Σ_u^0 and Λ_u^0 in terms of Σ^0 and Λ^0 (Eq. (8.124)), one obtains

$$\mu_p = \mu_{\Sigma^+} , \qquad \mu_{\Xi^-} = \mu_{\Sigma^-} , \qquad \mu_n = \mu_{\Xi^0} , \tag{8.166}$$

and

$$\mu_n = \tfrac{3}{2}\mu_{\Lambda^0} - \tfrac{1}{2}\mu_{\Sigma^0} , \qquad \mu_{\Sigma^0} - \mu_{\Lambda^0} = \tfrac{2}{\sqrt{3}}\mu_{\Sigma^0\Lambda^0} , \tag{8.167}$$

where $\mu_{\Sigma^0\Lambda^0}$ is the transition magnetic moment responsible for the electromagnetic decay $\Sigma^0 \to \Lambda^0\gamma$.

Finally, from Eq. (8.164), one can get two more relations

$$\mu_{\Lambda^0} = \tfrac{1}{2}\mu_n , \qquad \mu_p + \mu_n = -\mu_{\Sigma^-} . \tag{8.168}$$

All the magnetic moments of the $\frac{1}{2}^+$ baryon octet can then be expressed in terms of only μ_p and μ_n[20]. The experimental values are in reasonable agreement with these relations.

8.6 Beyond $SU(3)$

In this section we shall consider different kinds of extensions of the flavor $SU(3)$ symmetry. First of all, we shall discuss the problem of the quark statistics, i.e. the fact that the baryon states appear always symmetric with respect to the constituent quark exchange. In order to save the Fermi-Dirac statistics, hidden degrees of freedom were introduced in terms of a second independent $SU(3)$ group, so that the symmetry of the hadronic states was extended to the group $SU(3) \otimes SU(3)$.

A further insight in the structure of the hadronic states was obtained by combining the flavor $SU(3)$ group with the ordinary spin $SU(2)$ group, with

---

[20] S. Coleman, S.L. Glashow, Phys. Rev. Lett. 6, 423 (1961).

the introduction of the $SU(6)$ group which embeds the subgroup $SU(3) \otimes SU(2)$.

Finally, we shall mention the extension of the flavor $SU(3)$ group to the higher group $SU(4)$, that was introduced after the discovery of a new class of hadrons implying the existence a forth quark.

8.6.1 From flavor $SU(3)$ to color $SU(3)$

The introduction of quarks solved the puzzle of the lack of evidence of exotic hadrons that cannot be described as $(q\bar{q})$ and (qqq) states, and then with values of isospin and hypercharge higher than those included in octets for mesons and in octets and decuplets for baryons[21].

Mowever, a second puzzle appeared from a closer analysis of the baryon spectrum: how can there be three identical spin $\frac{1}{2}$ quarks in the same quantum state? In this situation the Pauli principle would be violated.

This question arose from the observation of the state Δ^{++} (with $S = \frac{3}{2}$, $S_z = \frac{3}{2}$ and positive parity) and similarly for Δ^- and Ω^-, which are components of the decuplet represented in Fig. 8.5, and which should be composed of three identical quarks of u, d and s type, respectively. Each of the three quarks has spin component $S_z = \frac{1}{2}$ and orbital momentum $\ell = 0$ (there are indeed other baryon multiplets with higher spin that require orbital momenta different from zero, but it is reasonable to assume $\ell = 0$ for the decuplet $\frac{3}{2}^+$, having the lowest mass values).

The solution of this puzzle is based on the hypothesis that each quark possesses extra quantum numbers, which correspond to those of another $SU(3)$. To make a distinction between the two symmetry groups, one uses whimsical names: *flavor* and *color*. With respect to the first group, denoted by $SU(3)_f$, quarks belong to a triplet with three different flavors: u, d, s. With respect to the second group, denoted by $SU(3)_c$, each quark belongs to another triplet, with three different colors: q_1, q_2, q_3 (they are named *red, white* and *blue*, and we use the notation $1 = red$, $2 = white$, $3 = blue$). Then the states quoted above become compatible with the Pauli principle:

$$\Delta^{++} \equiv (u_1, u_2, u_3), \quad \Delta^- \equiv (d_1, d_2, d_3), \quad \Omega^- \equiv (s_1, s_2, s_3), \quad (8.169)$$

---

[21] However, observation of new heavy mesons, with masses of the order of 3875 MeV, were observed a few years ago (S.-K. Choi *et al.* (Belle Collaboration), Phys. Rev. Lett. 91, 26200 (2003); B.Aubert *et al.* (BaBar Collaboration), Phys. Rev. D73: 011101 (2006)); they were interpreted as diquark-antidiquark states: $X(cu\bar{c}\bar{u})$ and $X(cd\bar{c}\bar{d})$. Recently also new mesons with non-zero electric charge and masses about 4430 MeV have been observed (S.-K. Choi *et al.*, Belle Collaboration, Phys. Rev. Lett. 100: 142001 (2008)) and interpreted as $Z(cu\bar{c}\bar{d})$. If such interpretations were well established, these results would demonstrate the existence of *exotic* states. For theoretical considerations and other possible interpretations see e.g.: L. Maiani, A.D. Polosa and V. Riquer, Phys. Rev. Lett. 99, 182003 (2007) and references therein.

provided they are completely antisymmetric in the (1,2,3) indeces.

In conclusion, one assumes that there are nine different quarks, which can be collected in a 3×3 matrix (rows refer to color and columns to flavor)

$$
q_i^\alpha = \begin{pmatrix} u_1 & u_2 & u_3 \\ d_1 & d_2 & d_3 \\ s_1 & s_2 & s_3 \end{pmatrix} ,
\tag{8.170}
$$

which corresponds to the $(3,3)$ IR of the group $SU(3)_f \otimes SU(3)_c$.

We remark that the symmetric quark model assumes the existence of only one set of the 3-quark states allowed with the introduction of color. In fact, in this model, a baryon can be defined as totally antisymmetric in the color indices i, j, k (singlet with respect to $SU(3)_c$), and then globally symmetric in all the other variables (flavor indices, spin, angular momentum) α, β, γ:

$$
B^{\alpha\beta\gamma} = \sum_{i,j,k} \epsilon^{ijk} q_i^\alpha q_j^\beta q_k^\gamma .
\tag{8.171}
$$

This situation corresponds to the so-called *quark statistics*.

For mesons, there would be 81 possible $q\bar{q}$ states; however, also in this case, one assumes that the physical states are color singlets, so that only 9 states are left

$$
M_\beta^\alpha = \sum_i \bar{q}_\beta^i q_i^\alpha .
\tag{8.172}
$$

In conclusion, we point out that the fundamental property of hadrons (including possible exotic states) is that of being color singlets, i.e. the $SU(3)_c$ quantum numbers remain always hidden.

Let us finally stress that the introduction of the group $SU(3)_c$ has been the starting point for the construction of the field theory of strong interactions, as it will be discussed in the next chapter.

8.6.2 The combination of internal symmetries with ordinary spin

The first idea of combining the internal symmetry of the isospin group $SU(2)_I$ with the ordinary spin group $SU(2)_S$ is due to Wigner[22], who introduced the group $SU(4)$ as an approximate symmetry of nuclei. The four nucleon states $p(+\frac{1}{2})$, $p(-\frac{1}{2})$, $n(+\frac{1}{2})$, $n(-\frac{1}{2})$ (we have indicated, in parentheses, the value of the spin component $S_z = \pm\frac{1}{2}$) are assigned to the fundamental IR of $SU(4)$, and the higher IR's provide a classification of the nuclear states[23].

In analogy with this application to nuclear physics, the group $SU(6)$, which contains the subgroup $SU(3)_f \otimes SU(2)_S$ was applied to particle

---

[22] E. Wigner, Phys. Rev. 51, 25 (1937).

[23] F.J. Dyson, *Symmetry Groups in Nuclear and Particle Physics*, Benjamin, New York (1966).

physics[24] in order to take into account also the ordinary spin in the classification of the lowest hadron states which do not require orbital excitations ($\ell = 0$).

In Table 8.10 we exhibit the lowest IR's and their content with respect to the subgroup $SU(3) \otimes SU(2)$.

Table 8.10. Decomposition of some IR's of $SU(6)$ in terms of $SU(3) \otimes SU(2)$.

$SU(6)$	$SU(3) \otimes SU(2)$
6	$(3, 2)$
$\bar{6}$	$(\bar{3}, 2)$
20	$(8, 2) \oplus (1, 4)$
35	$(8, 1) \oplus (8, 3) \oplus (1, 3)$
56	$(10, 4) \oplus (8, 2)$
70	$(10, 2) \oplus (8, 4) \oplus (8, 2) \oplus (1, 2)$

The classification of hadrons in multiplets of $SU(6)$ is very simple if one refers to the quark model. The six quark states $u(\pm\frac{1}{2})$, $d(\pm\frac{1}{2})$, $s(\pm\frac{1}{2})$ are assigned to the self-representation 6, and the basic vector is denoted by ζ^A with $A = (\alpha, a)$, where the two indeces α and a are relative to $SU(3)_f$ and $SU(2)_S$, respectively (see Appendix C).

In terms of the $SU(6)$ representations, one writes the analogues of the product decompositions $3 \otimes \bar{3}$ and $3 \otimes 3 \otimes 3$ of $SU(3)$:

$$6 \quad \otimes \quad \bar{6} \quad = \quad 1 \quad \oplus \quad 35 \tag{8.173}$$

and

$$6 \otimes 6 \otimes 6 = 56 \oplus 70 \oplus 70 \oplus 20 \tag{8.174}$$

---

[24] F. Gürsey, A. Pais and L. Radicati, Phys. Rev. Lett. 13, 173 (1964).

Under each IR we have drawn the corresponding Young tableau, in order to exhibit the symmetry properties of the $q\bar{q}$ and $3q$-states.

According to the above decompositions, one would expect that mesons are classified in the IR's 1 and 35 of $SU(6)$, and baryons in 56, 70 and 20. But, by looking at the decompositions of these representations in terms of those of the subgroup $SU(3)_f \otimes SU(2)_S$, one arrives at the following conclusions. The (lowest) meson states considered in connection with $SU(3)$ fit nicely in 35: in fact, this IR contains a spin 0 octet and a spin 1 nonet $(8 + 1)$. With respect to the baryon states, the spin $\frac{1}{2}$ octet and the $\frac{3}{2}$ decuplet fit nicely in 56. The interpretation given above for the lowest states in terms of S-wave $q\bar{q}$ and $3q$ systems accounts also for the right parity, which is odd for mesons, since the relative $q\bar{q}$ parity is odd $(J^P = 0^-, 1^-)$, and even for baryons $(J^P = \frac{1}{2}^+, \frac{3}{2}^+)$.

We would like to stress that, according to Eq. (8.171), only the 56 multiplet is allowed by the quark statistics (in the absence of orbital momentum), since it is completely symmetric in the flavor and spin degrees of freedom, and then it must be completely antisymmetric in color ($SU(3)_c$ singlet). This is a further confirmation of the quark statistics; moreover also the baryon octet, which has mixed symmetry with respect to $SU(3)_f$, becomes completely symmetric when combined with the mixed symmetric spin $\frac{1}{2}$ state. Other baryon multiplets, for which the spin-parity (J^P) values require the introduction of orbital momenta, can be assigned to the IR's 70 and 20 provided their states are globally symmetric in the flavor, spin and orbital momentum degrees of freedom. For instance, a set of $SU(3)_f$ multiplets with $J^P = \frac{1}{2}^-, \frac{3}{2}^-$ and $\frac{5}{2}^-$ can be fitted in the 70-supermultiplet with orbital momentum $\ell = 1$ and configuration $(1s)^2(2p)$.

Even if the $SU(6)$ model can provide only an approximate description of the hadronic states, since it is intrinsically non-relativistic, it gave rise to interesting relations among hadron masses and magnetic moments. The $SU(3)_f$ mass Eq. (8.146) can be extended with an additional spin term[25]:

$$m(I, Y, S) = a + bY + c\left[I(I+1) - \tfrac{1}{4}Y^2\right] + d\,S(S+1)\,. \qquad (8.175)$$

In particular, it gives rise to the relations

$$m_\Omega - m_{\Xi^*} = m_{\Xi^*} - m_{\Sigma^*} = m_{\Sigma^*} - m_\Delta = m_\Xi - m_\Sigma \qquad (8.176)$$

for baryon masses, and

$$m_{K^*}^2 - m_K^2 = m_\rho^2 - m_\pi^2 \qquad (8.177)$$

for meson masses. Both types of relations are well satisfied.

We refer to the original papers[26] for other interesting results obtained for the nucleon magnetic moments and for the ideal mixing between the vector mesons ω and ϕ.

---

[25] F. Gürsey, A. Pais and L. Radicati, Phys. Rev. Lett. 13, 173 (1964).

[26] See F. J. Dyson, *Symmetry Groups in Nuclear and Particle Physics*, Benjamin, New York (1966).

8.6.3 Extensions of flavor $SU(3)$

In the previous Sections, we made reference to the situation before 1974, until when there was evidence only for three flavor quarks (u, d, s). Two more types of quarks were discovered in the next few years, denoted by c for *charm* and b for *bottom*, and later on a sixth type of quark: t for *top*. The quark c was first introduced for theoretical reasons[27], to provide an explanation for a peculiar feature of the weak neutral current, which appeared to be flavor conserving, while the charged weak currents violate flavor.

A new type of hadron, requiring the introduction of a new quantum number called *charm*, was discovered in 1974. It was discovered independently in two different laboratories, Stanford[28] and Brookhaven[29], and it was observed immediately afterward also at Frascati[30]. It received two different names and since then it is denoted by J/ψ. All the experimental information is in agreement with the interpretation of this new hadron, which is a boson with $J^P = 1^-$, as a bound state of a quark-antiquark $c\bar{c}$ pair.

The existence of the b quark was inferred from the discovery in 1977, at the Fermi National Laboratory, of a new type of heavy meson[31], denoted by Υ, which had to carry a new quantum number, called *beauty*, and interpreted as a bound state of a quark-antiquark $b\bar{b}$ state.

The sixth quark, the top t, was discovered in 1995 at the Fermi National Laboratory[32] and it was found to be, as expected, much heavier that the other quarks. In Table 8.11 we give the values of the masses of the six quarks[33].

Due to the very large mass differences, it would be completely useless to extend the flavor $SU(3)_f$ group by including the complete set of the six quarks. However, the extension limited to the four lightest quarks (u, d, s, c) received some attention. The triplet given in Eq. (8.65) is replaced by

$$\xi = \begin{pmatrix} u \\ d \\ s \\ c \end{pmatrix}, \tag{8.178}$$

which can be assigned to the self-representation 4 of the flavor group $SU(4)$ This group was used in the literature for a classification of the new charmed hadrons, that were discovered after the appearance of the J/ψ boson. A new quantum number C was introduced to identify the new class of hadrons; the value $C = 1$ was assigned to the charm quark c.

[27] S.L. Glashow, J. Iliopoulos and L. Maiani, Phys. Rev. D2, 1285 (1970).

[28] J.J. Aubert *et al.*, Phys. Rev. Lett. 33, 1404 (1974).

[29] J.E. Augustin *et al.*, Phys. Rev. Lett. 33, 1406 (1974).

[30] C. Bacci *et al.*, Phys. Rev. Lett. 33, 1408 (1974).

[31] S.W. Herb *et al.*, Phys. Rev. Lett. 39, 252 (1977).

[32] F. Abe *et al.*, Phys. Rev. Lett. 74: 2626 (1995);
    S. Abachi *et al.*, Phys. Rev. Lett. 74: 2632 (1995).

[33] The Review of Particle Physics, C. Amsler et al., Phys. Lett. B 667, 1 (2008).

Table 8.11. The six quarks

Name	Symbol	Q	Mass (MeV)
up	u	$\frac{2}{3}$	$1.5 \div 3$
down	d	$-\frac{1}{3}$	$3 \div 7$
charmed	c	$\frac{2}{3}$	$\sim 1.25 \times 10^3$
strange	s	$-\frac{1}{3}$	~ 95
top	t	$\frac{2}{3}$	$\sim 1.7 \times 10^5$
bottom	b	$-\frac{1}{3}$	$\sim 4.2 \times 10^3$

In terms of the subgroup $SU(3)_f \otimes U(1)_C$ the quartet in Eq. (8.178) reduces to

$$4 = (3,0) \oplus (1,1), \tag{8.179}$$

where $(3,0)$ represents the quark triplet with $C = 0$ and $(1,1)$ the c-quark with $C = 1$. The electric charge of the charmed quark, which is equal to $\frac{2}{3}$, requires the modification of the Gell-Mann Nishijima relation in the form

$$Q = I_3 + \tfrac{1}{2}(B + S + C). \tag{8.180}$$

For the hypercharge, we make the choice $Y = 0$, which corresponds to

$$Y = B + S - \tfrac{1}{3}C, \tag{8.181}$$

so that Eq. (8.180) can be written in the form

$$Q = I_3 + \tfrac{1}{2}Y + \tfrac{2}{3}C. \tag{8.182}$$

The above choice for Y gives rise to charmed hadrons with fractional values of hypercharge, but it is convenient since it corresponds to regular weight diagrams in the (I_3, Y, C) space. For instance, the quark quartet can be represented by a regular tetrahedron as shown in Fig 8.12.

Extending to $SU(4)$ the hadron quark structure $q\bar{q}$ and qqq, mesons and baryons will be classified in terms of $4 \otimes \bar{4}$ and $4 \otimes 4 \otimes 4$, respectively[34]. According to the usual decompositions, one gets:

---

[34] For details on the hadron classification in $SU(4)$ see: M.K. Gaillard, B.W. Lee and J.L. Rosner, Rev. of Mod. Phys. 47, 277 (1975).

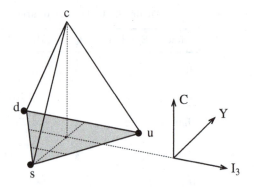

Fig. 8.12. Quark multiplet in $SU(4)$.

$$4 \quad \otimes \quad \bar{4} \quad = \quad 1 \quad \oplus \quad 15 \qquad\qquad (8.183)$$

and

$$4 \otimes 4 \otimes 4 = 20 \oplus 20' \oplus 20' \oplus \bar{4} \quad (8.184)$$

In terms of the subgroup $SU(3)_f \otimes SU(2)_c$, we have the following decompositions

$$15 = (8,0) \oplus (1,0) \oplus (\bar{3},1) \oplus (3,-1) \,,$$
$$20' = (8,0) \oplus (6,1) \oplus (\bar{3},1) \oplus (3,2) \,, \qquad (8.185)$$
$$20 = (10,0) \oplus (6,1) \oplus (3,2) \oplus (1,3) \,.$$

We see that the $SU(3)$ meson octets and nonets fit into 15, the baryon octets into 20' and the baryon decuplets into 20; these $SU(4)$ multiplets contain various charmed $SU(3)$ multiplets. In particular, the representations 20' of the $\frac{1}{2}^+$ baryon octet and 20 of the $\frac{3}{2}^+$ baryon decuplet are reported in Fig. 8.13 and Fig. 8.14, respectively.

Concerning mesons, we limit ourselves to illustrate the classification of mesons in the 15-multiplet. The lower 0^- meson states are indicated in Fig. 8.15, following the usual[35] nomenclature adopted for the charmed states.

---

[35] The Review of Particle Physics; C. Amsler et al., Phys. Lett. B 667, 1 (2008).

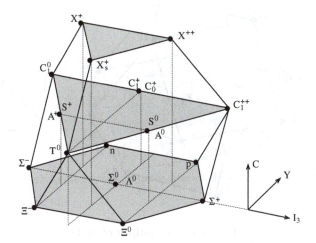

Fig. 8.13. Baryon $\frac{1}{2}^+$ 20'-multiplet of $SU(4)$.

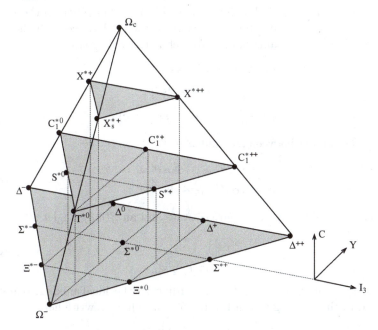

Fig. 8.14. Baryon $\frac{3}{2}^+$ 20-multiplet of $SU(4)$.

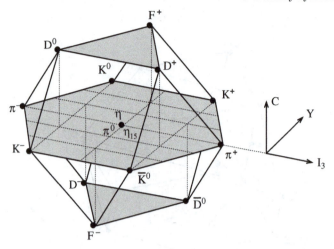

Fig. 8.15. Meson 15-multiplet of $SU(4)$.

The vector 1^- mesons, as the 0^- ones, can be fitted into the 15-multiplet. We denote by F^*, D^*, $\overline{F}^*$, $\overline{D}^*$ the vector counterparts of the pseudoscalar mesons F, D, $\overline{F}$, $\overline{D}$ indicated in figure. The $Q = Y = 0$ members of the $1 \oplus 15$ multiplet are of particular interest, because of their relevance with the J/ψ and the particles of the same family. We define the vector states

$$\omega_1' = \tfrac{1}{2}(u\bar{u} + d\bar{d} + s\bar{s} + c\bar{c}),$$
$$\omega_8 = \tfrac{1}{\sqrt{6}}(u\bar{u} + d\bar{d} - 2s\bar{s}),$$
$$\omega_{15} = \tfrac{1}{2\sqrt{3}}(u\bar{u} + d\bar{d} + s\bar{s} - 3c\bar{c}).$$
$$(8.186)$$

The following mixing between ω_{15} and ω_1'

$$\omega_1 = \sin\theta' \, \omega_{15} + \cos\theta' \, \omega_1',$$
$$\psi = \cos\theta' \, \omega_{15} - \sin\theta' \, \omega_1'$$
$$(8.187)$$

gives rise, with $\sin\theta' = \tfrac{1}{2}$, to a pure $c\bar{c}$ state and an $SU(3)$ singlet ω_1:

$$\psi = c\bar{c},$$
$$\omega_1 = \tfrac{1}{2}\omega_{15} + \tfrac{\sqrt{3}}{2}\omega_1'.$$
$$(8.188)$$

Moreover, by combining ω_1 and ω_8 according to the ideal mixing, one can finally obtain the states given in Eq. (8.158), which we rewrite here:

$$\omega = \tfrac{1}{\sqrt{2}}(u\bar{u} + d\bar{d}),$$
$$\phi = s\bar{s}.$$
$$(8.189)$$

One can say that the above situation corresponds to an ideal mixing also in $SU(4)$; the IR's 1 and 15 have to be considered together and one has a 16 multiplet of vector mesons. The new bosons of the J/ψ family are interpreted as bound $c\bar{c}$ states; since they have $C = 0$, one can refer to them as *hidden charm* states.

In conclusion, a simple look at Table 8.11 indicates that the $SU(2)$ symmetry, which involves the quarks u, d, is a good symmetry, only slightly broken by the small mass difference of the two quarks. The mass differences are greater in the case of the three quarks u, d, s involved in $SU(3)$, but, since they are smaller than the typical hadron scale, the symmetry is approximate but still useful in relating different properties of hadrons, as discussed in the previous sections.

Going beyond $SU(3)$ can be useful mainly for hadron classification, since the quark mass differences are very large and the hypothetical symmetries $SU(n)$ with $n > 3$ are badly broken. On one hand, investigations on hadron spectra are performed for classifying mesons and baryons with beauty (i.e. with a constituent bottom quark)[36]. On the other hand, several theoretical models, based on different groups, have been formulated with the attempt of reproducing the quark mass spectrum, which is part of the so-called *flavor problem*. However, up to now, the success is rather limited.

A more successful approach was obtained in the unification of the fundamental interactions, going from global to local gauge groups, in the frame of quantum field theory. The main result of this approach will be illustrated in the next Chapter.

Problems

8.1. Assuming charge independence, determine the ratio of the cross-sections relative to the two processes

$$p + d \to \pi^0 + {}^3He , \qquad\qquad p + d \to \pi^+ + {}^3H ,$$

where the deuteron d is taken as iso-singlet and $({}^3He, {}^3H)$ as iso-doublet.

8.2. Express the amplitudes for the processes

$$\pi^+\pi^+ \to \pi^+\pi^+ , \quad \pi^+\pi^- \to \pi^+\pi^- , \quad \pi^+\pi^- \to \pi^0\pi^0 , \quad \pi^0\pi^0 \to \pi^0\pi^0 ,$$

in terms of the indepemdent isospin amplitudes.

8.3. Find the expression of the $SU(2)$ generators in the three-dimensional IR starting from their expression in the two-dimensional IR ($I_i = \frac{1}{2}\sigma_i$) and making use of Eq. (8.37).

---

[36] The heavy bottom quark b has been included in an $SU(4)$ group based on the subset of quarks (u, d, s, b) for the classification of the $\frac{1}{2}^+$ baryons with beauty in a $20'$ multiplet. Recently, the Ω_b^- (ssb) state has been observed: V.M.Arbazov *et al.*, The DO Collaboration, Phys. Rev. Lett. 101, 232002 (2008).

8.4. From the usual form of the matrix elements of the $SU(2)$ generators in the three-dimensional representation

$$(I_i)_{jk} = -i\epsilon_{ijk}$$

obtain the basis in which I_3 is diagonal and the basic elements are identified with the pion states π^+, π^0, π^-. By constructing the raising and lowering operators $I_\pm$, show that

$$I_\pm \pi^0 = \sqrt{2}\pi^\pm.$$

8.5. Prove the decomposition into IR's of the direct products $8 \otimes 8$, $8 \otimes 10$, $10 \otimes \overline{10}$ given in Eq. (8.81), and give the $SU(2)_I \otimes U(1)$ content for each IR.

8.6. Introduce the shift operators

$$I_\pm = I_1 \pm iI_2 , \qquad U_\pm = U_1 \pm iU_2 , \qquad V_\pm = V_1 \pm iV_2 .$$

Find their commutation relations and their action on the relevant states. Use them to obtain Eq. (8.124), i.e. the combinations of Σ^0 and Λ^0 which correspond to Σ_u^0 and Λ_u^0.

8.7. From U-spin invariance derive the relation between the amplitudes of the decays

$$\pi^0 \to 2\gamma , \qquad\qquad \eta^0 \to 2\gamma .$$

8.8. Derive explicitly the $\omega - \phi$ mixing angle θ (Eq. (8.157)) and show that it can be related to the branching ratio of the decays

$$\omega \to e^+ e^- , \qquad\qquad \phi \to e^+ e^- .$$

8.9. Find explicitly, on the basis of U-spin invariance, the relations reported in Eq. (8.167), i.e.

$$\mu_n = \tfrac{3}{2}\mu_{\Lambda^0} - \tfrac{1}{2}\mu_{\Sigma^0} , \qquad\qquad \mu_{\Sigma^0} - \mu_{\Lambda^0} = \tfrac{2}{\sqrt{3}}\mu_{\Lambda^0 \Sigma^0}.$$

8.10. Find the eigenvalue of the Casimir operator $F^2 = \sum_i F_i F_i$ for the adjoint IR of $SU(3)$. Use this eigenvalue to verify Eq. (8.145).

8.11. From the definition of the F_k and D_k matrices (Eqs. (8.137), (8.138)), verify Eqs. (8.92), (8.139), (8.140).

8.12. A particular version of the quark model assumes a very strong binding in the symmetric S-wave qq (diquark) state. What are the consequences of this hypothesis on the spectrum of baryons in the frame of the $SU(6)$ model? Discuss also the meson states obtained by combining a diquark with an antidiquark.

8.13. Discuss the classification of the low-lying meson and baryon states in term of $SU(8)$, which includes $SU(4)$ and the spin group $SU(2)_S$.

8.14. Consider the group $SU(4)$ for the classification of the $\tfrac{1}{2}^+$ hadrons containing the quarks u, d, s, b and compare with what is described in Subsection 8.6.3 for charmed hadrons. How have Eqs. (8.180) and (8.181) to be modified?

9

Gauge symmetries

In this Chapter we examine the *gauge groups*, in which the parameters of the group transformations are continuous functions of the space-time coordinates; in other words, the global transformations considered up to now, in which the group parameters are space-time constants, are replaced by *local* ones. The gauge groups are extensively used in quantum field theory and, specifically, in the field theory of elementary particles.

9.1 Introduction

In the previous chapter we have considered in details some examples of unitary symmetries based on groups of global or rigid transformations: in this case the group parameters are constant in space-time and the transformations produce a rigid and simultaneous change in the whole space-time domain in which the physical system is defined. We showed that the use of this kind of transformation groups was extremely fruitful in the investigation of the symmetry properties of the fundamental interactions and in the classification of the elementary and compound particles. However, the invariance under such transformations is not sufficient to build a dynamical theory of elementary particles, in which one can evaluate cross sections for different processes and lifetimes for different decay modes. For this purpose, the introduction of a new kind of transformations, the so-called *gauge transformations*, was of fundamental importance.

The promotion of global to local transformations has deep implications, as we shall see in the following. Our considerations will be formulated in the frame of field theory, mainly at the classical level; specific problems require the use of quantum field theory, but we shall limit ourselves to a simple treatment, quoting convenient references for more details. The invariance under gauge transformations requires the introduction of gauge vector fields, which are interpreted as the quanta mediating the interactions among the fermions that are the fundamental constituents of matter. We know that there are four kinds

G. Costa and G. Fogli, *Symmetries and Group Theory in Particle Physics*,
Lecture Notes in Physics 823, DOI: 10.1007/978-3-642-15482-9_9,
© Springer-Verlag Berlin Heidelberg 2012

of fundamental interactions: electromagnetic, weak, strong and gravitational, but only the first three are relevant for particle physics, below the *Planck energy scale*[1].

As a first example of gauge field theory we shall consider the case of Quantum Electrodynamics (QED), which is based on the Abelian group $U(1)_Q$; it contains a single gauge vector field, which represents the photon. QED is a very successful theory, which has been tested with very high precision, since the theoretical calculations, based on higher order radiative corrections, have been matched by increasingly accurate experiments.

It is not surprising that a description of the other interactions of elementary particles requires to replace the group $U(1)$ by larger gauge groups which allow more vector particles. It is natural to go from the group $U(1)$ to non-Abelian groups, and this generalization produces profound changes. Among the non-Abelian gauge field theories, first introduced by Yang and Mills[2], we shall examine the case of Quantum Chromodynamics (QCD), which is based on the *gauged* version of the color $SU(3)_c$ group, considered in the previous chapter. QCD describes the strong interactions of *quarks*, mediated by eight vector bosons called *gluons*, which correspond to the eight generators of $SU(3)$. It is believed that these gauge symmetry is exact and remains unbroken at least up to extremely high energies.

Another kind of symmetry was discovered from the study of condensed matter and it was very important in the understanding of the nature of weak interactions. It is based on the mechanism of the so-called *spontaneous symmetry breaking* (SSB), an expression which refers to the following situation: the Lagrangian of the system is invariant under a specific gauge symmetry, but the solutions of the equations of motion (in particular, the lowest energy state, i.e. the vacuum) possess a lower symmetry. In this process, some of the fields, including the gauge vector bosons, which are massless in the Lagrangian, will acquire a mass.

The mechanism was applied to the process of unification of the fundamental interactions. In the so-called Standard Model, weak and electromagnetic interactions have been unified at a certain energy scale, called *Fermi scale*[3], of

---

[1] The Planck scale is related to the Newton gravitational constant G_N. It indicates the order of magnitude of the energy at which gravitational interactions become of order 1:

$$\frac{G_N M^2 / r}{Mc^2} \sim 1 . \tag{9.1}$$

This implies, assuming for r the natural unit of lenght, $r = \hbar/Mc$,

$$M_{\text{Planck}} = \left(\frac{G_N}{\hbar c}\right)^{-\frac{1}{2}} = 1.22 \times 10^{19} \text{ GeV} . \tag{9.2}$$

[2] C.N. Yang, R. Mills, Phys. Rev. 96, 191 (1954).

[3] The Fermi scale is related to the Fermi coupling constant of weak interactions: $(G_F)^{-1/2} = 3.4 \times 10^3$ GeV.

the order of 10^3 GeV, in a gauge field theory based on the group $SU(2) \otimes U(1)$. It contains four vector fields, three of which acquire mass through the SSB mechanism and they correspond to the vector bosons mediating the weak interactions; the fourth remains massless and describes the photon.

There are theoretical speculations and some experimental hints that the electroweak and the strong interactions may be unified at a much higher energy scale, and several Grand Unified Theories, based on larger gauge groups, have been proposed. After a discussion of the Standard Model, the simplest example of grand unification will be considered at the end of the Chapter.

9.2 Invariance under group transformations and conservation laws

Before going to the specific gauge theories which describe the different particle interactions, we want to re-examine in more detail the invariance properties of physical systems under group transformations and their connection with conservation laws, that we have already considered in Chapter 8.

The correspondence between conservation laws and symmetry properties represents one of the major outcomes of the applications of group theory to physics. The invariance of the laws of nature under space translations of the coordinate frame of reference leads to the conservation of momentum, while their invariance with respect to time translation leads to the conservation of energy. Invariance under rotation of the coordinate system about an arbitrary axis implies the conservation of angular momentum. Equivalence of left-handed and right-handed coordinate systems leads to conservation of parity. Equivalence of all frames of reference in uniform relative motion leads to the Lorentz transformations and to the laws of special relativity.

All these symmetries and invariance properties are of geometrical kind. In particle physics, the so-called *internal symmetries* cannot be described in geometrical terms, but they are extremely important in relating observed particle states to representations of certain Lie groups, and in connecting some dynamical aspects of particle interactions. The analysis of this kind of symmetries and their relevance for the study of elementary particles has been the main subject of Chapter 8.

The relationship between symmetries and conservation laws is expressed, in mathematical terms, by the well-known *Noether's theorem*. We reproduce the main steps of the proof of the theorem in classical field theory, in view of the applications considered in the next sections.

Let us consider a physical system described by a Lagrangian L; in a local field theory, L is expressed as a spatial integral of a Lagrangian density $\mathcal{L}(\phi, \partial_\mu \phi)$, which is a function of a single field[4] $\phi(x)$ and its derivatives $\partial_\mu \phi(x)$.

---

[4] The extension to the case of more fields is straightforward; see J.D. Bjorken, S.D. Drell, *Relativistic Quantum Fields*, McGraw-Hill Book Company (1965).

The fundamental quantity is the action[5], defined by

$$S = \int L dt = \int \mathcal{L}(\phi, \partial_\mu \phi) d^4 x \ . \tag{9.3}$$

According to the *principle of least action*, the evolution of the system between two times t_1 and t_2 corresponds to an extremum (normally a minimum), i.e.

$$\delta S = 0 \ . \tag{9.4}$$

This condition gives rise to the *Euler-Lagrange equation*, which is the equation of motion of the system:

$$\partial_\mu \left(\frac{\partial \mathcal{L}}{\partial(\partial_\mu \phi)} \right) - \frac{\partial \mathcal{L}}{\partial \phi} = 0 \ . \tag{9.5}$$

Noether's theorem is based on the analysis of the continuous transformation of the field $\phi(x)$; an infinitesimal variation can be written as

$$\phi(x) \rightarrow \phi'(x) = \phi(x) + \delta\phi(x) \ , \tag{9.6}$$

where $\delta\phi(x)$ is an infinitesimal deformation of the field. This transformation is a *symmetry* of the system if it leaves the action (9.3), and therefore the Euler-Lagrange equation, invariant. However, the principle of least action, Eq. (9.4), does not require $\delta\mathcal{L} = 0$ since a sufficient condition is that $\delta\mathcal{L}$ is equal to a four-divergence:

$$\delta\mathcal{L}(x) = \partial_\mu \mathcal{I}^\mu(x) \ . \tag{9.7}$$

In fact, the four-divergence gives rise, in the four-dimensional integral of Eq. (9.3), to a surface term which does not affect the derivation of the Euler-Lagrange equation[6]. Specifically, one gets

$$\delta\mathcal{L} = \frac{\partial \mathcal{L}}{\partial \phi} \delta\phi + \frac{\partial \mathcal{L}}{\partial(\partial_\mu \phi)} \delta(\partial_\mu \phi) = \left(\frac{\partial \mathcal{L}}{\delta\phi} - \partial_\mu \frac{\partial \mathcal{L}}{\partial(\partial_\mu \phi)} \right) \delta\phi + \partial_\mu \left(\frac{\partial \mathcal{L}}{\partial(\partial_\mu \phi)} \delta\phi \right) \ . \tag{9.8}$$

The first term in the above equation corresponds to the Euler-Lagrange equation and therefore is equal to zero. Then, taking into account Eq. (9.7), we can identify the second term with $\partial_\mu \mathcal{I}^\mu$ obtaining the *current*

$$j^\mu(x) = \frac{\partial \mathcal{L}}{\partial(\partial_\mu \phi)} \delta\phi - \mathcal{I}^\mu \ , \tag{9.9}$$

which satisfies the condition

---

[5] For a detailed discussion, we refer to: M.E. Peskin, D.V. Schröder, *An Introduction to Quantum Field Theory*, Addison-Wesley Publishing Company (1995).

[6] The surface contribution vanishes by appropriate boundary conditions of the field $\phi(x)$ and its derivatives. See e.g.: J.D. Bjorken and S.D. Drell, *Relativistic Quantum Fields*, McGraw-Hill Book Company (1965).

$$\partial_\mu j^\mu(x) = 0 \,. \tag{9.10}$$

The vanishing of the four divergence of j^μ leads to a *conservation law* in integral form, since the quantity

$$Q = \int j^0(x) d^3x \,, \tag{9.11}$$

where the integral is carried over all the space, is *constant in time*. In fact, by integration of Eq. (9.10) over all the space, one gets

$$\frac{dQ}{dt} = \int \partial_0 j^0(x) d^3x = -\int \nabla \cdot \mathbf{j}(x) d^3x = 0 \,. \tag{9.12}$$

It is useful to consider a couple of specific examples.

A - *Space-time translation*
In this case, the infinitesimal transformation is

$$x^\mu \rightarrow x^\mu + \epsilon^\mu \,, \tag{9.13}$$

where ϵ^μ is an infinitesimal constant four-vector. The corresponding variation of the field ϕ is given by

$$\delta\phi(x) = \phi(x + \epsilon) - \phi(x) = \epsilon^\mu \partial_\mu \phi(x) \,. \tag{9.14}$$

The Lagrangian density $\mathcal{L}$ is a scalar, so that one gets in a similar way

$$\delta\mathcal{L} = \epsilon^\mu \partial_\mu \mathcal{L} = \epsilon^\nu \partial_\mu (g^\mu_{\ \nu} \mathcal{L}) \,. \tag{9.15}$$

By inserting the two previous quantities into Eq. (9.8), we obtain:

$$\partial_\mu T^{\mu\nu} = 0 \,, \tag{9.16}$$

with

$$T^{\mu\nu} = \frac{\partial \mathcal{L}}{\partial(\partial_\mu \phi)} \partial^\nu \phi - g^{\mu\nu} \mathcal{L} \,. \tag{9.17}$$

This expression is called the *energy-momentum tensor*. One of the conserved quantity is the *Hamiltonian* H of the system ($\dot{\phi} \equiv \partial_0 \phi$):

$$H = \int T^{00} d^3x = \int \left(\frac{\partial \mathcal{L}}{\partial \dot{\phi}} \dot{\phi} - \mathcal{L} \right) d^3x \,, \tag{9.18}$$

which corresponds to the total energy. The other three conserved quantities ($i = 1, 2, 3$)

$$P^i = \int T^{0i} d^3x = -\int \frac{\partial \mathcal{L}}{\partial \dot{\phi}} \partial^i \phi \, d^3x \tag{9.19}$$

are interpreted as the components of the *linear momentum* $\mathbf{P}$.

B - *Phase transformation*

In the case of a complex field ϕ, the Lagrangian density is hermitian and then invariant under the *phase transformation*:

$$\phi \rightarrow e^{i\alpha}\phi, \quad \phi^* \rightarrow e^{-i\alpha}\phi^* \, , \tag{9.20}$$

where α is a real constant. For an infinitesimal value of α one has:

$$\delta\phi = i\alpha\phi, \quad \delta\phi^* = -i\alpha\phi^* \, . \tag{9.21}$$

The Lagrangian density of a complex scalar field is given by

$$\mathcal{L} = \partial_\mu\phi^*\partial^\mu\phi - \mu^2\phi^*\phi \, , \tag{9.22}$$

from which, making use of Eq. (9.5), one gets the Klein-Gordon equation (compare with Eq. (7.17)):

$$(\Box + \mu^2)\,\phi(x) = 0 \, . \tag{9.23}$$

One can easily check that the four-divergence of the current

$$j^\mu = i(\phi^*\partial^\mu\phi - \phi\,\partial^\mu\phi^*) \tag{9.24}$$

is zero, and the conserved "charge" (constant in time) is given by:

$$Q = i \int (\phi^*\dot\phi - \phi\dot\phi^*)d^3x \, . \tag{9.25}$$

It is also useful to consider the case of the spinor Lagrangian density

$$\mathcal{L} = \overline\psi(x)(i\gamma^\mu\partial_\mu - m)\psi(x) \, , \tag{9.26}$$

where $\overline\psi(x) = \psi^\dagger(x)\gamma^0$. It gives rise to the Dirac equation, already considered in Eq. (7.79):

$$(i\gamma^\mu\partial_\mu - m)\psi(x) = 0 \, , \tag{9.27}$$

and to the equation for the conjugate spinor

$$\overline\psi(x)(i\gamma^\mu\overleftarrow{\partial}_\mu + m) = 0 \, . \tag{9.28}$$

In this case the "conserved" current is

$$j^\mu(x) = \overline\psi(x)\gamma^\mu\psi(x) \, , \tag{9.29}$$

and the conserved charge is:

$$Q = \int \psi^\dagger(x)\gamma^\mu\psi(x)d^3x \, . \tag{9.30}$$

9.3 The gauge group $U(1)$ and Quantum Electrodynamics

In Section 8.3 we have examined a few examples of conserved quantum numbers based on the global symmetry $U(1)$. We want to discuss here the deep implications of promoting $U(1)$ to a local symmetry, i.e. to a continuous set of local phase transformations.

In order to explain the main difference in going from global to local transformations, we consider a simple example taken from quantum field theory. We start from the Lagrangian density (9.26)

$$\mathcal{L}_0(x) = \overline{\psi}(x)(i\gamma^\mu \partial_\mu - m)\psi(x) , \tag{9.31}$$

where $\psi(x)$ stands here for the free electron field. If α is a real constant, $\mathcal{L}_0(x)$ is invariant under the global phase transformation

$$\psi(x) \to e^{i\alpha}\psi(x) , \tag{9.32}$$

the conjugate transformation being applied to the conjugate field $\overline{\psi}(x) = \psi^\dagger(x)\gamma^0$. It is well known that this invariance leads to the conservation of an additive quantum number, which is identified with the total electric charge (9.30).

The invariance no longer holds if the constant α is replaced with a function $\alpha(x)$ depending on the space-time co-ordinate x. Invariance under the local transformation

$$\psi(x) \to e^{i\alpha(x)}\psi(x) , \tag{9.33}$$

would require a peculiar modification in Eq. (9.31): the usual derivative has to be replaced with the so-called *covariant derivative* D_μ

$$\partial_\mu \psi(x) \to D_\mu \psi(x) = [\partial_\mu - ieA_\mu(x)]\psi(x) , \tag{9.34}$$

where e is the absolute value of the electron charge and $A_\mu(x)$ a four-vector field which is assumed to transform according to

$$A_\mu(x) \to A_\mu(x) + \frac{1}{e}\partial_\mu \alpha(x) . \tag{9.35}$$

At the level of the Lagrangian, this corresponds to the addition of an interaction term,

$$e\overline{\psi}(x)\gamma^\mu \psi(x) A_\mu , \tag{9.36}$$

which couples the electron current to the four-vector field $A_\mu(x)$, to be interpreted as the electromagnetic potential. In order to complete the expression of the Lagrangian density, one has to add the free electromagnetic term

$$\mathcal{L}_0^{em} = -\tfrac{1}{4}F_{\mu\nu}(x)F^{\mu\nu}(x) , \tag{9.37}$$

where

$$F_{\mu\nu} = \partial_\mu A_\nu - \partial_\nu A_\mu \qquad (9.38)$$

is the electromagnetic field tensor, thus obtaining the complete Lagrangian density in the form

$$\mathcal{L}(x) = \overline{\psi}(x)(i\gamma^\mu \partial_\mu - m)\psi(x) - \tfrac{1}{4}F_{\mu\nu}(x)F^{\mu\nu}(x) + e\overline{\psi}(x)\gamma^\mu \psi(x)\, A^\mu. \quad (9.39)$$

This is nothing else that the well-known Lagrangian density of a charged spinor field interacting with the electromagnetic field, i.e. the Lagrangian of Quantum Electrodynamics (QED), which is clearly invariant under the gauge transformations (9.33) and (9.35). It is important to point out that a mass term of the type $m^2 A_\mu A^\mu$ would break the gauge symmetry and then it is not allowed, in agreement with the fact that the field A_μ represents the photon, which is massless.

In conclusion, the example of QED teaches us that, while the invariance under the global transformation (9.32) implies the conservation of the electric charge, the invariance under the local transformation (9.33) requires the introduction of a massless vector field which plays the role of carrier of the interaction. Let us note that this result,in the case of QED, is to some extent tautological, since it does not add anything new to the theory: the introduction of the covariant derivative of Eq. (9.34) is nothing else that the quantum form of the substitution $p_\mu \to p_\mu - eA_\mu$, which introduces the electromagnetic interaction at the classical level. But the relevant point is that the approach of requiring local gauge invariance is the general way of introducing the interaction in terms of new vector fields also in those cases in which a classical counterpart of the interaction does not exists. We will see the consequences of this approach in the next Sections.

9.4 The gauge group $SU(3)$ and Quantum Chromodynamics

The conclusion reached in the previous Section can be extended to a more general case. Let us consider a quantum field theory which is invariant under a gauge group $\mathcal{G}$, i.e. the group transformations are imposed locally. It can be shown that, if the group $\mathcal{G}$ is of order r, i.e. if it has r generators, then the invariance under the local transformations of $\mathcal{G}$ requires the introduction of r massless vector fields, transforming as the adjoint representation of $\mathcal{G}$. We shall not reproduce the proof of this theorem here, but we refer to specific textbooks for details[7]. In general, the group $\mathcal{G}$ is non-Abelian and this fact implies profound changes in the corresponding field theory, with respect to the Abelian case of QED. Non-Abelian gauge theories, the first of which was

---

[7] See e.g. M.E. Peskin, D.V. Schröder, *An Introduction to Quantum Field Theory*, Addison-Wesley Publishing Company (1995).

built by Yang and Mills[8], have been extensively used in particle physics, as we shall see in the following.

In this Section we apply the above rule to the color group $SU(3)_c$. We saw in Subection 8.6.1 that the introduction of the color group solves the puzzle of quark statistics, but it does not solve a further puzzle: why all hadron states are color singlets, and there is no evidence of hadrons with open color? Or, in other words, why quarks are always bound in colorless hadrons states and do not appear as free particles? The quantum field theory based on the color group $SU(3)_c$, and called for this reason Quantum Chromodynamics (QCD), provides a reasonable solution to this puzzle, even if it cannot be considered yet as a rigorous proof [9].

In the following, we summarize some of the main properties of QCD[10], which is the present theory of strong interactions. The Lagrangian density for the free quark fields is a simple generalization of Eq. (9.31):

$$\mathcal{L}_0(x) = \sum_j \bar{q}^j(x)(i\gamma^\mu \partial_\mu - m_j)q^j(x) , \tag{9.40}$$

where the sum is over the quark flavors ($j = 1, 2, ...6$) and q denotes a triplet of $SU(3)_c$

$$q = \begin{pmatrix} q_1 \\ q_2 \\ q_3 \end{pmatrix} . \tag{9.41}$$

$\mathcal{L}_0$ is invariant under the global transformation

$$q(x) \rightarrow Uq(x) , \tag{9.42}$$

where U is an element (3×3 matrix) of $SU(3)_c$.

However, if one imposes that $SU(3)_c$ is a group of local gauge transformations

$$q(x) \rightarrow U(x)q(x) , \tag{9.43}$$

the group elements become functions of x, and they can be written explicitly in terms of the λ matrices (8.85) and of 8 real functions $\alpha_k(x)$

$$U(x) = \exp\left\{ i \sum_k \alpha_k(x)\lambda_k \right\} . \tag{9.44}$$

The invariance under these local transformations requires the introduction of 8 real fields called *gluons* and denoted by $G_\mu^\alpha(x)$, where μ is the Lorentz index

---

[8] C.N. Yang and R.Mills, Phys.Rev.96, 191 (1954).

[9] M.E. Peskin, D.V. Schröder, quoted ref.; S.Weinberg, *The Quantum Theory of Fields - Vol. II. Modern Applications*, Cambridge University Press, (1996).

[10] D.J. Gross and F. Wilczek, Phys.Rev. D8, 3633 (1973); H. Frizsch and M. Gell-Mann, Proceed. of the XVI Intern. Conf. on High Energy Physics, Chicago (1972).

and α specifies the elements of an octet of $SU(3)_c$. They can be included in a 3×3 Hermitian traceless matrix, defined by

$$G_\mu(x) = \sum_\alpha G_\mu^\alpha(x) \tfrac{1}{2}\lambda_\alpha = G_\mu^\dagger(x) \,. \qquad (9.45)$$

The gauge invariant Lagrangian density is given by the following expression

$$\mathcal{L}(x) = \sum_j \bar{q}^j(x)(i\gamma^\mu D_\mu - m_j)q^j(x) - \tfrac{1}{2}\mathrm{Tr}(G_{\mu\nu}G^{\mu\nu}) \,, \qquad (9.46)$$

where D_μ is the *covariant derivative*

$$\partial_\mu \rightarrow D_\mu = \partial_\mu + ig_s G_\mu(x) \,, \qquad (9.47)$$

and g_s denotes the *strong coupling constant* between quarks and gluons. The last term in Eq. (9.46) contains the *field strength* $G_{\mu\nu}$, given by

$$G_{\mu\nu} = \partial_\mu G_\nu - \partial_\nu G_\mu + ig_s[G_\mu, G_\nu] \,, \qquad (9.48)$$

where the presence of the commutator is a consequence of the non-Abelian character of the gauge group. One can show that the above expression is invariant under the transformations

$$q^j(x) \rightarrow U(x)q^j(x) \,, \qquad (9.49)$$

and

$$G_\mu(x) \rightarrow U(x)G_\mu(x)U^\dagger(x) + \frac{i}{g_s}(\partial_\mu U(x))U^\dagger(x) \,. \qquad (9.50)$$

It is instructive to make a comparison between QCD, which is a non-Abelian gauge field theory based on the group $SU(3)$, and QED, Abelian gauge theory based on the group $U(1)$. For this comparison, it is useful to re-write Eqs. (9.46) and (9.48) in a more explicit form:

$$\mathcal{L}(x) = \sum_j \bar{q}^j(x) \left[i\gamma^\mu(\partial_\mu + ig_s G_\mu^\alpha(x)\tfrac{1}{2}\lambda_\alpha) - m_j\right] q^j(x) - \tfrac{1}{4}G_{\mu\nu}^\alpha(x)G_\alpha^{\mu\nu}(x) \,,$$
$$\qquad (9.51)$$

and

$$G_{\mu\nu}^\alpha = \partial_\mu G_\nu^\alpha(x) - \partial_\nu G_\mu^\alpha(x) - g_s f_{\alpha\beta\gamma}G_\mu^\beta(x)G_\nu^\gamma(x) \,. \qquad (9.52)$$

The main difference is the following: while photons are electrically neutral and therefore there is no direct coupling among themselves in the Lagrangian, gluons are color octets, they carry color charges and there are direct couplings among them in the Lagrangian. The quark-gluon and gluon-gluon couplings are represented in Fig. 9.1; we note that only the first coupling has a QED analogue, while there are no analogues for the two others. This feature, which is a consequence of the different properties of the two gauge groups, makes

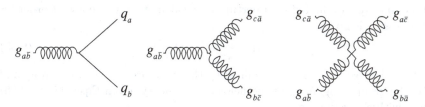

Fig. 9.1. *Gluon vertices in QCD. Since gluons are color-anticolor pairs, also three-gluons and four-gluons vertices are present, besides the qqg vertex, analogous to the qqγ vertex in QED. Color indices are also indicated.*

a big difference between electron and quark interactions in their high energy behavior.

In order to understand this fact, one should analyze the renormalization properties of the two couplings: the electric charge e and the strong coupling g_s. We shall not perform here a detailed analysis, which is outside the scope of this book, but limit ourselves to a simple semi-qualitative discussion. It is well known that the bare coupling constants which appear in the Lagrangians are renormalized by higher order corrections, so that one has to deal with effective couplings which depend on the energy scale, or better on the momentum transfer $q^2 = -\mu^2$. Then the coupling constants are replaced by *running couplings*[11]. While a generic coupling $g(\mu^2)$, which is a function of μ^2, is logarithmically divergent when evaluated in perturbation theory, the difference of its values at two different scales is finite:

$$g(\mu^2) - g(\mu'^2) = \text{finite} . \tag{9.53}$$

A change in scale is called *scale or conformal transformation*; the set of these transformations is called *renormalization group*[12]. It was first realized by Gell-Mann and Low[13] that, by changing the effective energy scale from a value μ, one can define the theory at another scale, and it is a similar replica of itself. One obtains the so-called renormalization group equation[14]

$$\frac{dg}{d\ln\mu} = \beta(g), \tag{9.54}$$

and the important point is that the function $\beta(g)$ depends only on the parameters of the theory, such as the couplings, but not on the scale μ. The

---

[11] See e.g. F. Halzen, A.D. Martin, *Quarks and Leptons: An Introductory Course in Modern Particle Physics*, John Wiley & Sons (1984); M.E. Peskin, D.V. Schröder, quoted ref.; S. Weinberg, quoted ref.

[12] N.N. Bogoliubov and D.V. Shirkov, *The Theory of Quantized Fields*, Interscience, New York (1959).

[13] M. Gell-Mann and F. Low, Phys. Rev. 95, 1330 (1954).

[14] H. Georgi, H.R. Quinn, S. Weinberg, Phys. Rev. Lett. 33, 451 (1974).

function $\beta(g)$, evaluated by a perturbative calculation at the lowest order for $g(\mu^2) \ll 1$, is given by

$$\beta(g) = \frac{1}{4\pi}bg^3 + O(g^5), \tag{9.55}$$

where the coefficient b is determined by the structure of the gauge group and by the group assignment of the fields of the theory. From Eqs. (9.54), (9.55) and with the definition

$$\alpha(\mu^2) = \frac{g^2(\mu^2)}{4\pi} , \tag{9.56}$$

we obtain the equation which gives the variation of the coupling with the squared momentum

$$\frac{1}{\alpha(Q^2)} = \frac{1}{\alpha(\mu^2)} - b \ln \frac{Q^2}{\mu^2} , \tag{9.57}$$

where, for the sake of convenience, we have introduced the positive quantity $Q^2 = -q^2$.

Let us consider a fixed value of μ^2 while varying Q^2. In the case $b > 0$, the coupling $\alpha(Q^2)$ increases logarithmically with Q^2. Specifically, in the case of QED, one gets for the electromagnetic coupling α_{em}:

$$\frac{1}{\alpha_{\text{em}}(Q^2)} = \frac{1}{\alpha_{\text{em}}(\mu^2)} - \frac{1}{3\pi} \ln \frac{Q^2}{\mu^2} . \tag{9.58}$$

The effective electric charge increases with Q^2, i.e. with decreasing distance. This result can be interpreted in the following way: the *bare* charge is screened by the presence of virtual e^+e^- pairs, but going to smaller distances one penetrates the polarization cloud and the screening effect is reduced. The variation of α_{em} with Q^2 is very small: from the low-energy value of the fine-structure constant $\alpha \approx \frac{1}{137}$, one gets the value $\approx \frac{1}{128}$ at $Q \approx 250$ GeV; this behavior has been tested experimentally.

In the case $b < 0$ we get instead

$$\lim_{Q^2 \to \infty} \alpha(Q^2) \to 0 , \tag{9.59}$$

which means that, by increasing Q^2, and correspondingly the energy, the coupling decreases and tends to zero for $Q^2 \to \infty$. This situation is called *asymptotic freedom*, and it is what happens in the case of QCD[15]. In this case, the coefficient b in Eq. (9.57) is given by $b = 7/4\pi$, so that one obtains for the strong interaction coupling $\alpha_s = g_s^2/4\pi$:

$$\frac{1}{\alpha_s(Q^2)} = \frac{1}{\alpha_s(\mu^2)} + \frac{7}{4\pi} \ln \frac{Q^2}{\mu^2} . \tag{9.60}$$

---

[15] D.J. Gross and F. Wilczek, Phys. Rev. Lett. 30, 1343 (1973); H.D. Politzer, Phys. Rev. Lett. 30, 1346 (1973).

In conclusion, the coupling $\alpha_s(Q^2)$ decreases with increasing Q^2, so that, for high Q^2 (i.e. short distances), perturbation theory is valid. This peculiar feature has been confirmed by the deep inelastic scattering of high-energy electrons and neutrinos by protons. At high energy, the proton reveals its structure: it is made of free point-like constituents, called *partons*, which are *valence* quarks, virtual quark-antiquark pairs and gluons (for a review see for instance ref[16]).

On the other hand, for very small Q^2 values (i.e large distances), the coupling α_s becomes very strong, and this explains the impossibility of pulling quarks apart. This behavior would explain the *confinement* of colored quarks inside hadrons; perturbation theory cannot be used, but this conjecture is supported by lattice calculations. The Q^2 dependence of α_s has been checked experimentally.

9.5 The mechanism of spontaneous symmetry breaking

The approach followed for the construction of the field theory of strong interactions cannot be applied, even with the appropriate modifications, to the case of weak interactions.

It was well known that the original Fermi theory of weak interactions is not renormalizable. Even if one assumes that the weak interactions are mediated by massive intermediate vector bosons, the theories one can build are, in general, not renormalizable. The condition for building a gauge theory, as in the cases of QED and QCD, is that the gauge vector bosons are massless, but this appears to be in contrast with the short range nature of weak interactions which requires the exchange of heavy bosons.

It was only by means of the so-called mechanism of *spontaneous symmetry breaking* (SSB) that the solution of the problem was found. This mechanism, first introduced by Higgs[17] for explaining specific phenomena occurring in condensed matter physics, was the new key ingredient for building a renormalizable field theory of weak interactions.

Before applying this mechanism to the realistic case, we prefer, for didactic reasons, to consider two simple examples in order to illustrate the central point of the mechanism without being involved with other, non essential, technicalities.

9.5.1 Spontaneous symmetry breaking of a discrete symmetry

Let us consider the Lagrangian density of a real scalar field,

$$\mathcal{L} = T - V = \tfrac{1}{2}\partial_\mu\phi\partial^\mu\phi - V(\phi) , \qquad (9.61)$$

---

[16] G. Sterman *et al.*, Rev. Mod. Phys. 67, 157 (1995)

[17] P.W. Higgs, Phys. Rev. Lett. 13, 508 (1964); Phys. Rev. 145, 1156 (1966).

where $V(\phi)$, usually called potential, contains the mass and the self-interaction terms. In its simplest form, it is given by

$$V(\phi) = \tfrac{1}{2}\mu^2\phi^2 + \tfrac{1}{4}\lambda\phi^4 , \qquad (9.62)$$

with $\lambda > 0$ to ensure the existence of a spectrum of stable bound states. Due to the form of the potential, the Lagragian is seen to satisfy the reflection symmetry $\phi \to -\phi$. In general, we can assume this symmetry property for the Lagrangian and consider the potential (9.62) as the first two terms of a power expansion of a generic $V(\phi)$.

We want to analyze the properties of the ground state, i.e. the lowest energy state of the system, which corresponds to the vacuum state.

In the case $\mu^2 > 0$, the behavior of the potential is given in Fig. 9.2: the Lagrangian (9.61) describes a scalar field of mass μ, with the ground state identified by $\phi = 0$. The excited levels can be obtained through a perturbative expansion around $\phi = 0$, and are related to the self-interaction term $\tfrac{1}{4}\lambda\phi^4$, which corresponds to the minimal coupling, represented by a four-particle vertex. The reflection symmetry is manifest.

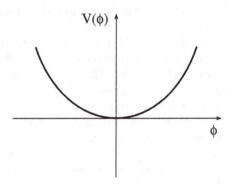

Fig. 9.2. *Case of a discrete symmetry: the potential $V(\phi)$ for $\mu^2 > 0$.*

The situation is, however, completely different if we assume $\mu^2 < 0$. The term $\tfrac{1}{2}\mu^2$, in fact, cannot be interpreted as a mass term, and the potential, on the basis of the minimum condition

$$\frac{\partial V(\phi)}{\partial \phi} = \phi(\mu^2 + \lambda\phi^2) = 0 , \qquad (9.63)$$

exhibits now two minima, located at

$$\phi_{\text{min}} = \pm v \qquad \text{with} \qquad v = \sqrt{-\frac{\mu^2}{\lambda}} , \qquad (9.64)$$

whereas $\phi = 0$ is now a relative maximum. The behavior of $V(\phi)$ for $\mu^2 < 0$ is reported in Fig. 9.3.

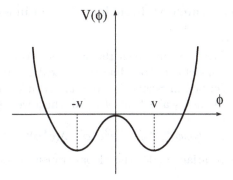

Fig. 9.3. *Case of a discrete symmetry: the potential $V(\phi)$ for $\mu^2 < 0$.*

If now we want to describe the spectrum of the system, we have to choose one of the two minima as ground state and then perturb the system starting from the minimum we have chosen. Let us choose for example $\phi = +v$ as ground state and shift the field $\phi(x)$ in such a way that the ground state occurs at the zero of the new field $\eta(x)$:

$$\phi(x) \rightarrow \eta(x) + v . \tag{9.65}$$

By applying the shift (9.65) to the Lagrangian density (9.61), we find for it the new form

$$\mathcal{L}' = \tfrac{1}{2}\partial_\mu\eta\partial^\mu\eta - v^2\lambda\eta^2 - v\lambda\eta^3 - \tfrac{1}{4}\lambda\eta^4 + \text{constant terms} . \tag{9.66}$$

Considering now $\mathcal{L}'$ as the Lagrangian which describes the system, one can see that the new field $\eta(x)$ has the mass

$$m_\eta = \sqrt{2v^2\lambda} = \sqrt{-2\mu^2} . \tag{9.67}$$

since the term η^2 exhibits the right sign. Moreover, it is possible, at least in principle, to derive the spectrum of the physical states through a perturbative expansion, taking $\eta(x) = 0$ as the physical ground state (vacuum) of the system.

Seemingly, the reflection symmetry is lost in $\mathcal{L}'$, since it contains a term η^3, which changes its sign under the reflection $\eta(x) \rightarrow -\eta(x)$. However, the change of variables from $\phi(x)$ to $\eta(x)$ has no physical relevance, but the initial symmetry of the Lagrangian density $\mathcal{L}$ is no longer manifest in $\mathcal{L}'$. It is usual to say that the symmetry is *hidden*; its only manifestation is the relation among the coefficients of the three terms of the potential in $\mathcal{L}'$, which depend only on the two parameters μ^2 and λ of the potential in Eq. (9.61). In fact, the Lagrangian densities $\mathcal{L}$ and $\mathcal{L}'$ are completely equivalent; it is only the specific choice of the ground state that breaks the symmetry, and this is called the mechanism of spontaneous symmetry breaking. For a clarification of this point we refer to the next subsection, where the present situation is generalized to the case of a continuous symmetry.

9.5.2 Spontaneous symmetry breaking of a continuous global symmetry

Let us consider now the case in which the symmetry is a continuous global symmetry, as in the model corresponding to the scalar version of the spinor electrodynamics, described in Sect. 9.3. We start then with the Lagrangian density (9.22) of a complex scalar field $\phi(x)$, to which we add a self-coupling term:

$$\mathcal{L} = (\partial_\mu \phi)^* (\partial^\mu \phi) - \mu^2 \phi^* \phi - \lambda (\phi^* \phi)^2 \ . \tag{9.68}$$

This Lagrangian is invariant under the global transformations of the group $U(1)$

$$\phi(x) \ \rightarrow \ e^{i\alpha} \, \phi(x) \ , \quad \phi^*(x) \ \rightarrow \ e^{-i\alpha} \phi^*(x) \ , \tag{9.69}$$

and we know that this invariance property gives rise to an additive conserved quantity, usually interpreted as the electric charge of the system.

As in the case of a discrete symmetry, it is useful to define the last two term in Eq. (9.68) as the *potential* of the system:

$$V(\phi) = \mu^2 |\phi|^2 + \lambda |\phi|^4 \ . \tag{9.70}$$

In the case $\mu^2 > 0$, the usual minimum of the potential, obtained from the equation

$$\frac{dV}{d|\phi|} = 2\mu^2 |\phi| + 4\lambda |\phi|^3 = 0 \ , \tag{9.71}$$

is given by $|\phi|_{\min} = 0$, as it is shown in Fig. 9.4.

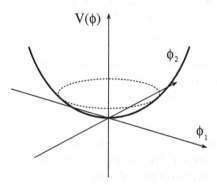

Fig. 9.4. *The potential $V(\phi)$ for $\mu^2 > 0$.*

However, if the peculiar case $\mu^2 < 0$ were occurring, $|\phi| = 0$ would become a relative maximum and the minimum would be obtained for

$$|\phi|^2_{\min} = -\frac{\mu^2}{2\lambda} \ . \tag{9.72}$$

In fact, as it is shown in Fig. 9.5, the above equation represents a continuum set of solutions, since the phase of ϕ at the minimun is completely arbitrary:

$$\phi_{\min} = |\phi|_{\min} e^{i\gamma} . \tag{9.73}$$

The case $\mu^2 < 0$ is not unrealistic, since the renormalization procedure replaces the constant μ^2 appearing in the Lagrangian density (9.68) by a function $\mu^2(Q^2)$, which can change its sign at a certain energy scale.

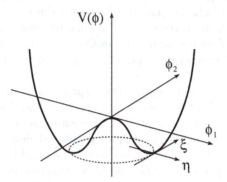

Fig. 9.5. *The potential $V(\phi)$ for $\mu^2 < 0$.*

The solution given by Eq. (9.73) represents an infinite set of degenerate vacuum states: the choice of a particular vacuum state gives rise to *spontaneous symmetry breaking*. Since the Lagrangian (9.68) is invariant under the phase transformations (9.69), any specific value of γ is equally good and, for the sake of convenience, we choose the value $\gamma = 0$. Then, by writing

$$\phi = \frac{1}{\sqrt{2}} \{\phi_1(x) + i\phi_2(x)\} , \tag{9.74}$$

the minimum of the potential is given by

$$(\phi_1)_{\min} = \sqrt{\frac{-\mu^2}{\lambda}} \equiv v . \tag{9.75}$$

The quantity v is called *vacuum expectation value* (v.e.v.) of the scalar field. As in the case of the discrete symmetry, we shift the two fields ϕ_1, ϕ_2 according to our choice of the minimum:

$$\phi_1(x) = \eta(x) + v , $$
$$\phi_2(x) \equiv \xi(x) , \tag{9.76}$$

and the Lagrangian density $\mathcal{L}$ of Eq. (9.68) takes the form

$$\mathcal{L}' = \tfrac{1}{2}\partial_\mu\eta\partial^\mu\eta + \tfrac{1}{2}\partial_\mu\xi\partial^\mu\xi - \tfrac{1}{2}(-2\mu^2)\eta^2 - \tfrac{1}{4}\lambda(\eta^2+\xi^2)^2 - \lambda v(\eta^2+\xi^2)\eta \ . \quad (9.77)$$

The Lagrangian densities $\mathcal{L}$ and $\mathcal{L}'$ given in Eqs. (9.68) and (9.77) are expressed in terms of different variables, but they are completely equivalent. In principle, the use of either $\mathcal{L}$ or $\mathcal{L}'$ should give the same physical results; however, this would be true only if one could get an *exact* solution of the theory. In fact, in general, since exact solutions are not available, one has to use perturbation theory, and the situation for approximate solutions is rather different. In the case $\mu^2 < 0$, one cannot start from the Lagrangian density $\mathcal{L}$ and perform a perturbation expansion about the value $\phi(x) = 0$ (as in the case $\mu^2 > 0$), since this value corresponds to an unstable configuration and perturbation theory would give meaningless results. Instead, starting from $\mathcal{L}'$, one can treat the interaction terms by perturbation about the stable configuration $\phi(x) = (\phi_1)_{min}$.

Let us compare more closely the two expressions of $\mathcal{L}$ and $\mathcal{L}'$. The first expression, considered in the case $\mu^2 > 0$, contains a complex scalar field ϕ (or, equivalently, two real fields ϕ_1 and ϕ_2) with squared mass μ^2; the second one, to be used in the case $\mu^2 < 0$, contains a real scalar field $\eta(x)$ with squared mass $m_\eta^2 = -2\mu^2 = 2\lambda v^2$, and a *massless* real scalar field ξ. The shifted field $\eta(x)$ represents the quantum excitations, above the constant background value v, along the radial direction; the field $\xi(x)$ represents a *massless mode*, and it corresponds to excitations occurring along the flat direction of the potential.

If we consider the case in which the parameter μ^2 changes continuously with the energy scale from positive to negative values, we can describe the change of the behavior of the system as a *phase transition*. The transition occurs from a state which is invariant under the transformations (9.69) of the group $U(1)$ to a new state in which this invariance is lost; as a consequence, the additive quantity, previously interpreted as electric charge, looses its meaning and it is no longer conserved. This is a simple example of a general situation in which the *spontaneous symmetry breaking* (SSB) of a continuous symmetry occurs in a transition accompanied by the appearance of massless scalar bosons called *Goldstone bosons*.

The appearance of Goldstone bosons in the SSB of a continuous symmetry is a consequence of the so-called Goldstone theorem[18]. At first sight, the presence of massless bosons makes it difficult to apply the SSB mechanism to a realistic theory of weak interactions mediated by massive vector bosons. Then one has to find a way that allows to evade the Goldstone theorem. Indeed, this way exists and it is realized in the case in which the global symmetry is promoted to a *gauge* symmetry.

---

[18] J. Goldstone, Nuovo Cimento 19, 154 (1961); Y. Nambu, Phys. Rev. Lett. 4, 380 (1960).

9.5.3 Spontaneous symmetry breaking of a gauge symmetry: the Higgs mechanism

As we will see in the following, the Goldstone theorem is evaded in the case of gauge theories, in which the scalar fields interact with gauge vector fields. This is the famous *Higgs mechanism*, which we illustrate in the frame of the simple model, based on the group $U(1)$, considered above. We start from the Lagrangian (9.68), but replace the Eqs. (9.69) by the local transformations

$$\phi(x) \; \rightarrow \; e^{i\alpha(x)} \, \phi(x) \; , \quad \phi^*(x) \; \rightarrow \; e^{-i\alpha(x)} \phi^*(x) \; . \tag{9.78}$$

The invariance under these transformations requires the introduction of the *covariant derivative*

$$\partial_\mu \phi(x) \rightarrow D_\mu \phi(x) = [\partial_\mu + ie A_\mu(x)] \phi(x) \; , \tag{9.79}$$

where $A_\mu(x)$ is the vector field with transformation property defined in Eq. (9.35)[19]. Then Eq. (9.68) is replaced by

$$\mathcal{L}(\phi, A_\mu) = (D_\mu \phi)^* D^\mu \phi - V(\phi) - \tfrac{1}{4} F_{\mu\nu} F^{\mu\nu} \; , \tag{9.80}$$

where, as usual, $F_{\mu\nu} = \partial_\mu A_\nu - \partial_\nu A_\mu$. Eq. (9.80) can be rewritten in the form

$$\mathcal{L}(\phi, A_\mu) = \partial_\mu \phi^* \partial^\mu \phi - V(\phi) - \tfrac{1}{4} F_{\mu\nu} F^{\mu\nu} - e j^\mu A_\mu \; , \tag{9.81}$$

in terms of the current

$$j^\mu = i(\phi^* \, \partial^\mu \phi - \phi \, \partial^\mu \phi^*) \; . \tag{9.82}$$

Eqs. (9.80), (9.81) correspond, for $\mu^2 > 0$, to the Lagrangian density of the scalar electrodynamics, i.e. of a charged scalar field coupled with the electromagnetic field. Since the first two terms in Eq. (9.81) are the same as those in Eq. (9.68), the mechanism of SSB can occur also in the present case for $\mu^2 < 0$. Making use of the notation given in Eq. (9.76) and with $m_\eta^2 = -2\mu^2$, Eq. (9.81) can be expressed in the form

$$\mathcal{L} = \tfrac{1}{2} \partial_\mu \eta \partial^\mu \eta + \tfrac{1}{2} \partial_\mu \xi \partial^\mu \xi - \tfrac{1}{2} m_\eta^2 \eta^2 - \tfrac{1}{4} \lambda (\eta^2 + \xi^2)^2 \tag{9.83}$$

$$- \lambda v(\eta^2 + \xi^2)\eta - \tfrac{1}{4} F_{\mu\nu} F^{\mu\nu} + \tfrac{1}{2} e^2 v^2 A_\mu A^\mu$$

$$+ \tfrac{1}{2} e^2 A_\mu A^\mu (\eta^2 + \xi^2 + 2v\eta) - e(\eta \partial_\mu \xi + \xi \partial_\mu \eta) A^\mu + e v \partial_\mu \xi A^\mu \; .$$

We point out an important result: the vector field A_μ, which is massless before the SSB, acquires a mass different from zero: this is a simple example of the Higgs mechanism. The value of the mass is determined by

---

[19] We note that, in the present case, we have changed the sign of the charge e with respect to that of the spinor electrodynamics.

$$m_A = ev \, , \tag{9.84}$$

i.e. by the coupling constant times the v.e.v. of the scalar field.

However, one has to check that the number of the degrees of freedom remains the same before and after the transition. In the first phase the field A_μ is massless and transversal (two degrees of freedom), while, in the second phase, it becomes massive and therefore it acquires also a longitudinal component (three degrees of freedom). This additional degree corresponds to the ξ field, which becomes unphysical and it is "eaten" by the vector field: in other words, the Goldstone boson disappears and it is replaced by the longitudinal component of the vector field. On the other hand, the η field is physical and massive: it is called *Higgs boson*.

The unphysical field can be eliminated by adopting a different parametrization for the scalar field $\phi(x)$:

$$\phi(x) = \frac{1}{\sqrt{2}} e^{i\xi(x)/v} \left[\eta(x) + v \right] \, , \tag{9.85}$$

with $\eta(x)$ and $\xi(x)$ real scalar fields. By applying the local gauge transformations

$$\phi(x) \quad \rightarrow \quad \phi'(x) = e^{-i\xi(x)/v}\phi(x) = \frac{1}{\sqrt{2}}\left[\eta(x) + v\right] \, ,$$

$$A_\mu(x) \rightarrow A'_\mu(x) = A_\mu(x) + \frac{1}{ev} \, \partial_\mu \xi(x) \, , \tag{9.86}$$

the Lagrangian (9.83) takes the form

$$\mathcal{L}'' = \tfrac{1}{2}\partial_\mu\eta\partial^\mu\eta - \tfrac{1}{2}m_\eta^2\eta^2 - \tfrac{1}{4}F_{\mu\nu}F^{\mu\nu} + \tfrac{1}{2}m_A^2 A_\mu A^\mu + \tfrac{1}{2}e^2 A_\mu A^\mu \eta^2$$
$$+ e^2 v A_\mu A^\mu \eta - \tfrac{1}{4}\lambda\eta^4 - \lambda v\eta^3 \, , \tag{9.87}$$

in which the field $\xi(x)$ has disappeared.

In conclusion, we would like to formulate the Goldstone theorem and the Higgs mechanism for a general case, without giving the proof[20].

A system, for which the Lagrangian is invariant under the transformations of a group $\mathcal{G}$, can have eigenstates which posses a lower symmetry, corresponding to that of a subgroup $\mathcal{G}'$ of $\mathcal{G}$. In general, if n and n' are the numbers of generators of $\mathcal{G}$ and $\mathcal{G}'$, respectively, the phenomenon of SSB is characterized by the appearance of $n - n'$ Goldstone bosons. This is the content of the general Goldstone theorem.

In the case of a local gauge theory, in which the Lagrangian is invariant under the local gauge group $\mathcal{G}$, there are n massless gauge vector fields. In the process of SSB in which the unbroken symmetry is $\mathcal{G}'$, the $n - n'$ Goldstone bosons are "eaten" by $n - n'$ vector fields which become massive, while the remaining n', corresponding to the generators of the subgroup $\mathcal{G}'$, remain massless.

---

[20] J. Goldstone, A. Salam and S. Weinberg, Phys. Rev. 127, 965 (1962).

Before applying this mechanism to the theory of weak interactions, we would like to make a digression on the approximate symmetries considered in Chapter 9, which can be obtained by the QCD Lagrangian making use of the SSB mechanism.

9.6 Spontaneous breaking of the chiral symmetry of QCD

As an example of spontaneous symmetry breaking it interesting to examine the approximate chiral symmetry of QCD. Let us consider the Lagrangian density of QCD, Eq. (9.46), taking into account only the terms containing the color triplets with *up* and *down* flavors

$$\mathcal{L} = \bar{u}(x)(i\gamma^\mu D_\mu - m_u)u(x) + \bar{d}(x)(i\gamma^\mu D_\mu - m_d)d(x) + ... \qquad (9.88)$$

Since the masses of the u and d are very small in comparison with the other heavier quarks, as shown in Table 8.11, it is reasonable to consider the approximate case in which the two lightest quarks are massless. Then we can rewrite Eq. (9.88) in the form

$$\mathcal{L} = \left(\bar{u}(x), \bar{d}(x)\right)(i\gamma^\mu D_\mu)\begin{pmatrix} u(x) \\ d(x) \end{pmatrix} ... = \bar{q}(x)(i\gamma^\mu D_\mu)q(x)... , \qquad (9.89)$$

where q stands here for the quark (flavor) doublet

$$q = \begin{pmatrix} u \\ d \end{pmatrix} . \qquad (9.90)$$

The Lagrangian (9.89) is clearly invariant under the isospin group $SU(2)_I$, with the generators $I_i = \frac{1}{2}\sigma_i$, defined in Subsection 8.4.1. Since the mass terms have been eliminated, there are no couplings between left-handed and right-handed quarks

$$q_L = \tfrac{1}{2}(1 - \gamma_5)q , \qquad q_R = \tfrac{1}{2}(1 + \gamma_5)q . \qquad (9.91)$$

Therefore the Lagrangian (9.89) is symmetric under the separate *chiral* isospin transformations of the direct product of two commuting $SU(2)$ groups

$$SU(2)_L \otimes SU(2)_R . \qquad (9.92)$$

The corresponding generators

$$I_{Li} = \tfrac{1}{2}(1 - \gamma_5)I_i , \qquad I_{Ri} = \tfrac{1}{2}(1 + \gamma_5)I_i , \qquad (9.93)$$

which act respectively on q_L and q_R, satisfy the commutation relations

$$[I_{Li}, I_{Lj}] = i\epsilon_{ijk}I_{Lk} \ ,$$
$$[I_{Ri}, I_{Rj}] = i\epsilon_{ijk}I_{Rk} \ , \tag{9.94}$$
$$[I_{Li}, I_{Rj}] = 0 \ .$$

The group (9.92) has an $SU(2)$ subgroup with generators

$$I_i = I_{Ri} + I_{Li} \ , \tag{9.95}$$

which is the usual isospin group $SU(2)_I$. Moreover, it has another $SU(2)$ subgroup, which we denote by $SU(2)_A$, with generators

$$I_{Ai} = I_{Ri} - I_{Li} \ . \tag{9.96}$$

The generators of $SU(2)_I$ and $SU(2)_A$ satisfy the commutation relations

$$\begin{aligned}
[I_i, I_j] \ \ &= i\epsilon_{ijk}I_k \ , \\
[I_i, I_{Aj}] \ \ &= i\epsilon_{ijk}I_{Ak} \ , \\
[I_{Ai}, I_{Aj}] &= i\epsilon_{ijk}I_k \ .
\end{aligned} \tag{9.97}$$

From the Lagrangian (9.89), according to Noether's theorem (Section 9.2), one can derive two conserved currents:

$$\mathbf{j}_L^\mu = i\bar{q}_L\gamma^\mu\mathbf{I}_L q_L \ , \qquad \mathbf{j}_R^\mu = i\bar{q}_R\gamma^\mu\mathbf{I}_R q_R \ , \tag{9.98}$$

where $\mathbf{I}_L$, $\mathbf{I}_R$ are three-vectors with components (9.93). The sum of $\mathbf{j}_L$ and $\mathbf{j}_R$ gives the *vector* isospin current

$$\mathbf{j}_V^\mu = i\bar{q}\gamma^\mu\mathbf{I}q \ , \tag{9.99}$$

while the difference gives the *axial-vector* isospin current

$$\mathbf{j}_A^\mu = i\bar{q}\gamma^\mu\gamma_5\mathbf{I}q = i\bar{q}\gamma^\mu\mathbf{I}_A q \ , \tag{9.100}$$

where $\mathbf{I} = \mathbf{I}_R + \mathbf{I}_L$ and $\mathbf{I}_A = \mathbf{I}_R - \mathbf{I}_L$. The currents of Eqs. (9.99) and (9.100) are conserved, i.e. they satisfy the conditions

$$\partial_\mu \mathbf{j}_V^\mu = 0 \ , \qquad \partial_\mu \mathbf{j}_A^\mu = 0 \ . \tag{9.101}$$

The charges associated to these currents are defined by the operators

$$\mathcal{I} = \int d^3x \mathbf{j}_V^0 \qquad \mathcal{I}_A = \int d^3x \mathbf{j}_A^0 \ , \tag{9.102}$$

that, with the chosen normalization of the currents, satisfy the same commutation relations (9.97) of the matrices I_i, I_{Ai}.

However, while the transformations generated by the isospin $\mathcal{I}$ operator are manifest symmetries of the strong interactions, those generated by the axial-vector charge $\mathcal{I}_A$ do not correspond to any symmetry of hadrons. In fact, such a symmetry would require that, for any one-hadron state $|h>$, there should be a degenerate state $\mathcal{I}_A|h>$ with the same spin and internal quantum numbers, but with opposite parity. The hadron spectrum does not show the existence of parity doubling. While $SU(2)_I$ is a good symmetry, the $SU(2)_A$ symmetry appears to be broken.

In 1960, Nambu and Jona-Lasinio[21] formulated the hypothesis that the chiral symmetry is spontaneously broken. Let us suppose that the chiral symmetry $SU(2)_L \otimes SU(2)_R$ is spontaneously broken down to $SU(2)_I$ and examine the consequences of this hypothesis. The breaking is expected to occur through *condensate* of quark-antiquark pairs, characterized by a nonzero vacuum expectation value of the scalar operator

$$< 0 \,|\, \bar{q}q \,|\, 0 > \; = \; < 0 \,|\, \bar{q}_L q_R + \bar{q}_R q_L \,|\, 0 > \; \neq \; 0 \,. \qquad (9.103)$$

The v.e.v. indicates that the two quark helicities get mixed, and so the u and d quarks acquire effective masses as they move through the vacuum.

By Goldstone's theorem, the breaking of the symmetry associated to the axial-vector charges leads to three *massless* Goldstone bosons, with the same quantum numbers of the broken generators $\mathcal{I}_A$. Then they have spin zero, isospin $I = 1$, negative parity.

The fact that the up and down quarks have non-zero masses indicates that the chiral symmetry is only approximate, so that its breaking entails the existence of approximately massless Goldstone bosons. The real spectrum of hadrons does not contain massless particles, but there is an isospin triplet of relatively light mesons, the *pions*, which would correspond to the Goldstone bosons.

In conclusion, we have shown that the isospin symmetry of strong interactions can be derived from the QCD Lagrangian: it is not a fundamental symmetry linking *up* and *down* quarks, but it is an approximate symmetry related to the fact that the masses m_u and m_d in (9.88) are small compared with the effective scale of hadrons.

Replacing in the Lagrangian (9.89) the quark doublet (9.90) by the triplet

$$q = \begin{pmatrix} u \\ d \\ s \end{pmatrix} , \qquad (9.104)$$

one gets a chiral $SU(3)_L \otimes SU(3)_R$ symmetry; this symmetry would be spontaneously broken down to the flavor $SU(3)_f$ with the appearance of an octet of pseudoscalar Goldstone bosons. They would be interpreted as the meson octet (8.100), containing, besides the π-meson, the K-meson and the η. The interpretation of the flavor symmetry is similar to that given for the isospin $SU(2)_I$ symmetry. However, the fact that the mass m_s of the strange quark s is much bigger than m_u and m_d, even if smaller than the strong interaction

---

[21] Y. Nambu and G. Jona-Lasinio, Phys.Rev.122, 345 (1960)

scale, indicates that the approximate $SU(3)_f$ symmetry is less good than the isospin symmetry.

9.7 The group $SU(2) \otimes U(1)$ and the electroweak interactions

In Section 9.5 we mentioned that the present theory of weak interactions is a gauge field theory based on the mechanism of spontaneous symmetry breaking. In this Section we develop this subject by considering the main properties of weak interactions, their connection to the electromagnetic ones and the possibility of describing the two kinds of interactions in terms of a unified gauge theory, based on the group $SU(2) \otimes U(1)$. The spontaneous breaking of the gauge symmetry through the Higgs mechanism is assumed to take place. In this way one is able to reproduce not only the right V-A structure of the weak charged interactions mediated by massive vector bosons and the vector structure of the electromagnetic interactions mediated by a massless photon, but also to describe the weak neutral currents in a consistent way with the experimental observations. Moreover, the same mechanism of SSB plays the fundamental role of giving mass to the fermions fields. They are massless in the starting Lagrangian, owing to their group assignment and to the gauge symmetry, and their masses are generated through their couplings with the Higgs fields.

9.7.1 Toward the unification of weak and electromagnetic interactions

Before examining the structure of the gauge group, the gauge vector bosons and their couplings to leptons and quarks, we need a few preliminary considerations.

The fundamental fermions are the quarks and the leptons. There are six quarks (u, d, s, c, b, t), which are listed in Table 8.11. One assigns to each of them a different flavor, as well as one of the three different colors. All quarks participate in strong, weak and electromagnetic interactions.

The leptons do not carry color, since they do not participate in strong interactions, but only in weak and electromagnetic interactions. There are three charged leptons (e^-, μ^-, τ^-) and three neutral ones, the neutrinos $(\nu_e, \nu_\mu, \nu_\tau)$, which, being neutral, interact only weakly. Leptons can be arranged in three pairs, each containing a charged lepton and a neutrino, identified by a specific lepton number (L_e, L_μ, L_τ). We list the six leptons in Table 9.1[22].

---

[22] In the Table there is no indication about neutrino masses. Let us remind that, in the Standard Model, neutrinos are assumed to be massless, this choice being mainly dictated by the appearance in the experiments of only left-handed neutrinos (and right-handed anti-neutrinos). However, the observed phenomenon of

Table 9.1. The six leptons

Name	Symbol	Q	Mass (MeV)
e-neutrino	ν_e	0	
electron	e	-1	0.511
μ-neutrino	ν_μ	0	
muon	μ	-1	105.66
τ-neutrino	ν_τ	0	
tau	τ	-1	1776.84

As it is well known, the first phenomenological model of weak interactions was formulated by Fermi[23] in order to explain the nuclear β-decay; it consists of a four-fermion contact interaction. Later on, the current-current interaction Lagrangian

$$\mathcal{L} = \tfrac{4}{\sqrt{2}} G_F j^\mu(x) j^\dagger_\mu(x) \tag{9.105}$$

was proposed[24]. In the above equation, G_F is the Fermi coupling constant[25] and the current $j^\mu(x)$ is of $V - A$ type, i.e. it is the difference between a *vector* and an *axial-vector* term[26].

The Lagrangian (9.105) describes *the charged current* interactions as the product of a charge-raising with a charge-lowering current and it applies to all hadronic and leptonic processes: it exhibits the *universality* of the weak interactions. In order to clarify its implications, let us consider the specific case of the lepton current, limiting ourselves here to the (ν_e, e^-) pair (the explicit expressions for the lepton and quark currents will be examined later, in the Subsection 9.7.3).

The electron currents can be written in the form

---

neutrino oscillations indicates indirectly that neutrinos are massive particles. This feature will be briefly discussed in Subsection 9.7.3.

[23] E. Fermi, Z. Phys. 88, 161 (1934).

[24] R.P. Feynman and M. Gell-Mann, Phys. Rev. 109, 193 (1958); E.C.G. Sudarshan and R.E. Marshak, Phys.Rev. 109, 1860 (1958).

[25] The coupling constant G_F has the dimension of a square mass and its value is given approximately by $G_F m_p^2 \simeq 10^5$, where m_p is the proton mass.

[26] The numerical factor $\frac{4}{\sqrt{2}}$ contains the factor $\frac{1}{\sqrt{2}}$ of historical origin, introduced as a normalization factor when the axial term γ_5 was added. The factor 4 is generally introduced in such a way as to express the charged currents (9.106) in terms of the helicity projection operator $\frac{1}{2}(1 - \gamma_5)$.

$$j_\mu(x) = \bar{\nu}_e(x)\gamma_\mu \tfrac{1}{2}(1 - \gamma_5)e(x) = \bar{\nu}_{e_L}(x)\gamma_\mu e_L(x) \,,$$

$$j_\mu^\dagger(x) = \bar{e}(x)\gamma_\mu \tfrac{1}{2}(1 - \gamma_5)\nu_e(x) = \bar{e}_L(x)\gamma_\mu \nu_{e_L}(x) \;:. \tag{9.106}$$

They correspond to a change of the electric charge $\Delta Q = \pm 1$ and contain only the left-handed fields, $e_L = \tfrac{1}{2}(1 - \gamma_5)e$ and $\nu_{e_L} = \tfrac{1}{2}(1 - \gamma_5)\nu_e$.

By introducing E_L as a doublet of the two left-handed fields,

$$E_L = \begin{pmatrix} \nu_e \\ e^- \end{pmatrix}_L , \tag{9.107}$$

one can rewrite the currents (9.106) in the form

$$j_\mu^{(\pm)}(x) = \overline{E}_L(x)\gamma_\mu I_\pm E_L(x) \,, \tag{9.108}$$

making use of the isospin shift operators $I_\pm = I_1 \pm i I_2$, where I_i $(i = 1, 2, 3)$ stand for the isospin generators. In the present case, they are expressed in the self-representation and are identified by $I_i = \tfrac{1}{2}\sigma_i$ in terms of the usual Pauli matrices; they are referred to as the components of a *weak isospin*. In this way, we have implicitly introduced an $SU(2)$ group for the left-handed leptons, which we denote by $SU(2)_L$.

The assumed $SU(2)$ structure, however, requires in principle also the presence of a neutral component, related to the third generator I_3. Indeed, this appeared to be the case, since it turned out that the charged currents are not sufficient to describe all the weak interaction reactions, since it was discovered that also weak *neutral currents* $(\Delta Q = 0)$ exist. Then it seems natural to add a neutral partner to the charged ones:

$$j_\mu^{(3)}(x) = \overline{E}_L(x)\gamma_\mu I_3 E_L(x) \,, \tag{9.109}$$

by completing in this way a *weak isospin* triplet of weak currents:

$$j_\mu^{(i)}(x) = \overline{E}_L(x)\gamma_\mu I_i E_L(x) \,. \tag{9.110}$$

The phenomenological analysis, however, revealed that the structure of the physical neutral current is not so simple, since it must contain also the right-handed fields, specifically the right-handed component of the electron field $e_R = \tfrac{1}{2}(1 + \gamma_5)e$ (whereas from the experimental evidence, we are induced to assume that neutrinos are left-handed only).

Since we have introduced three currents, there must be three vector bosons coupled to them, and the minimal gauge theory should be based on the group $SU(2)_L$. However, the presence of right-handed fields in the neutral current indicates that, in order to describe also the couplings of the right-handed fields, one has to enlarge the gauge symmetry beyond $SU(2)_L$. At this point it is useful to introduce also the electromagnetic current (9.29), which contains both left- and right-handed fields. In the case of the electron, it can be written as follows in terms of the electric charge operator Q:

$$j_\mu^{em} = \bar{e}(x)\gamma^\mu Q e(x) = \bar{e}_L(x)\gamma_\mu Q e_L(x) + \bar{e}_R(x)\gamma_\mu Q e_R(x) \ . \tag{9.111}$$

with the convention that the eigenvalues of Q are expressed in units of the elementary charge e taken with positive sign (so that $Q = -1$ in the case of the electron).[27]

If now we express the electric charge Q by the same relation (8.45) used in the formalism of the isotopic spin (in other words, we impose the same relation between Q and *weak* isospin used for the description of the *strong* isospin),

$$Q = I_3 + \tfrac{1}{2}Y \ , \tag{9.112}$$

we complete the list of the operators that describe both the electromagnetic and the weak interactions with the introduction of a *weak hypercharge* Y, whose eigenvalues are chosen in such a way to get the right values for Q. Going from the operators to the corresponding currents, Eq. (9.111) can be written in the form

$$j_\mu^{em} = j_\mu^{(3)} + \tfrac{1}{2}j_\mu^Y \ , \tag{9.113}$$

with the introduction of the *hypercharge weak current*, which, for the (ν_e, e^-) pair, can be written as

$$j_\mu^Y = \overline{E}_L(x)\gamma_\mu Y E_L(x) + \bar{e}_R(x)\gamma_\mu Y e_R(x) \ , \tag{9.114}$$

with the assignment $Y = -1$ for the left-handed isospin doublet E_L and $Y = -2$ and for the right-handed isospin singlet e_R.

The electromagnetic current appears in this way a combination of the neutral component of the weak isospin current and of the hypercharge current. This feature suggests that also the physical weak neutral current will appear as a linear combination of the same two currents, $j_\mu^{(3)}$ and j_μ^Y. We will see that this is indeed the case in the next Subsections.

In conclusion, the above considerations lead to base the gauge theory on the group

$$\mathcal{G} = SU(2)_L \otimes U(1)_Y \ , \tag{9.115}$$

the gauge group $U(1)_{em}$, which describes the electromagnetic interactions, being properly included in $\mathcal{G}$. For the sake of completeness, we report here the commutation rules of the generators of $\mathcal{G}$,

$$[I_i, I_j] = i\epsilon_{ijk}I_k \ ,$$
$$[I_i, Y] = 0 \ , \tag{9.116}$$

---

[27] This means that in order to write the electromagnetic current with its numerical value, we have to multiply j_μ^{em} by the coupling constant e, identified with the absolute numerical value of the electric charge. In a similar way, the charged weak currents need to be multiplied by suitable weak couplings g and g', as we will see later.

which are obviously the same of the analogous generators used in the strong interaction case.

We are now in the position of specifying the ingredient of the gauge theory. Since the group $\mathcal{G}$ of Eq. (9.115) has four generators, two of which appear both in the weak and in the electromagnetic currents, one can argue that the gauge theory will include both interactions. Weak and electromagnetic interactions are unified within the so-called *electroweak theory*.

9.7.2 Properties of the gauge bosons

We denote by $A_\mu^i (i = 1, 2, 3)$ and B_μ the gauge fields corresponding, respectively, to the generators I_i and Y. The short range nature of the interactions requires the vector bosons to be *massive* and this fact, as already mentioned in Section 9.5, prevents the renormalization of the field theory. The solution of the problem is based on the concept of *spontaneous symmetry breaking* (SSB) and on the Higgs mechanism. The successful gauge field theory was formulated independently by Glashow, Salam and Weinberg[28]; it is called the *Standard Model* of electroweak interactions. n fact, a gauge theory, based on the Higgs mechanism, which is consistent with all phenomenological requirements, can be built only by including both weak and electromagnetic currents. Few years later, it was proved that the theory is indeed renormalizable[29].

Out of the four gauge fields A_μ^i, B_μ, three of them (or rather linear combinations of them) acquire mass and can be identified with the physical bosons $W^\pm$ and Z^0. The remaining fourth combination, representing the photon, must remain massless, so that the SSB must keep unbroken a $U(1)$ subgroup of $\mathcal{G}$, which turns out to be the subgroup $U(1)_Q$.

In order to implement the Higgs mechanism, it is necessary to introduce a set of scalar fields. The minimum choice consists in a doublet of complex fields

$$\phi = \begin{pmatrix} \phi^+ \\ \phi^0 \end{pmatrix} , \tag{9.117}$$

with $I = \frac{1}{2}$ and $Y = 1$. This choice not only breaks both $SU(2)_L$ and $U(1)_Y$, but allows to couple in a simple way Higgs and fermion fields through Yukawa-type couplings, as we will see in the next Subsection.

The terms of the Lagrangian which contain the gauge vectors and the scalar fields are the following

$$\mathcal{L}_{\text{gauge}} = (D_\mu \phi)^\dagger D^\mu \phi - V(\phi^\dagger \phi) - \tfrac{1}{4} F_{\mu\nu}^i F^{i,\mu\nu} - \tfrac{1}{4} B_{\mu\nu} B^{\mu\nu} , \tag{9.118}$$

where

---

[28] S.L. Glashow, Nucl. Phys. 22, 579 (1961); S. Weinberg, Phys. Rev. Lett. 19, 1269 (1967); A. Salam, Proceedings of the VIII Nobel Symposium, ed. N. Svartholm; Almquist and Wiksells (1968), p.367.

[29] G. 't Hooft, Nucl. Phys. B3, 167 (1971); G. 't Hooft and M. Veltman, Nucl. Phys. B44, 189 (1972).

$$V(\phi^\dagger \phi) = \mu^2 \phi^\dagger \phi + \lambda (\phi^\dagger \phi)^2 , \qquad (9.119)$$

$$F^i_{\mu\nu} = \partial_\mu A^i_\nu - \partial_\nu A^i_\mu + g\epsilon_{ijk} A^j_\mu A^k_\nu , \qquad (9.120)$$

$$B_{\mu\nu} = \partial_\mu B_\nu - \partial_\nu B_\mu \qquad (9.121)$$

and

$$D_\mu = \partial_\mu - igI_i A^i_\mu - ig'\tfrac{1}{2}YB_\mu . \qquad (9.122)$$

The constants g and g' specify the couplings of the gauge fields of the groups $SU(2)_L$ and $U(1)_Y$, respectively, to all the other fields of the theory.

Generalizing the procedure of the Higgs mechanism described in Subsection 9.5.3, we summarize here the main steps. The minimum of the potential (9.119), for $\mu^2 < 0$, is given by

$$(\phi^\dagger \phi)_{\min} = -\frac{\mu^2}{2\lambda} \equiv \tfrac{1}{2}v^2 . \qquad (9.123)$$

However, since the electric charge Q is conserved, only the neutral component ϕ^0 can have non-vanishing v.e.v. It is convenient to rewrite Eq. (9.117) in the form

$$\phi = \frac{1}{\sqrt{2}} \begin{pmatrix} \phi_1^+ + i\phi_2^+ \\ \phi_1^0 + i\phi_2^0 \end{pmatrix} \qquad (9.124)$$

and make the following choice:

$$\phi_{\min} = \frac{1}{\sqrt{2}} \begin{pmatrix} 0 \\ v \end{pmatrix} , \qquad (9.125)$$

while the other components of ϕ have vanishing v.e.v.'s. The Higgs mechanism assures that the three Goldstone bosons (corresponding to the fields ϕ_1^+, ϕ_2^+ and ϕ_2^0) are absorbed by the three gauge fields that become massive. With a convenient choice of the gauge, that is the so-called *unitary* gauge[30], one can get rid of the unphysical fields, and only one physical field $h(x)$ is left:

$$\phi(x) \to \phi'(x) = \frac{1}{\sqrt{2}} \begin{pmatrix} 0 \\ v + h(x) \end{pmatrix} . \qquad (9.126)$$

Finally, the Lagrangian (9.118) reduces to:

$$\mathcal{L}_{\text{gauge}} = \tfrac{1}{2}\partial_\mu h \partial^\mu h - \tfrac{1}{2}m^2 h^2 - \tfrac{1}{4}F^i_{\mu\nu}F^{i,\mu\nu} - \tfrac{1}{4}B_{\mu\nu}B^{\mu\nu} +$$

$$+ \tfrac{1}{8}v^2 g^2 \left(A^{(1)}_\mu A^{(1),\mu} + A^{(2)}_\mu A^{(2),\mu} \right) + \qquad (9.127)$$

$$+ \tfrac{1}{8}v^2 \left(g^2 A^{(3)}_\mu A^{(3),\mu} + g'^2 B_\mu B^\mu - 2gg' A^{(3)}_\mu B^\mu \right) + \text{interaction terms} .$$

---

[30] M.E. Peskin and D.V. Schroeder, quoted ref.

We see that mass terms have been originated for the gauge fields. Let us consider separately charged and neutral fields. For the first ones, it is convenient to define new fields which carry definite charge, i.e.

$$W_\mu^{(\pm)} = \tfrac{1}{\sqrt{2}}\left(A_\mu^{(1)} \mp iA_\mu^{(2)}\right) . \tag{9.128}$$

Taking into account the squared mass term expected for a charged vector field $M_W^2 W_\mu^{(+)} W^{(-)\mu}$, we find the mass

$$M_W = \tfrac{1}{2}vg . \tag{9.129}$$

The last explicit term in Eq. (9.127) refers to the two neutral vector bosons. One sees that the fields $A_\mu^{(3)}$ and B_μ get mixed, so that, in order to identify the physical fields, one needs to diagonalize the squared mass matrix

$$\tfrac{1}{4}v^2 \begin{pmatrix} g^2 & -gg' \\ -gg' & g'^2 \end{pmatrix} . \tag{9.130}$$

One eigenvalue is equal to zero, while the other, by comparing it with the squared mass term expected for a neutral vector field $M_Z^2 Z_\mu Z^\mu$, gives

$$M_Z = \tfrac{1}{2}v\sqrt{g^2 + g'^2} , \tag{9.131}$$

which represents the mass of the neutral vector boson. The eigenvectors corresponding to the zero and M_Z eigenvalues can be written in the form

$$\begin{aligned} A_\mu &= \cos\theta_w B_\mu + \sin\theta_w A_\mu^{(3)} , \\ Z_\mu &= -\sin\theta_w B_\mu + \cos\theta_w A_\mu^{(3)} . \end{aligned} \tag{9.132}$$

where we have introduced the *weak (Weinberg) mixing angle*, which can be taken as a free parameter of the theory and is related to the couplings by the relation

$$\cos\theta_w = \frac{g}{\sqrt{g^2 + g'^2}} . \tag{9.133}$$

We see that the scheme we have followed predicts $M_W < M_Z$, since, by comparing Eqs. (9.129) and (9.131), one finds

$$\frac{M_W}{M_Z} = \cos\theta_w . \tag{9.134}$$

However, the ratio M_W/M_Z depends on the specific choice made for the Higgs fields. Eq. (9.134) corresponds to the case of the *minimal* choice, i.e. the doublet of Eq. (9.124). In general, one gets

$$\frac{M_W^2}{M_Z^2} = \rho\cos^2\theta_w , \tag{9.135}$$

where ρ is a phenomenological parameter, depending on the representations of the Higgs fields one adopts (compare with Problem 9.5). At present the experimental data prefer $\rho \simeq 1$, which is an indication in favor of the Standard Model in its *minimal* form (only one or more Higgs doublets). But we cannot exclude contributions from Higgs multiplets other than doublets with small relative weight due to higher masses.

In conclusion, in the minimal version of the Standard Model the sector of the Lagrangian (9.118), which refers to the masses and couplings of the gauge fields, contains three free parameters: g, g' and v or, equivalently, g, v and θ_w. These parameters can be determined from experimental quantities, as it will be indicated in the next Subsection. Another important prediction, not yet confirmed, is the existence of the scalar boson h, which is called *Higgs boson*. Its squared mass is given by

$$m_h^2 = 2v^2\lambda \,, \tag{9.136}$$

but its value cannot be determined since the parameter λ is completely unknown. Experimentally, there is the lower limit $m_h > 114$ GeV, found in the LEP collider experiments at CERN[31], and the physicist community expects that the Higgs boson will be discovered soon at LHC.

9.7.3 The fermion sector of the Standard Model

The inclusion of quarks and leptons in the Standard Model requires their specific assignment to the IR's of the group $SU(2)_L \otimes U(1)_Y$. For both leptons and quarks, the left-handed fields are assigned to doublets of $SU(2)_L$, and the right-handed to singlets. In Tables 9.2 and 9.3 we list the quantum numbers (weak I_3, weak Y and Q) of a lepton pair and of a quark pair. There are three *generations* of leptons and quarks, and the classification is the same for all of them.

The inclusion of all these fermions requires the addition of several terms in the electroweak Lagrangian. In the following, we give the expressions for a generic fermion field $\psi(x)$.

First of all, one needs a kinetic term which includes the coupling of the fermion field with the gauge fields

$$\mathcal{L}_f = \overline{\psi}_L i\gamma^\mu D_\mu^L \psi_L + \overline{\psi}_R i\gamma^\mu D_\mu^R \psi_R \,, \tag{9.137}$$

where ψ_L and ψ_R stand for an $SU(2)$ doublet and a singlet, respectively, and the covariant derivatives are given by

---

[31] Direct searches set the lower limit $m_h > 114.4$ GeV at 95% C.L.; combining this limit with the precision electroweak measurements, the upper bound $m_h < 186$ GeV at 95% C.L. was obtained (The LEP Working Group for Higgs Boson Searches, Phys. Lett. B565, 61 (2003)). Searches for Higgs bosons are under way at the Tevatron at Fermilab (see e.g. T. Aaltonen et al., CDF Collaboration, Phys. Rev. Lett. 104, 061802 (2010)).

Table 9.2. Lepton classification

Symbol	I	I_3	Q	Y
ν_e	$\frac{1}{2}$	$\frac{1}{2}$	0	-1
e_L^-	$\frac{1}{2}$	$-\frac{1}{2}$	-1	-1
e_R^-	0	0	-1	-2

Table 9.3. Quark classification

Symbol	I	I_3	Q	Y
u_L	$\frac{1}{2}$	$\frac{1}{2}$	$\frac{2}{3}$	$\frac{1}{3}$
d_L	$\frac{1}{2}$	$-\frac{1}{2}$	$-\frac{1}{3}$	$\frac{1}{3}$
u_R	0	0	$\frac{2}{3}$	$\frac{4}{3}$
d_R	0	0	$-\frac{1}{3}$	$-\frac{2}{3}$

$$D_\mu^L = \partial_\mu - igI_i A_\mu^i(x) - ig'\tfrac{1}{2}Y B_\mu(x) \tag{9.138}$$

and

$$D_\mu^R = \partial_\mu - ig'\tfrac{1}{2}Y B_\mu(x) \,. \tag{9.139}$$

Making use of the expressions (9.128) and (9.132), one can obtain, from the above equations, the couplings of the fermion fields with the physical gauge vector fields $W_\mu^{(\pm)}$, Z_μ^0 and A_μ, which can be written in the following form

$$\mathcal{L}_{\text{coupl}} = \frac{g}{\sqrt{2}}\left(j_w^{(+)\mu}W_\mu^{(+)} + j_w^{(-)\mu}W_\mu^{(-)}\right) + \frac{g}{\cos\theta_w}j_{NC}^\mu Z_\mu^0 + ej_{em}^\mu A_\mu \,, \tag{9.140}$$

where we have identified the electric charge in terms of the couplings g and g', or equivalently in terms of one of them and the mixing angle θ_w:

$$e = \frac{gg'}{\sqrt{g^2 + g'^2}} = g\sin\theta_W = g'\cos\theta_W \,. \tag{9.141}$$

The Lagrangian (9.140) contains the weak charged current

$$j_w^{(\pm)\mu} = \overline{\psi}_L \gamma^\mu I_\pm \psi_L \tag{9.142}$$

and the electromagnetic current j_{em}^μ defined in Eq. (9.111). The weak neutral current j_μ^{NC} is coupled to the massive neutral vector boson Z_μ^0 with the coupling

$$g_N = \frac{g}{\cos\theta_w} = \frac{e}{\sin\theta_w \cos\theta_w} \,; \tag{9.143}$$

it is called *physical neutral current* and it is given by

$$j_\mu^{NC} = j_\mu^{(3)} - \sin^2\theta_w j_\mu^{em} . \tag{9.144}$$

The above relation solves the problem of finding the structure of the weak neutral current and allows to identify the couplings of a generic fermion field $\psi(x)$ to the vector boson Z_μ^0 in terms of its third isospin component I_3 and its charge Q, according to:

$$j_\mu^{NC} = \overline{\psi}_L \gamma_\mu I_3 \psi_L - \sin^2\theta_w \,\overline{\psi}\gamma_\mu Q\psi . \tag{9.145}$$

By comparing with the expected $V - A$ structure expressed in terms of two generic couplings c_V and c_A

$$j_\mu^{NC} = \overline{\psi}\gamma_\mu \tfrac{1}{2}(c_V - c_A\gamma_5)\psi , \tag{9.146}$$

one obtains the couplings c_V, c_A of each fermion f in terms of the eigenvalues of I_3 and Q according to:

$$c_V = I_3 - 2\sin^2\theta_w Q , \tag{9.147}$$

$$c_A = I_3 . \tag{9.148}$$

The extraordinary agreement between the experimental estimate of the fermion couplings in the different neutral current interactions and the prediction of the Standard Model decreed the success of the theory.

It is useful to give the explicit expressions of the weak currents for the first generation of leptons and quarks (similar expression can be written for the fermions of the other two generations):

$$j_w^{(+)\mu} = \overline{\nu}_e\gamma^\mu \tfrac{1}{2}(1 - \gamma_5)e + \overline{u}\gamma^\mu \tfrac{1}{2}(1 - \gamma_5)d , \tag{9.149}$$

$$j_w^{(-)\mu} = \overline{e}\gamma^\mu \tfrac{1}{2}(1 - \gamma_5)\nu_e + \overline{d}\gamma^\mu \tfrac{1}{2}(1 - \gamma_5)u , \tag{9.150}$$

$$
\begin{aligned}
j_Z^\mu = {} & \overline{\nu}_e\gamma^\mu \tfrac{1}{2}(1 - \gamma_5)\nu_e - \overline{e}\gamma^\mu \tfrac{1}{2}(1 - \gamma_5)e + \\
& + \overline{u}\gamma^\mu \tfrac{1}{2}(1 - \gamma_5)u - \overline{d}\gamma^\mu \tfrac{1}{2}(1 - \gamma_5)d + \\
& - 2\sin^2\theta_w\left(\tfrac{2}{3}\overline{u}\gamma^\mu u - \tfrac{1}{3}\overline{d}\gamma^\mu d - \overline{e}\gamma^\mu e\right) .
\end{aligned}
\tag{9.151}
$$

In the Lagrangian (9.140) the coupling of the $W^\pm$ bosons with the charged currents is simply given by $\frac{1}{\sqrt{2}}g$. If now we introduce explicitly the W propagator, we can compare the low energy limit of the charged current interaction described by the Lagrangian (9.140) with the current-current interaction described by the Lagrangian (9.105): in this way the Fermi coupling constant G_F can be related to the constant g and to the W mass M_W by the relation

$$G_F = \frac{\sqrt{2}g^2}{8M_W^2} = \frac{1}{\sqrt{2}\,v^2} . \tag{9.152}$$

From the experimental values of e, G_F and $\sin^2 \theta_w$ (the last quantity obtained from the analysis of the weak neutral current at low energy), one can determine the three parameters of the theory, and finally the values of the masses (9.129) and (9.131). With $v \simeq 250$ GeV and $\sin^2 \theta_W \simeq 0.23$, one finds $M_W \simeq 80$ GeV and $M_Z \simeq 91$ GeV. The discovery[32] of the two bosons $W^\pm$ and Z_0 at CERN in 1983 confirmed these predictions, and this completed the success of the Standard Model.

Finally, in order to complete the Lagrangian of the Standard Model, one has to add the terms which couple the fermion fields with the scalar doublet, which are usually referred to as *Yukawa terms*. We show here only the structure of a typical term; limiting to the first generation (q_L stands for the u_L-d_L doublet), one has

$$\mathcal{L}_{\text{Yuk}} = f_q(\bar{q}_L \phi d_R) + f_q'(\bar{q}_L \tilde{\phi} u_R) \,, \tag{9.153}$$

where

$$\tilde{\phi} = i\tau_2 \phi^* = \begin{pmatrix} \phi^0 \\ -\phi^- \end{pmatrix} \,, \tag{9.154}$$

and f_q, f_q' are two dimensionless coupling constants. We remark that no mass term is present in the Lagrangian, owing to the gauge invariance and to the different assignments of left-handed and right-handed fermions. The fermion masses are generated by the v.e.v. of the scalar field and they are given by an expression of the type

$$m_q = \tfrac{1}{2} f_q v \quad \text{or} \quad m_q' = \tfrac{1}{2} f_q' v \,. \tag{9.155}$$

The complete expression of the Yukawa Lagrangian density is given by

$$\mathcal{L}_{\text{Yuk}} = \sum_{i,j} \left[f_{ij}(\bar{q}_L^i \phi \, d_R^j) + f_{ij}'(\bar{q}_L^i \tilde{\phi} \, u_R^j) \right] \,, \tag{9.156}$$

where each of the indices (i, j) refers to one of the three quark generations. We note that there is a different coupling constant for each pair of quarks, and these constants are free parameters, not determined by the theory. Therefore one gets two non-diagonal mass matrices for the *up*- and *d*-type quarks, which we denote by $M^{(u)}$ and $M^{(d)}$. These matrices can be diagonalized by means of two unitary matrices $U_{L,R}$ and $D_{L,R}$, as follows

$$M_{\text{diag}}^{(u)} = U_L^\dagger M^{(u)} U_R \quad , \quad M_{\text{diag}}^{(d)} = D_L^\dagger M^{(d)} D_R \,. \tag{9.157}$$

and Eq. (9.156) can be expressed in terms of the mass eigenstates with the substitutions

$$u_{L,R}^i \to U_{L,R}^{ij} u_{L,R}^j \,, \tag{9.158}$$

$$d_{L,R}^i \to D_{L,R}^{ij} d_{L,R}^j \,. \tag{9.159}$$

[32] G. Arnison et al., Phys. Lett. B122, 103 (1983); Phys. Lett. B126, 398 (1983); M. Banner et al., Phys. Lett. B122, 476 (1983); P. Bagnaia et al., Phys. Lett. B129, 130 (1983).

Let us now introduce these substitutions in the quark terms of the weak charged current (9.142), obtaining

$$j_{\text{quark}}^{(+)\mu} = \overline{u}_L^i \gamma^\mu (U_L^\dagger D_L)^{ij} d_L^j \ . \tag{9.160}$$

The unitary matrix $V = U_L^\dagger D_L$ is the well-known CKM (Cabibbo, Kobayashi, Maskawa) mixing matrix[33], which contains three angles and a CP-violating phase. The effect of this matrix is to mix the flavors in the weak charged current that exhibits explicitly flavor-changing terms. On the other hand, one can easily check, due to the unitarity of the U and D matrices, that the quark terms of the weak neutral current are flavor-conserving.

It is useful to consider the simple case of two generations $((u,d)$ and $(c,s))$. In this case the matrix V contains only one free parameter, the so-called *Cabibbo angle* θ_c $(\sin \theta_c \simeq 0.22)$ and it is usually written as

$$V = \begin{pmatrix} \cos \theta_c & \sin \theta_c \\ -\sin \theta_c & \cos \theta_c \end{pmatrix} \ . \tag{9.161}$$

Then the explicit expression for the weak charged current follows from Eq. (9.160):

$$j_{\text{quark}}^{(+)\mu} = \overline{u}_L \gamma^\mu (\cos \theta_c d_L + \sin \theta_c s_L) + \overline{c}_L \gamma^\mu (-\sin \theta_c d_L + \cos \theta_c s_L) \ . \tag{9.162}$$

In the case of leptons, since the Standard Model contains only left-handed neutrinos, mass terms are generated only for charged leptons; they are of the type

$$\mathcal{L}_{\text{Yuk}} = f_\ell (\overline{\ell}_L \phi e_R) \ , \tag{9.163}$$

with additional coupling constants f_ℓ, while neutrinos remain massless. However, the rather recently observed phenomenon of neutrino oscillations[34], experimentally verified at a high level of accuracy, implies that neutrinos are massive particles. More precisely, neutrinos of given flavor, ν_e, ν_μ, ν_τ behave as linear combinations of three eigenstates of mass, ν_1, ν_2, ν_3, whose different evolution in time, due to the mass differences, give rise to oscillations. Neutrino mass eigenstates and flavor eigenstates are related by a mixing matrix,

---

[33] N. Cabibbo, Phys. Rev. Lett. 10, 531 (1963); M. Kobayashi and K. Maskawa, Progr. Theor. Phys. 49, 652 (1973).

[34] The first evidence of neutrino oscillations has been obtained by the Superkamiokande Collaboration from the observation of the zenith angle dependence of their atmospheric neutrino data: Super-Kamiokande Collaboration, Y. Fukuda et al., Phys. Rev. Lett. 81, 1562 (1998). Neutrino oscillations have been observed also for solar neutrinos, as the result of the interpretation of the solar neutrino data related to the so-called "Solar neutrino problem". Experiments performed with "terrestrial" neutrinos, i.e neutrinos coming from artificial sources, have confirmed the oscillation phenomenon. For a report on these subjects see for example G.L. Fogli et al., *Global analysis of three-flavor neutrino masses and mixings*, Progr. Part. Nucl. Phys. 57, 742 (2006).

in analogy with the quark mixing case. At present, we know with a good precision the two mass differences, but still ignore the absolute neutrino masses. We know only that these values are very small, much smaller than the masses of the charged leptons. In particular, there is an upper limit[35] for the mass of the ν_e of the order of 2 eV.

It is clear that, in order to accommodate massive neutrinos in the theory, the Standard Model has to be extended either including right-handed neutrinos and/or extra scalar Higgs fields. Different mechanisms have been proposed which justify the absence, at present energies, of right-handed neutrinos and give possible explanation of the smallness of the neutrino masses. Moreover, it remains still open the possibility that neutrinos behave as Majorana particles, i.e. that they are coincident with their own antiparticles. We do not enter here in a detailed descriptions of these topics, that are discussed in specialized textbooks and review papers[36].

Before leaving this Subsection we would like to discuss briefly the problem of renormalization, mentioned in Section 9.5. In general, this problem arises in the quantization of a classical field theory. In quantum field theories, one makes use of perturbation expansion that allows to evaluate higher order corrections[37]. In general, these corrections are invalidated by the appearance of infinities which, however, can be hidden in the re-definition of the parameters defined in the Lagrangian (such as masses, coupling constants, etc.). Then the final results are finite and can be compared with the experimental data. This situation occurs in the so-called *renormalizable* theories, such as QED, QCD and the electroweak standard theory.

Other theories are not renormalizable, since one cannot eliminate all the divergent terms, thus preventing the possibility of obtaining finite results. This may happen in gauge theories which contain both vector and axial-vector currents, such as in the case of the electroweak Standard Model. In this kind of

---

[35] Two experiments, the Mainz and Troitsk experiments, have proved that if the neutrino rest mass is non-zero, it is, respectively, less than 2.3 eV (C. Kraus et al., Eur. Phys. J. C 40, 447 (2005)) and 2.5 eV (V.M. Lobashev, Nucl. Phys. A 719, 153c (2003)). Further decrease of this limit is out of the possibilities of those experiments. There is however a project, KATRIN, that, with an analogical spectrometer of considerably larger dimensions, will be presumably able to determine the upper limit of the neutrino mass at the level of 0.3 eV (see for example C. Weinheimer, *Neutrino mass from β decay*, Proc. of the Neutrino Oscillation Workshop (NOW 2006), Conca Specchiulla, Otranto, Italy, 9−15 September 2006. P. Bernardini, G.L. Fogli, E. Lisi eds., Nucl. Phys. Proc. Suppl. 168, 1 (2007)).

[36] R.N. Mohapatra and P.B. Pal, *Massive Neutrinos in Physics and Astrophysics*, World Scientific (2003); M. Fukugita and T. Yanagida, *Physics of Neutrinos and Applications to Astrophysics*, Springer-Verlag (2003); C. Giunti and C.W. Kim, *Fundamentals of Neutrino Physics and Astrophysics*, Oxford University Press (2007).

[37] See e.g. M.E. Peskin, D.V. Schroeder, *An Introduction to Quantum Field Theory*, Addison-Wesley Publishing Company (1995); S. Weinberg, *The Quantum Theory of Fields - Vol. II. Modern Applications*, Cambridge University Press, 1996.

theories, it was shown that some *anomalies* appear, which spoil the gauge symmetry of the Lagrangian thus preventing the application of the renormalization procedure. Specifically, it is the case of the so-called *ABJ* anomaly[38], also called *chiral* anomaly, which depends on the set of fermion fields that are present in the Lagrangian, Therefore the only safe and acceptable theories are those that are anomaly free, in which the anomalies are absent or the anomalous contributions are cancelled among themselves. This requirement imposes severe constraints in the construction of gauge theories of elementary particles.

At one loop level, the Feynman diagrams that can give rise to anomalies are of the type represented in Fig. 9.6, where the triangle fermion loop is coupled at its vertices, through the gauge bosons, either to one axial current and two vector currents, or to three axial currents. It is remarkable that the condition for anomaly cancellation depends only on the group structure of the theory and on the fermion assignment.

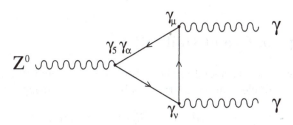

Fig. 9.6. *Example of Feynman graph giving rise to an anomalous contribution. Two vector currents (two photons) are coupled through a triangular fermion loop to an axial current (the axial part of the coupling of a Z^0 vector boson).*

In fact, the condition for absence of anomalies is given by[39]:

$$\text{Tr}\left(\gamma_5\{T_\alpha, T_\beta\}T_\gamma\right) = 0 , \tag{9.164}$$

where T_α, T_β, T_γ are generators of the Lie algebra of the gauge group, which couple the gauge bosons with the fermion fields. The γ_5 matrix in Eq. (9.164) shows that the anomaly is associated with chiral currents: it gives a factor -1 for left-handed fermion and $+1$ for right-handed ones. The trace is taken over the group representations which include all the fermions of the theory.

In the case of the electroweak Standard Model, the generators are the isospin components I_i and the hypercharge Y, and the fermions are assigned to isospin doublets ($I = \frac{1}{2}$) and singlets ($I = 0$). Since, in the case of doublets, one has

[38] S. Adler, Phys. Rev. 177, 2426 (1969); J.S. Bell, R. Jackiw, Nuovo Cimento 60A, 47 (1969).
[39] H. Georgi and S.L. Glashow, Phys. Rev. D6, 429 (1972).

$$I_i = \tfrac{1}{2}\sigma_i \quad , \qquad \text{Tr}(I_i) = 0 \quad , \qquad \{I_i, I_j\} = \tfrac{1}{2}\delta_{ij} \ , \tag{9.165}$$

the only non trivial conditions imposed by Eq. (9.164) are:

$$\text{Tr}(Y) = 0 \quad \text{and} \quad \text{Tr}(Y^3) = 0 \ , \tag{9.166}$$

where the first trace is taken over the fermion isospin doublets only, and the second over both isospin singlets and doublets. Looking at the Tables 9.2 and 9.3 and taking into account that there are three colored quarks for each flavor, one can check that the above conditions are satisfied. The electroweak Standard Model is anomaly free and one realizes that color is a necessary ingredient for satisfying this requirement.

Finally, we want to mention that the inclusion of QCD, with the extension of the electroweak gauge group to $SU(3)_c \otimes SU(2)_L \otimes U(1)_Y$, could generate anomalous triangle graphs with one or two vertices coupled to gluons, but it is easy to show that the conditions (9.166) are sufficient to avoid all anomalies also in this case.

9.8 Groups of Grand Unification

Combining what we discussed in the Subsections (9.4) and (9.7), we see that the field theory of elementary particles is based on the group

$$G_{\text{SM}} = SU(3)_c \otimes SU(2)_L \otimes U(1)_Y \ . \tag{9.167}$$

The group $SU(3)_c$ describes the strong interactions of quarks, which interact with the 8 colored gluons; the group $SU(2)_L \otimes U(1)_Y$ describes the electroweak interactions of quarks and leptons, which generate, through the Higgs mechanism, the masses of quarks and charged leptons.

In spite of the success of both QCD and the electroweak theory, the model presents still some unsatisfactory features:

- The unification is incomplete: there are three independent couplings (g_s, g, g') and also the electroweak unification is not complete, since the ratio g'/g is a free parameter.
- The electric charge is not a generator of the group G_{SM}, and there is no theoretical motivation for the fact that the charge of the proton is exactly equal to the charge of the positron.
- The theory has too many free parameters; e.g. there are no relations among the fermion masses.

On the other hand, all quarks and leptons can be grouped in three generations (or families) which show the same structure. Each generation contains a charged lepton and its neutrino partner; three colored quarks of *up* type and three of *down* type. These considerations lead to the idea of the existence of a higher symmetry, which should be valid at extremely high energies and should unify all the fundamental interactions (except gravity); it would be broken down at lower energies to the symmetry described by the group G_{SM}. On this idea, several Grand Unified Theories (GUT's) were proposed. The simplest one is based on the unitary group $SU(5)$, which contains the group G_{SM} of Eq.(9.167) as a subgroup[40].

In order to classify the fundamental fermions in the IR's of $SU(5)$, it is sufficient to consider only the left-handed fields ψ_L (the right-handed fields ψ_R are assigned to the conjugate representations). Each generation of quarks and leptons contains 15 independent fields, that can be assigned to the representation

$$15 = \bar{5} + 10 . \tag{9.168}$$

In fact, writing explicitly the decompositions of $\bar{5}$ and 10 in terms of the subgroup $SU(3) \otimes SU(2)$,

$$\bar{5} = (1,2) + (\bar{3},1) \tag{9.169}$$

$$10 = (3,2) + (\bar{3},1) + (1,1) , \tag{9.170}$$

it is easy to verify that the two IR's can accommodate the fermions of each generation, by identifying as follows their respective content in terms of left-handed fields:

$$(\nu_e, e^-)_L , \quad (\bar{d}_1, \bar{d}_2, \bar{d}_3)_L , \tag{9.171}$$

$$(u_1, u_2, u_3; d_1, d_2, d_3)_L , \quad (\bar{u}_1, \bar{u}_2, \bar{u}_3)_L , \quad (e^+)_L . \tag{9.172}$$

The advantages of the model based on the unification group $SU(5)$ are summarized in the following.

First, the electric charge is a generator of the group, and it is represented by a traceless matrix. In the case of the representation $\bar{5}$, one gets the relations

$$Q_{e^-} + 3Q_{\bar{d}} = 0 \quad \rightarrow \quad Q_d = \tfrac{1}{3}Q_{e^-} = -\tfrac{1}{3} , \tag{9.173}$$

and, since the up quark is the isospin ($I_3 = +\tfrac{1}{2}$) partner of the $down$ quark,

$$Q_u = Q_d + 1 = \tfrac{2}{3} . \tag{9.174}$$

The fractional electric charges of quarks ($\tfrac{2}{3}$ and $-\tfrac{1}{3}$) result as a consequence of the existence of three colors. The above relations explain the identity of the electric charge of the proton (uud bound system) and the charge of the positron.

Second, the behavior with the energy scale of the three effective couplings (g_s, g, g') indicate that they converge, at very high energy, towards the same region. In order to clarify this point, we consider again Eq. (9.57), which we rewrite here for α_i (i=1,2,3):

---

[40] H. Georgi and S.L. Glashow, Phys. Rev. Lett. **32**, 438 (1974); H. Georgi, H.R. Quinn and S. Weinberg, Phys. Rev. Lett. **32**, 451 (1974).

$$\frac{1}{\alpha_i(Q^2)} = \frac{1}{\alpha_i(\mu^2)} - b_i \, \ln \frac{Q^2}{\mu^2} \, . \tag{9.175}$$

The gauge couplings $\alpha_1, \alpha_2, \alpha_3$ are relative to the subgroups $U(1)$, $SU(2)$ and $SU(3)$, respectively:

$$\alpha_1 = \frac{5}{3} \frac{g'^2}{4\pi} \, , \quad \alpha_2 = \frac{g^2}{4\pi} \, , \quad \alpha_3 = \frac{g_s^2}{4\pi} \, . \tag{9.176}$$

The factor $\frac{5}{3}$ in the definition of α_1 is introduced for normalizing the three couplings in a consistent way for grand unification[41]. We point out that the fine structure constant α_{em} of QED is given by

$$\frac{1}{\alpha_{em}} = \frac{1}{\alpha_2} + \frac{5}{3} \frac{1}{\alpha_1} \, . \tag{9.177}$$

The values of the coefficients b_i relative to the three subgroups are given by[42]:

$$b_i = \begin{pmatrix} b_1 \\ b_2 \\ b_3 \end{pmatrix} = -\frac{1}{4\pi} \begin{pmatrix} 0 \\ \frac{22}{3} \\ 11 \end{pmatrix} + \frac{N_f}{6\pi} \begin{pmatrix} 1 \\ 1 \\ 1 \end{pmatrix} + \frac{N_H}{4\pi} \begin{pmatrix} \frac{1}{10} \\ \frac{1}{6} \\ 0 \end{pmatrix} \, , \tag{9.178}$$

where N_f is the number of quark flavors and N_H the number of scalar (Higgs) doublets ($N_H = 1$ in the minimal SM). The values of the three couplings are known at low energies; by extrapolation from these values to higher and higher energies, one obtains the straight lines represented in the left panel of Fig. 9.7. We see that there are two different points of interception, so that the three couplings, even if they are getting close to each other, do not unify at a single point.

A great improvement of the situation can be obtained by replacing the SM that we have considered with a *supersymmetric* version. *Supersymmetry* was introduced about 30 years ago; it is now a general ingredient of all present particle theories, including *string theories*. It is based on an extension of the Lie algebra of a gauge group, such as G_{SM} and $SU(5)$, to a "graduate" Lie algebra which contains, besides the usual generators, also *spinor* generators, which transform as spinors with respect to the Poincaré group. The analysis of supersymmetry is beyond the scope of this book, and we indicate, instead, a few references in which the subject is described in great detail[43].

Here we limit ourselves to a few qualitative considerations. Supersymmetry implies that all known particles possess supersymmetric partners having opposite statistics. In the supersymmetric (SUSY) version of the SM, all particles

[41] H. Georgi, H.R. Quinn and S. Weinberg, Phys. Rev. Lett. 32, 451 (1974).

[42] M.B. Einhorn and D.R. Jones, Nucl. Phys. B196, 475 (1982).

[43] D. Bailin, A. Love, *Supersymmetric Gauge Field Theory and String Theory*, Institute of Physics Publishing, London (1994); S. Weinberg, *The Quantum Theory of Fields - Vol.III, Supersymmetry*, Cambridge University Press (2000).

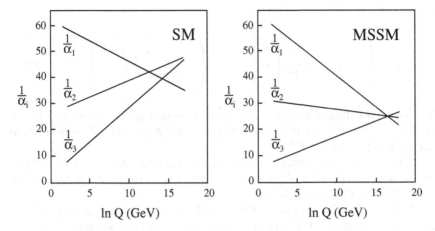

Fig. 9.7. *Behavior of the inverse of the three running coupling constants α_i in terms of $\ln(Q) \approx \ln E$. On the left panel the evolution is estimated within the Standard Model: the unification of the three couplings at a well-defined energy is not reached. The unification is instead obtained in the Minimal Supersymmetric Standard Model, as it is reported on the right panel.*

are doubled, in the sense that each multiplet is replaced by a supermultiplet which contains both fermion and bosons whose spin differs by $\frac{1}{2}$. Specifically, quarks and leptons ($S = \frac{1}{2}$) have scalar ($S = 0$) partners (squarks and sleptons); gauge bosons ($S = 1$) have fermionic partners ($S = \frac{1}{2}$), called gauginos; finally, the Higgs boson ($S = 0$) has a ($S = \frac{1}{2}$) partner, the Higgsino. Supersymmetry would require that a particle and its partner have the same mass; however, since no supersymmetric partner has been discovered so far, supersymmetry has to be broken, and one expects that its breaking occurs at not too high energy (around or slightly above the Fermi energy scale, of the order of a few hundreds GeV).

Even if supersymmetry has not been established by experiments, it is a constant ingredient of the present theories; in particular, the SUSY version of the SM leads to some advantages and theoretical improvements. First, supersymmetry provides a solution to the so-called *problem of hierarchy*, i.e. the problem of understanding how the electroweak scale, not protected by a symmetry, can be stable in front of the quadratic higher order corrections, which would take the scalar particle mass to the extremely high scale energy of grand unification. In supersymmetric theories one gets a cancellation among higher order corrections, and the divergence is reduced to a logarithmic behavior.

A second interesting feature is that the behavior of the running couplings is modified by the presence of the supersymmetric partners and it does indicate a unification of the three couplings. In the MSSM, the minimal supersymmetric version of the SM (which contains the particles listed above and one additional Higgs doublet), the values of the b_i coefficients given in Eq. (9.178) are replaced by:

$$b_i = \begin{pmatrix} b_1 \\ b_2 \\ b_3 \end{pmatrix} = -\frac{1}{4\pi}\begin{pmatrix} 0 \\ 6 \\ 9 \end{pmatrix} + \frac{N_f}{4\pi}\begin{pmatrix} 1 \\ 1 \\ 1 \end{pmatrix} + \frac{N_H}{4\pi}\begin{pmatrix} \frac{3}{10} \\ \frac{1}{2} \\ 0 \end{pmatrix}. \qquad (9.179)$$

The corresponding behavior of the running couplings is represented in the right panel of Fig. 9.7 and it shows the interesting property of unification, as first pointed out in 1991[44].

Whether supersymmetry exists in Nature and possesses the expected properties, will be hopefully discovered by the forthcoming experiments at the Large Hadron Collider (LHC) at CERN.

Among other consequences of the $SU(5)$ model, we quote a general characteristic feature of GUT's: the fact that quarks and leptons are in the same group representation indicates that they can be interchanged among each other, and this leads to the prediction that the proton is no longer stable, but it can decay, fortunately with a very long life-time. The experimental searches of this effect in the big undergound laboratories may confirm, in the future, this kind of prediction.

We remark that the GUT based on $SU(5)$ contains only massless neutrino, as in the Standard Model. To implement non-vanishing neutrino masses, one has to extend the theory, including more fields: either other fermion fields, like right-handed neutrinos, or other Higgs, like scalar iso-triplets.

Other GUT's have been proposed, based on higher groups, such as $SO(10)$ and E_6. In particular, models based on the orthogonal group $SO(10)$ can accommodate quarks and leptons of each generation in the single IR 16, and this representation can accommodate also a right-handed neutrino ν_R. It is worthwhile to mention that the gauge theories based on $SO(10)$ and E_6 are automatically anomaly free.

In this chapter, we have not taken into account the gravitational interactions. This is due to the fact that they play practically no role, in particle physics, in the region of accessible energies. In fact, they become important only at the Planck scale, around 10^{19} GeV. A complete theory of particles would require also the unification of gravity with the other interactions. Unfortunately, one is faced with a severe theoretical difficulty, that of combining, in the frame of quantum field theory, general relativity with quantum mechanics. All the attempts to find a solution in this direction failed; gauge field theories, which include gravitational interactions, were constructed, but they are non-renormalizable.

The only real solution was found going beyond the usual field theory, in which particles are point-like, to the *string theories*, in which the fundamental constituents are one-dimensional vibrating strings. In these theories, a necessary ingredient is supersymmetry, which leads to the formulation of *superstring*

---

[44] U. Amaldi, W. de Boer, H. Fuerstennau, Phys. Lett. B260, 447 (1991).

theory[45]. A necessary tool is provided by group theory; the elimination of all possible anomalies requires that superstrings live in a ten-dimensional space-time, and that the internal symmetry groups are either $SO(32)$ or $E_8 \otimes E_8$. Even if no realistic model, which could be confronted with experimental information, has been produced up to now, superstring theory is widely and deeply investigated, since it has the attractive feature of unifying all the fundamental interactions.

Problems

9.1. Give the explicit proof of the invariance of the QCD Lagrangian (9.46) under the gauge transformations (9.49) and (9.50).

9.2. The Lagrangian density

$$\mathcal{L} = \sum_i \{ \tfrac{1}{2} \partial_\mu \phi_i \partial^\mu \phi_i - \tfrac{1}{2} \mu^2 \phi_i^2 - \tfrac{1}{4} \lambda (\phi_i \phi_i)^2 \} ,$$

where the $\phi_i(x)$'s stand for a set of N real fields, is invariant under the orthogonal group $O(N)$. Considering the case $\mu^2 < 0$ in which the symmetry can be spontaneously broken, determine the symmetry of the ground state and the number of Goldstone bosons.

9.3. Evaluate the leading contribution to the low energy ν_e-electron scattering amplitude due to the $W^\pm$ exchange in the Standard Model, and compare it with the amplitude obtained from the effective Lagrangian (9.105); derive the relation (9.152) between the Fermi coupling constant G_F and the gauge coupling constant g. For the same process, derive also the contribution to the scattering amplitude due to the Z^0 exchange.

9.4. Consider the electroweak Standard Model based on the gauge group $SU(2)_L \otimes U(1)_Y$ and include, besides the usual complex scalar doublet ϕ of Eq. (9.117), a real scalar triplet Ψ with $I = 1$, $Y = 0$ and v.e.v. $< \Psi^0 >_0 = V$.

Determine the mass ratio M_W / M_Z and compare it with Eq. (9.134). Could the scalar triplet alone produce the required breaking $SU(2)_I \otimes U(1)_Y \to U(1)_Q$?

9.5. Consider the relation (9.135), in which the ratio of the two squared vector boson masses depends also on the phenomenological parameter ρ. Assuming that several Higgs fields $h_\ell(x)$ belonging to different representations of the gauge group contribute to the SSB, show that ρ can be expressed as a function of the weak isospin I_ℓ and hypercharge Y_ℓ of the Higgs fields in the form

$$\rho = \frac{\sum_\ell v_\ell^2 [I_\ell(I_\ell + 1) - \tfrac{1}{4} Y_\ell^2]}{\tfrac{1}{2} \sum_\ell v_\ell^2 Y_\ell^2} .$$

---

[45] See e.g. M.B. Green, J.H. Schwarz & E. Witten, *Superstring theory - Volume I, Introduction*, Cambridge University Press (1987).

9.6. The so-called Left-Right Symmetric Model is based on the gauge group $SU(2)_L \otimes SU(2)_R \otimes U(1)_Y$, with generators I_L, I_R and Y, and electric charge defined by $Q = I_{3L} + I_{3R} + \frac{1}{2}Y$. It contains an extra triplet of vector gauge fields $(A^i_\mu)_R = (1,3,0)$ with the same coupling g, in addition to the vector gauge fields of the Standard Model: $(A^i_\mu)_L = (3,1,0)$ and $B_\mu = (1,1,0)$. Left-handed and right-handed fermions are assigned symmetrically to doublets of $SU(2)_L$ and $SU(2)_R$, respectively.

Consider the simple version of the scalar sector containing two complex scalar doublets $\phi_L = (2,1,1)$ and $\phi_R = (1,2,1)$ and suppose that the scalar potential has a minimum for $< \phi^0_L >_0 = v_L$ and $< \phi^0_R >_0 = v_R$. Evaluate the masses of the vector bosons with the hypothesis $v_R \gg v_L$.

9.7. In the $SU(5)$ GUT model, the gauge vector bosons are assigned to the adjoint representation 24. Decompose this representation into the IR's of the subgroup $G_{SM} = SU(3)_c \otimes SU(2)_L \otimes U(1)_Y$ and identify the vector gauge multiplets corresponding to this subgroup. Exhibit the quantum numbers (electric charge Q and hypercharge Y) of the extra multiplets contained in 24.

9.8. In the $SU(5)$ GUT model the SSB from $SU(5)$ to $G_{SM} = SU(3)_c \otimes SU(2)_L \otimes U(1)_Y$ is accomplished by a scalar potential which contains a scalar field Ψ in the adjoint representation 24 and which has a minimum for a v.e.v. $< \Psi >_0 \neq 0$.

What is the form of $< \Psi >_0$ required to produce the desired breaking, if Ψ is written as a 5×5 traceless tensor? What is the spectrum of the vector gauge bosons after the SSB?

In order to implement a second stage of breaking from G_{SM} to $SU(3)_c \otimes U(1)_Q$ it is necessary to introduce another scalar field with v.e.v. different from zero. Which is the minimal choice?

A

Rotation matrices and Clebsch-Gordan coefficients

In this Appendix we collect some formulae that are useful for specific applications of the rotation group discussed in Chapter 2. First we give the explicit expressions of the functions $d^j_{m'm}$ and of the spherical harmonics for the lowest angular momentum values. Then we report a few tables of Clebsch-Gordan coefficients which are used in the addition of angular momenta.

A.1 Reduced rotation matrices and spherical harmonics

The reduced rotation matrix $d^{(j)}_{m'm}(\beta)$ enter the rotation matrix $D^{(j)}(\alpha, \beta, \gamma)$ according to (see Eq. (2.63))

$$D^{(j)}_{m'm}(\alpha, \beta, \gamma) = \langle j, m'|e^{-i\alpha J_z}e^{-i\beta J_y}e^{-i\gamma J_z}|j, m\rangle = e^{-i\alpha m'}d^{(j)}_{m'm}(\beta)e^{-i\gamma m} ,$$
(A.1)

where one defines:

$$d^{(j)}_{m'm}(\beta) = \langle j, m'|e^{-i\beta J_y}|j, m\rangle .$$
(A.2)

The general expression proposed by Wigner for the d-functions[1] is given in Eq. (2.65), that we report again here for the sake of completeness:

$$d^{(j)}_{m'm}(\beta) = \sum_s \frac{(-)^s[(j+m)!(j-m)!(j+m')!(j-m')!]^{1/2}}{s!(j-s-m')!(j+m-s)!(m'+s-m)!} \times$$
$$\times \left(\cos\frac{\beta}{2}\right)^{2j+m-m'-2s} \left(-\sin\frac{\beta}{2}\right)^{m'-m+2s} ,$$
(A.3)

where the sum is over the values of the integer s for which the factorial arguments are equal or greater than zero.

---

[1] See e.g. M.E. Rose, *Elementary Theory of Angular Momentum*, John Wiley and Sons, 1957.

G. Costa and G. Fogli, *Symmetries and Group Theory in Particle Physics*,
Lecture Notes in Physics 823, DOI: 10.1007/978-3-642-15482-9,
© Springer-Verlag Berlin Heidelberg 2012

We give here a few general properties:

$$d^j_{m'm}(0) = \delta_{m',m} , \qquad\qquad (A.4)$$

$$d^j_{m'm}(\pi) = (-)^{j-m}\delta_{m',-m} , \qquad\qquad (A.5)$$

$$d^j_{m'm}(\beta) = d^j_{mm'}(-\beta) , \qquad\qquad (A.6)$$

$$d^j_{m'm}(-\beta) = (-)^{m'-m}d^j_{m'm}(\beta) , \qquad\qquad (A.7)$$

and list the explicit expressions of the matrices for the lowest angular momenta[2] $d^{\frac{1}{2}}$ and d^1:

$$d^{\frac{1}{2}} = \begin{pmatrix} \cos\frac{\beta}{2} & -\sin\frac{\beta}{2} \\[2mm] \sin\frac{\beta}{2} & \cos\frac{\beta}{2} \end{pmatrix} , \qquad\qquad (A.8)$$

$$d^1 = \begin{pmatrix} \frac{1}{2}(1+\cos\beta) & -\frac{1}{\sqrt{2}}\sin\beta & \frac{1}{2}(1-\cos\beta) \\[2mm] \frac{1}{\sqrt{2}}\sin\beta & \cos\beta & -\frac{1}{\sqrt{2}}\sin\beta \\[2mm] \frac{1}{2}(1-\cos\beta) & \frac{1}{\sqrt{2}}\sin\beta & \frac{1}{2}(1+\cos\beta) \end{pmatrix} , \qquad (A.9)$$

together with the most relevant matrix elements of the $d^{\frac{3}{2}}$ (the other matrix elements being easily deduced on the basis of the symmetry properties given above):

$$
\begin{aligned}
d^{3/2}_{3/2,3/2} &= \tfrac{1}{2}(1+\cos\beta)\cos\frac{\beta}{2} , & d^{3/2}_{3/2,1/2} &= -\tfrac{\sqrt{3}}{2}(1+\cos\beta)\sin\frac{\beta}{2} , \\[2mm]
d^{3/2}_{3/2,-1/2} &= \tfrac{\sqrt{3}}{2}(1-\cos\beta)\cos\frac{\beta}{2} , & d^{3/2}_{3/2,-3/2} &= -\tfrac{1}{2}(1-\cos\beta)\sin\frac{\beta}{2} , \\[2mm]
d^{3/2}_{1/2,1/2} &= \tfrac{1}{2}(3\cos\beta-1)\cos\frac{\beta}{2} , & d^{3/2}_{1/2,-1/2} &= -\tfrac{1}{2}(3\cos\beta+1)\sin\frac{\beta}{2} .
\end{aligned}
$$
$$(A.10)$$

Finally we give the expressions of the spherical harmonics Y_ℓ^m limiting ourselves to the case $\ell = 1$[3]:

$$Y_1^0 = \sqrt{\frac{3}{4\pi}}\cos\theta , \quad Y_1^1 = -\sqrt{\frac{3}{8\pi}}\sin\theta e^{i\phi} , \quad Y_1^{-1} = \sqrt{\frac{3}{8\pi}}\sin\theta e^{-i\phi} . \quad (A.11)$$

---

[2] A more complete list can be found in: *The Review of Particle Physics*, C. Amsler et al., Physics Letters B 667, 1 (2008).

[3] For $\ell = 2$ see the reference quoted above.

A.2 Clebsch-Gordan coefficients

We report in the following the values of the Clebsch-Gordan coefficients $C(j_1, j_2, j; m_1, m_2, m)$ defined in Eq. (2.72), where j is the eigenvalue of the total angular momentum $\mathbf{J} = \mathbf{J_1} + \mathbf{J_2}$ and $m = m_1 + m_2$ the corresponding component along the x^3-axis. We limit ourselves only to the cases $j_2 = \frac{1}{2}$, very often useful in the calculations. The coefficients for higher values of j_2 can be found in the quoted references.

In Table A.1 we report the coefficients $C(j_1, j_2, j; m_1, m_2, m)$, for any value of j_1 and $j_2 = \frac{1}{2}$.

Table A.1. Clebsch-Gordan coefficients $C(j_1, \frac{1}{2}, j; m_1, m_2, m)$.

j	$m_2 = \frac{1}{2}$	$m_2 = -\frac{1}{2}$
$j_1 + \frac{1}{2}$	$\sqrt{\dfrac{j_1 + m_1 + 1}{2j_1 + 1}}$	$\sqrt{\dfrac{j_1 - m_1 + 1}{2j_1 + 1}}$
$j_1 - \frac{1}{2}$	$-\sqrt{\dfrac{j_1 - m_1}{2j_1 + 1}}$	$\sqrt{\dfrac{j_1 + m_1}{2j_1 + 1}}$

The specific cases $(j_1 = \frac{1}{2}, j_2 = \frac{1}{2})$ and $(j_1 = 1, j_2 = \frac{1}{2})$ are given in the Tables A.2 and A.3, respectively.

Table A.2. Clebsch-Gordan coefficients $C(\frac{1}{2}, \frac{1}{2}, j; m_1, m_2, m)$.

j	m_2 / m_1	$\frac{1}{2}$	$-\frac{1}{2}$
1	$\frac{1}{2}$	1	$\frac{1}{\sqrt{2}}$
1	$-\frac{1}{2}$	$\frac{1}{\sqrt{2}}$	1
0	$\frac{1}{2}$	0	$\frac{1}{\sqrt{2}}$
0	$-\frac{1}{2}$	$-\frac{1}{\sqrt{2}}$	0

Table A.3. Clebsch-Gordan coefficients $C(1, \frac{1}{2}, j; m_1, m_2, m)$.

j	m_1 \ m_2	$\frac{1}{2}$	$-\frac{1}{2}$
$\frac{3}{2}$	1	1	$\sqrt{\frac{1}{3}}$
	0	$\sqrt{\frac{2}{3}}$	$\sqrt{\frac{2}{3}}$
	-1	$\sqrt{\frac{1}{3}}$	1
$\frac{1}{2}$	1	0	$\sqrt{\frac{2}{3}}$
	0	$-\sqrt{\frac{1}{3}}$	$\sqrt{\frac{1}{3}}$
	-1	$-\sqrt{\frac{2}{3}}$	0

In Table A.4 we report the coefficients $C(j_1, 1, j; m_1, m_2, m)$, for any value of j_1 and $j_2 = 1$.

Table A.4. Clebsch-Gordan coefficients $C(j_1, 1, j; m_1, m_2, m)$.

j	$m_2 = 1$	$m_2 = 0$	$m_2 = -1$
$j_1 + 1$	$\sqrt{\dfrac{(j_1+m_1+1)(j_1+m_1+2)}{(2j_1+1)(2j_1+2)}}$	$\sqrt{\dfrac{(j_1-m_1+1)(j_1+m_1+1)}{(2j_1+1)(j_1+1)}}$	$\sqrt{\dfrac{(j_1-m_1+1)(j_1-m_1+2)}{(2j_1+1)(2j_1+2)}}$
j_1	$-\sqrt{\dfrac{(j_1+m_1+1)(j_1-m_1)}{2j_1(j_1+1)}}$	$\dfrac{m_1}{\sqrt{j_1(j_1+1)}}$	$\sqrt{\dfrac{(j_1-m_1+1)(j_1+m_1)}{2j_1(j_1+1)}}$
$j_1 - 1$	$\sqrt{\dfrac{(j_1-m_1-1)(j_1-m_1)}{2j_1(2j_1+1)}}$	$-\sqrt{\dfrac{(j_1-m_1)(j_1+m_1)}{j_1(2j_1+1)}}$	$\sqrt{\dfrac{(j_1+m_1)(j_1+m_1-1)}{2j_1(2j_1+1)}}$

Finally, the specific case $(j_1 = 1, j_2 = 1)$ is given in Tables A.5.

Table A.5. Clebsch-Gordan coefficients $C(1,1,j;m_1,m_2,m)$.

j	m_1 \ m_2	1	0	-1
2	1	1	$\sqrt{\tfrac{1}{2}}$	$\sqrt{\tfrac{1}{6}}$
	0	$\sqrt{\tfrac{1}{2}}$	$\sqrt{\tfrac{2}{3}}$	$\sqrt{\tfrac{1}{2}}$
	-1	$\sqrt{\tfrac{1}{6}}$	$\sqrt{\tfrac{1}{2}}$	1
1	1	0	$\sqrt{\tfrac{1}{2}}$	$\sqrt{\tfrac{1}{2}}$
	0	$-\sqrt{\tfrac{1}{2}}$	0	$\sqrt{\tfrac{1}{2}}$
	-1	$-\sqrt{\tfrac{1}{2}}$	$-\sqrt{\tfrac{1}{2}}$	0
0	1	0	0	$\sqrt{\tfrac{1}{3}}$
	0	0	$-\sqrt{\tfrac{1}{3}}$	0
	-1	$\sqrt{\tfrac{1}{3}}$	0	0

B

Symmetric group and identical particles

In this Appendix we examine briefly the symmetry properties of identical particles, which have to be taken into account when dealing with more particle states. This symmetry appears only in quantum mechanics, and it is related to the fact that particles of the same kind are to be considered absolutely *indistinguishable* from one another.

B.1 Identical particles

Suppose we have a system of n particles of the same kind, e.g. protons, not interacting among each other. A state of the system could be represented in terms of one-particle states:

$$|a_1 b_2 c_3 \ldots z_n> = |a_1> |b_2> |c_3> \ldots |z_n>, \qquad (B.1)$$

where a_1, b_2, ... correspond to the dynamical variables of the first, second, ... particle, respectively. The fact that the n particles are identical has as a consequence that a transition from the above state to the state e.g.

$$|a_2 b_1 c_3 \ldots z_n> = |a_2> |b_1> |c_3> \ldots |z_n>, \qquad (B.2)$$

obtained by interchanging particles 1 and 2, could not be observable by any means. It is then necessary to describe the situation represented by Eqs. (B.1) and (B.2) in terms of the *same state*.

More generally, we want to have a situation in which *any permutation* (built by repeated interchanges of two particles) among the n identical particles leads essentially to the same state of the system. The process of permuting the particles will be represented by a linear operator P in the Hilbert space.

Since the n particles are indistinguishable, the Hamiltonian H of the system will be a *symmetrical function* of the dynamical variables of the n particles, and then it will commute with the permutation operator

G. Costa and G. Fogli, *Symmetries and Group Theory in Particle Physics*,
Lecture Notes in Physics 823, DOI: 10.1007/978-3-642-15482-9,
© Springer-Verlag Berlin Heidelberg 2012

$$[P, H] = 0 .\qquad\qquad(B.3)$$

This means that a state which has initially some symmetry property (e.g. it is totally symmetric under P) will always conserve this symmetry.

B.2 Symmetric group and Young tableaux

In order to study in general the possible symmetry properties (under P) of the quantum mechanical states, it is convenient to make use of the *symmetric group*. We summarize briefly here the relevant properties of this group.

As defined in Section 1.1, the symmetric group $\mathcal{S}_n$ is the group of permutations of n objects. Each permutation can be decomposed into a product of transpositions (i.e. permutations in which only two elements are interchanged).

The order of the group is $n!$. We recall a useful theorem of finite groups: the sum of the squares of the dimensions of the IR's equals the order of the group.

We want now to make a correspondence between the IR's of the group $\mathcal{S}_n$ and the so-called *Young tableaux*.

Let us start with the states of two identical particles. The only possible permutation is the transposition P_{12} and one can build two independent states, respectively *symmetrical* and *antisymmetrical*[1]:

$$|\Phi_s> = |12> + |21> = (1 + P_{12})|12> = S_{12}|12> ,$$
$$|\Phi_a> = |12> - |21> = (1 - P_{12})|12> = A_{12}|12> ,\qquad(B.4)$$

where we have introduced the symmetrizing and antisymmetrizing operators

$$S_{12} = (1 + P_{12}) ,$$
$$A_{12} = (1 - P_{12}) .\qquad\qquad(B.5)$$

We can take $|\Phi_s>$ and $|\Phi_a>$ as bases of the two non equivalent (one-dimensional) IR's of $\mathcal{S}_2$, which are $(1, 1)$ and $(1, -1)$. A convenient notation is given in terms of the diagrams

for $(1, 1)$,   for $(1, -1)$,

which are particular Young tableaux, whose general definition will be given later.

Before giving the rules for the general case, let us consider explicitly the case of three identical particles. In principle we have 6 different states

---

[1] In the following the compact notation $|123...> = |a_1 b_2 c_3 ...>$ will be used.

$$|\Phi_1> = S_{123}|123>, \qquad\qquad |\Phi_4> = A_{23}S_{12}|123>,$$
$$|\Phi_2> = A_{123}|123>, \qquad\qquad |\Phi_5> = A_{23}S_{13}|123>, \qquad\text{(B.6)}$$
$$|\Phi_3> = A_{13}S_{12}|123>, \qquad\qquad |\Phi_6> = A_{12}S_{13}|123>,$$

where
$$S_{123} = 1 + P_{12} + P_{13} + P_{23} + P_{13}P_{12} + P_{12}P_{13},$$
$$A_{123} = 1 - P_{12} - P_{13} - P_{23} + P_{13}P_{12} + P_{12}P_{13}. \qquad\text{(B.7)}$$

In fact, there are only 4 independent states: one is totally symmetric, one totally antisymmetric, and two have mixed symmetry. They correspond to the (3 box) Young tableaux in Table B.1. To each Young tableau one can associate an IR of S_3. Moreover, one can also know the dimension of each IR, looking at the "standard arrangement of a tableau" or "standard Young tableau". Let us label each box with the numbers 1, 2, 3 in such a way that the numbers increase in a row from left to right and in a column from top to bottom. We see from the Table that there is only a standard tableau for each of the first two patterns, and two different standard tableaux for the last one. The number of different standard arrangements for a given pattern gives the dimension of the corresponding IR.

Table B.1. Young tableaux and IR's of S_3.

Young tableaux	Symmetry	Standard tableaux	IR dimension	Young operator
⬚⬚⬚	totally symmetric	1 2 3	1	S_{123}
⬚/⬚/⬚	totally antisymmetric	1/2/3	1	A_{123}
⬚⬚/⬚	mixed symmetry	1 2 / 3 1 3 / 2	2	$A_{13}S_{12}, A_{12}S_{13}$

To each standard tableau we can associate one of the symmetry operators (or *Young operators*) used in Eqs. (B.6), (B.7). We note that we have one-dimensional IR's for the totally symmetric and antisymmetric cases, which means that the corresponding states are *singlets*, and a two-dimensional IR in the case of mixed symmetry, which corresponds to a *doublet* of degenerate states[2]. One can check that this result agrees with the rule of finite groups mentioned above: in fact $3! = 1 + 1 + 2^2$.

The above example can be easily extended to the case of n identical particles. In this case the group is $\mathcal{S}_n$ and one has to draw all Young tableaux consisting of n boxes. Each tableau is identified with a given partition $(\lambda_1, \lambda_2, \ldots, \lambda_n)$ of the number n (i.e. $\sum_i \lambda_i = n$), ordering the λ_i in such a way that

$$\lambda_1 \geq \lambda_2 \geq \ldots \geq \lambda_n . \tag{B.8}$$

The corresponding Young tableau

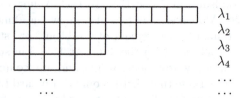

has λ_i boxes in the i-th row. Then for each tableau one finds all standard arrangements. Each Young tableau corresponds to an IR of $\mathcal{S}_n$, and the dimension of the IR is equal to the number of different standard arrangements.

To each IR there corresponds a n-particle state with a different symmetry property, and the dimension of the IR gives the degeneracy of the state.

In order to obtain the state with the required symmetry property from the state (B.1), we introduce a *Young operator* for each standard Young tableau τ:

$$Y_\tau = \left(\sum_{\text{col}} A_\nu \right) \left(\sum_{\text{rows}} S_\lambda \right) , \tag{B.9}$$

where S_λ is the symmetrizer corresponding to λ boxes in a row and A_ν the antisymmetrizer corresponding to ν boxes in a column, the two sums being taken for each tableau over *all* rows and columns, respectively.

The explicit expressions for S_λ and A_ν have already been given by (B.5) and (B.7) in the simplest cases $\lambda \leq 3$, $\nu \leq 3$. In general, they are defined by

$$S_\lambda = \sum_p \begin{pmatrix} 1 & 2 & \ldots & \lambda \\ p_1 & p_2 & \ldots & p_\lambda \end{pmatrix} , \tag{B.10}$$

$$A_\nu = \sum_p \epsilon_p \begin{pmatrix} 1 & 2 & \ldots & \nu \\ p_1 & p_2 & \ldots & p_\nu \end{pmatrix} , \tag{B.11}$$

---

[2] According to Eq. (B.6), one would expect two distinct two-dimensional IR's, corresponding to the mixed symmetry states $|\Phi_3\rangle$, $|\Phi_4\rangle$, $|\Phi_5\rangle$, $|\Phi_6\rangle$. However, the two IR's can be related to one another by a similarity transformations, i.e. they are equivalent.

where $\sum_p$ indicates the sum over all permutations and $\epsilon_p = +$ or $-$ in correspondence to an *even* or *odd* permutation (a permutation is even or odd if it contains an even or odd number of transpositions).

The state with the symmetry of the τ standard Young tableau is obtained by means of

$$|\Phi_\tau\rangle = Y_\tau |a_1 b_2 \ldots z_n\rangle . \tag{B.12}$$

Particular importance have the states which are totally symmetrical or totally antisymmetrical, because they are those occurring in nature. They correspond to the Young tableaux consisting of only one row or one column, so that they are simply given by

$$|\Phi_s\rangle = S_n |a_1 b_2 \ldots z_n\rangle , \tag{B.13}$$

$$|\Phi_a\rangle = A_n |a_1 b_2 \ldots z_n\rangle . \tag{B.14}$$

An assembly of particles which occurs only in symmetrical states (B.13) is described by the *Bose-Einstein statistics*. Such particles occur in nature and are called *bosons*. A comparison between the Bose statistics and the usual classical (Boltzmann) statistics (which considers particles distinguishable) shows e.g. that the probability of two particles being in the same state is greater in Bose statistics than in the classical one.

Let us examine the antisymmetrical state (B.14). It can be written in the form of a determinant

$$\begin{pmatrix} a_1 & a_2 & \ldots & a_n \\ b_1 & b_2 & \ldots & b_n \\ \ldots & \ldots & \ldots & \ldots \\ z_1 & z_2 & \ldots & z_n \end{pmatrix} . \tag{B.15}$$

It then appears that if two of the states $|a\rangle, |b\rangle, \ldots, |z_n\rangle$ are the same, the state (B.14) vanishes. This means that *two particles cannot occupy the same state*: this rule corresponds to the Pauli exclusion principle. In general, the occupied states must be all independent, otherwise (B.15) vanishes. An assembly of particles which occur only in the antisymmetric states (B.14) is described by the *Fermi-Dirac statistics*. Such particles occur in nature: they are called *fermions*.

Other quantum statistics exist which allow states with more complicated symmetries than the complete symmetry or antisymmetry. Until now, however, all experimental evidence indicates that only the Bose and the Fermi statistics occur in nature; moreover, systems with integer spin obey the Bose statistics, and systems with half-integer spin the Fermi statistics. This connection of spin with statistics is shown to hold in quantum field theory (spin-statistics theorem), in which bosons and fermions are described by fields which commute and anticommute, respectively, for space-like separations.

Then, unless new particles obeying new statistics are experimentally detected, all known particles states belong to the one-dimensional representation of the symmetric group. We note, however, that the states of n-particles have to be singlet (totally symmetric or totally antisymmetric) under permutations of *all* variables (space-time coordinates, spin, internal quantum numbers); if one takes into account only a subset of variables, e.g. the internal ones, a state can have mixed symmetry. In this case, one has to make use of the other – in general higher dimensional – IR's of the symmetric group $\mathcal{S}_n$.

C

Young tableaux and irreducible representations of the unitary groups

In this Appendix we illustrate the use of the Young tableaux in the construction of tensors which are irreducible with respect to the group $U(n)$ of unitary transformations.

The main property on which the method is based, and which will not be demonstrated here, is that the irreducible tensors can be put in one-to-one correspondence with the IR's of the symmetric group; these are associated, as pointed out in Appendix B, to the different Young tableaux.

On the other hand, since the irreducible tensors are bases of the IR's of the group $U(n)$, one can go from the Young tableaux to the IR's of the group $U(n)$ itself.

The method can be used, in general, for the group $GL(n, C)$ of linear transformations, but we shall limit here to the groups $U(n)$ and $SU(n)$ which are those considered in Chapter 8[1]. In the following, the main properties will be illustrated by examples which have physical interest.

C.1 Irreducible tensors with respect to $U(n)$

Let us consider a n-dimensional linear vector space; its basic elements are *controvariants vectors* defind by sets of n complex numbers ξ^α ($\alpha = 1, 2, \ldots, n$) and denoted by:

$$\xi \equiv \begin{pmatrix} \xi^1 \\ \xi^2 \\ \vdots \\ \xi^n \end{pmatrix}, \tag{C.1}$$

The group $U(n)$ can be defined in terms of unitary transformations

---

[1] For the extension of the method to $GL(n, C)$ see e.g. M. Hamermesh *"Group theory and its applications to physical problems"*, Addison-Wesley (1954).

G. Costa and G. Fogli, *Symmetries and Group Theory in Particle Physics*,
Lecture Notes in Physics 823, DOI: 10.1007/978-3-642-15482-9,
© Springer-Verlag Berlin Heidelberg 2012

$$\xi \to \xi' = U\xi \qquad \text{i.e.} \qquad \xi'^{\alpha} = U^{\alpha}{}_{\beta}\xi^{\beta} , \tag{C.2}$$

where

$$U^{\dagger}U = I \qquad \text{i.e.} \qquad (U^{\dagger})^{\alpha}{}_{\beta}U^{\beta}{}_{\gamma} = \delta^{\alpha}{}_{\gamma} . \tag{C.3}$$

We define *covariant vector* the set of n complex numbers η_{α} ($\alpha = 1, 2, \ldots, n$)

$$\eta \equiv (\eta_1 \ \eta_2 \ \cdots \ \eta_n) , \tag{C.4}$$

which transforms according to

$$\eta \to \eta' = \eta U^{\dagger} \qquad \text{i.e.} \qquad \eta'_{\alpha} = \eta_{\beta}(U^{\dagger})^{\beta}{}_{\alpha} . \tag{C.5}$$

It follows that the quantity

$$\xi^{\dagger} = (\xi_1 \ \xi_2 \ \cdots \ \xi_n) , \tag{C.6}$$

where

$$\xi_{\alpha} \equiv (\xi^{\alpha})^{*} , \tag{C.7}$$

transforms as a covariant vector.

It is also immediate to check that the *scalar product*

$$\eta\xi = \eta_{\alpha}\xi^{a} \tag{C.8}$$

is *invariant* under the group transformations.

A general mixed tensor $\zeta^{\alpha\beta\cdots}_{\mu\nu\cdots}$ where α, β, ... are controvariant and μ, ν, ... covariant indeces, is defined by the transformation property

$$\zeta'^{\,\alpha\beta\cdots}_{\mu\nu\cdots} = U^{\alpha}{}_{\alpha'}U^{\beta}{}_{\beta'}\cdots\zeta^{\alpha'\beta'\cdots}_{\mu'\nu'\cdots}(U^{\dagger})^{\mu'}{}_{\mu}(U^{\dagger})^{\nu'}{}_{\nu}\cdots . \tag{C.9}$$

We note, in passing, that the tensor $\delta^{\alpha}{}_{\beta}$ is *invariant* under $U(n)$.

A general tensor of the type given above is, in general, *reducible*. Let us first consider, for the sake of simplicity, a tensor of the type $\zeta^{\alpha\beta\gamma\cdots}$ with r indeces of controvariant type only: clearly, it corresponds to the basis of the direct product representation

$$U \otimes U \otimes \ldots \otimes U , \tag{C.10}$$

where U appears r times. In order to reduce $\zeta^{\alpha\beta\gamma\cdots}$ into the irreducible tensors which are bases of the IR's of $U(n)$ we make use of the Young tableaux.

We build all the different *standard Young tableaux* with r boxes, as defined in Appendix B. To each standard tableau τ we associate a Young operator Y_{τ} as defined in (B.9).

By applying the operator Y_{τ} to the tensor $\zeta = \zeta^{\alpha\beta\cdots}$, one gets a tensor $\theta = \theta^{\alpha\beta\cdots}$ which has the permutation symmetry of the corresponding standard tableau:

$$\theta = Y_{\tau}\zeta . \tag{C.11}$$

One can show, in general, that the Young operators Y_τ commute with the transformations of the group $U(n)$. Then one can write

$$(U \otimes U \otimes \ldots)\theta = (U \otimes U \otimes \ldots)Y_\tau\zeta = Y_\tau(U \otimes U \otimes \ldots)\zeta = Y_\tau\zeta' = \theta' . \quad \text{(C.12)}$$

This result can be understood if one considers a tensor as product of basic vectors $\zeta^{\alpha\beta\ldots} = \xi^\alpha\xi^\beta \ldots$. Now the transformations of the product do not depend on the order in which the vectors are taken. This means that the transformations of $U(n)$ commute with the operations of permutations of the individual vectors, and therefore that they do not change the symmetry character of a tensor.

The meaning of Eq. (C.12) is that the subspace spanned by the tensor θ is invariant under the transformations of $U(n)$. Therefore, the tensor θ can be taken as the basis of a IR of the group $U(n)$.

Taking into account all the possible Young operators Y_τ, i.e. all possible standard tableaux, one can then decompose the tensor ζ into the irreducible tensors $Y_\tau\zeta$ and, therefore, the reducible representation (C.10) into the IR's contained in it, the dimension of each IR being equal to the number of independent components of the irreducible tensor.

Example 1

Let us consider the third order tensor

$$\zeta^{\alpha_1\alpha_2\alpha_3} = \xi^{\alpha_1}\xi^{\alpha_2}\xi^{\alpha_3} , \quad \text{(C.13)}$$

where ξ^{α_i} is a generic component of the vector ξ defined by Eq. (C.2). Taking into account the standard Young tableaux with three boxes (see Appendix B), we obtain easily the following decomposition

$$\zeta^{\alpha_1\alpha_2\alpha_3} = \zeta^{\{\alpha_1\alpha_2\alpha_3\}} \oplus \zeta^{\{\alpha_1\alpha_2\}\alpha_3} \oplus \zeta^{\{\alpha_1\alpha_3\}\alpha_2} \oplus \zeta^{[\alpha_1\alpha_2\alpha_3]} , \quad \text{(C.14)}$$

i.e.

Each index α_i goes from 1 to n; the number of components of a tensor corresponds to the different ways of taking all the independent sets $(\alpha_1, \alpha_2, \alpha_3)$ with given permutation properties. With the help of a little bit of combinatorics, one can determine the dimension N of the different kinds of tensors in (C.14):

$$\zeta^{\{\alpha\beta\gamma\}} \qquad N = \binom{n+3-1}{3} = \frac{1}{6}n(n+1)(n+2) , \quad \text{(C.15)}$$

$$\zeta^{\{\alpha\beta\}\gamma} \qquad N = \frac{1}{3}n(n^2-1) , \quad \text{(C.16)}$$

$$\zeta^{[\alpha\beta\gamma]} \qquad N = \binom{n}{3} = \frac{1}{6}n(n-1)(n-2) \, . \qquad (C.17)$$

Let us now consider a mixed tensor, such as $\zeta_{\mu\nu...}^{\alpha\beta...}$ defined by Eq. (C.9). In this case one has to apply the above procedure *independently* to the upper (controvariant) and lower (covariant) indeces. In other words, one has to symmetrize the tensor, according to all possible Young standard tableaux, in both the upper and lower indeces, independently. Moreover, one has to take into account that a further reduction occurs, for a mixed tensor, due to the invariance of the tensor δ^{α}_{β}. In fact, a tensor $\zeta_{\mu\nu...}^{\alpha\beta...}$, already symmetrized, can be further reduced according to

$$\delta^{\mu}_{\alpha}\zeta_{\mu\nu...}^{\alpha\beta...} = \zeta_{\alpha\nu...}^{\alpha\beta...} \, , \qquad (C.18)$$

and in order to get irreducible tensors, one has to *contract* all possible pairs of upper and lower indeces, and separate the "trace" from the traceless tensors. An example will suffice, for our purpose, to clarify the situation.

Example 2
Let us consider the $(n \times n)$-component mixed tensor $\zeta^{\alpha}_{\beta} = \xi^{\alpha}\xi_{\beta}$. It can be decomposed in the form

$$\zeta^{\alpha}_{\beta} = \hat{\zeta}^{\alpha}_{\beta} + \frac{1}{n}\delta^{\alpha}_{\beta}\xi^{\gamma}\xi_{\gamma} \, , \qquad (C.19)$$

i.e. in the trace $\mathrm{Tr}\zeta = \xi^{\gamma}\xi_{\gamma}$ and the traceless tensor

$$\hat{\zeta}^{\alpha}_{\beta} = \zeta^{\alpha}_{\beta} - \frac{1}{n}\delta^{\alpha}_{\beta}\xi^{\gamma}\xi_{\gamma} \, , \qquad (C.20)$$

which has $n^2 - 1$ components.

C.2 Irreducible tensors with respect to $SU(n)$

The matrices U of the group $SU(n)$ satisfy the further condition

$$\det U = 1 \, , \qquad (C.21)$$

which can be written also in the form

$$\epsilon_{\beta_1...\beta_n}U^{\beta_1}_{\alpha_1} \ldots U^{\beta_n}_{\alpha_n} = \epsilon_{\alpha_1...\alpha_n} \, , \qquad (C.22)$$

with the introduction of the completely antisymmetric tensor

$$\epsilon_{\alpha_1\alpha_2...\alpha_n} = \epsilon^{\alpha_1\alpha_2...\alpha_n} \, , \qquad (C.23)$$

the only components different from zero being those obtained by permutations from $\epsilon_{12...n} = +1$, equal to $+1$ or -1 for even and odd permutations, respectively.

Eq. (C.22) shows that the tensor ϵ is invariant under the group $SU(n)$. This means that the corresponding IR is the *identity* (one-dimensional) representation. It is interesting to note that the corresponding standard Young tableau consists of a column of n boxes:

$$\epsilon_{\alpha_1 \alpha_2 ... \alpha_n} \longrightarrow \boxed{} \; = \; \bullet$$

From this fact, useful important properties are derived for the irreducible tensors of $SU(n)$ (which hold, in general, for the unimodular group $SL(n, C)$). Making use of the antisymmetric tensor ϵ, one can transform covariant indeces into controvariant ones, and viceversa; for example one can write

$$\epsilon^{\alpha_1 \alpha_2 ... \alpha_n} \hat{\zeta}^{\beta_1}_{\alpha_1} = \theta^{\beta_1 [\alpha_2 ... \alpha_n]} . \tag{C.24}$$

In this way, one can transform all mixed tensors into one kind of tensors, say controvariant. In fact, one can show that *all* the IR's of $SU(n)$ can be obtained starting only from one kind of tensors, e.g. $\zeta^{\alpha\beta...}$.

In particular, the covariant vector ξ_α is transformed through

$$\xi_{\alpha_1} = \epsilon_{\alpha_1 \alpha_2 ... \alpha_n} \zeta^{[\alpha_2 ... \alpha_n]} , \tag{C.25}$$

i.e. into the $(n-1)$-component tensor $\zeta^{[\alpha_2 ... \alpha_n]}$, which corresponds to a one-column Young tableau with $(n-1)$ boxes.

A Young tableau (with r boxes) employed for the IR's of $U(n)$ or $SU(n)$ is identified by a set of n integer $(\lambda_1, \lambda_2 ... \lambda_n)$ with the conditions $\lambda_1 \geq \lambda_2 ... \geq \lambda_n$, $\sum_{i=1}^{n} \lambda_i = r$.

In fact, the maximum number of boxes in a column is n, since column means complete antisymmetrization and each index α of a tensor can go from 1 to n.

However, in the case of $SU(n)$ the use of the invariant tensor $\epsilon_{\alpha_1 \alpha_2 ... \alpha_n}$ shows that, from the point of view of the IR's and irreducible tensors, each Young tableau can be replaced by one in which all complete (n boxes) columns are erased.

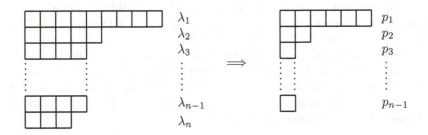

Each IR of $SU(n)$ is then specified by a set of $(n-1)$ integers $(p_1, p_2 \ldots p_{n-1})$, the relation between the two sets λ_i and p_i being

$$
\begin{aligned}
p_1 &= \lambda_1 - \lambda_2 , \\
p_2 &= \lambda_2 - \lambda_3 , \\
&\vdots \\
p_{n-1} &= \lambda_{n-1} - \lambda_n .
\end{aligned} \tag{C.26}
$$

It can be shown that the dimension of the IR of $SU(n)$ identified by the set p_i is given by

$$
\begin{aligned}
N = \frac{1}{1!2!\ldots(n-1)!} & (p_1+1)(p_1+p_2+2)\ldots(p_1+\ldots+p_{n-1}+n-1)\cdot \\
& \cdot(p_2+1)(p_2+p_3+2)\ldots(p_2+\ldots+p_{n-1}+n-2)\cdot \\
& \cdots\cdots\cdots \\
& \cdot(p_{n-2}+1)(p_{n-2}+p_{n-1}+2)\cdot \\
& \cdot(p_{n-1}+1) .
\end{aligned} \tag{C.27}
$$

In particular, one gets for $SU(2)$

$$
N = p_1 + 1 , \tag{C.28}
$$

and for $SU(3)$

$$
N = \tfrac{1}{2}(p_1+1)(p_1+p_2+2)(p_2+1) . \tag{C.29}
$$

We point out that in the case of $U(n)$, while the dimension of any IR is still given by Eq. (C.27), the set $(p_1, p_2 \ldots p_{n-1})$ is no longer sufficient to identify completely the IR. For $U(n)$ one needs to know also the number of boxes of the tableau corresponding to the set $(\lambda_1, \lambda_2 \ldots \lambda_n)$, i.e. the integer $r = \sum_i \lambda_i$. This fact can be interpreted in the following way: the groups $U(n)$ and $SU(n)$ are related by

$$
U(n) = SU(n) \otimes U(1) , \tag{C.30}
$$

and one can associate to the group $U(1)$ an additive quantum number. Fixing this quantum number for the vector ξ^α to a given value, say ρ, the value for a controvariant tensor $\xi^{\alpha_1 \alpha_2 \ldots \alpha_r}$ is $r\rho$. The value for the covariant vector, according to Eq. (C.7) is $-\rho$.

Moreover, in the case of $U(n)$, controvariant and covariant indeces cannot be transformed into each others, and, in general, irreducible tensors which are equivalent with respect to $SU(n)$ will not be equivalent with respect to $U(n)$. The same will occur for the corresponding IR's.

Given a Young tableau defined by the set $(p_1, p_2 \ldots p_{n-1})$, it is useful to introduce the *adjoint* Young tableau defined by $(p_{n-1}, p_{n-2} \ldots p_1)$: it corresponds to the pattern of boxes which, together with the pattern of the first tableau, form a rectangle of n rows. For example, for $SU(3)$, the following tableaux are adjoint to one another since

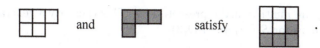

Given an IR of $SU(n)$ corresponding to a Young tableau, the adjoint IR is identified by the adjoint tableau; the two IR's have the same dimensionality (see Eq. (C.27)), but are in general *inequivalent* (they are equivalent only for $SU(2)$).

A Young tableau is *self-adjoint* if it coincides with its adjoint, and the corresponding IR is called also *self-adjoint*. For $SU(3)$, e.g., the octet, corresponding to the tableau

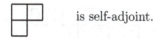

is self-adjoint.

Fundamental representations are said the IR's which correspond to set $(p_1, p_2 \ldots p_{n-1})$ of the type

$$\begin{cases} p_i = 1 \\ p_j = 0 \quad (j \neq i) \end{cases} . \tag{C.31}$$

$SU(n)$ then admits $n - 1$ fundamental representations, identified by the tableaux with only one column and number of boxes going from 1 to $n - 1$.

It is useful to distinguish different classes of IR's for the group $SU(n)$. In the case of $SU(2)$, as we know already from the study of the rotation group in Chapter 2, there are two kinds of IR's, the *integral* and *half-integral* representations, corresponding to even and odd values of p_1 (p_1 gives, in this case, just the number of boxes in the Young tableaux; in terms of the index j of the $D^{(j)}$ it is $p_1 = 2j$).

For $n > 2$, given the total number r of boxes in a tableau, we define the number k:

$$k = r - \ell n , \tag{C.32}$$

where ℓ and k are non-negative integers such that

$$0 \leq k \leq n - 1 . \tag{C.33}$$

For $SU(n)$, one can then distinguish n classes of IR's, corresponding to the values $k = 0, 1, \ldots n - 1$. Each fundamental IR identifies a different class.

It is interesting to note that the direct product of two IR's of classes k_1 and k_2 can be decomposed into IR's which belong all to the class $k = k_1 + k_2$ (modulo n). In particular, product representations of class $k = 0$ contain only IR's of the same class $k = 0$.

This fact is related to the followinge circumstance: the group $SU(n)$ contains as invariant subgroup the abelian group Z_n (of order n) which consists in the n-th roots of unity. With respect to the factor group $SU(n)/Z(n)$, the only IR's which are single-valued are those of class $k = 0$; the other IR's of $SU(n)$

(class $k \neq 0$) are multi-valued representations. This is clearly a generalization of what happens for $SU(2)$ and $SO(3)$.

C.3 Reduction of products of irreducible representations

It is useful to give a general recipe for the reduction of the direct product of IR's of $SU(n)$. A simple example is provided already by Eq. (C.14).

The recipe is based on the analysis of the construction of irreducible tensors and can be expressed as follows. Given two Young tableaux, insert in one of them the integer k, $(k = 0, 1, \ldots n - 1)$ in all the boxes of the k-th row, e.g.

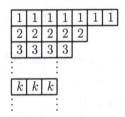

Then enlarge the other tableau by attaching in all "allowed" ways successively the boxes of type 1, those of type 2, etc. In each step, the following conditions have to be fulfilled:

a) Each tableau must be a *proper* tableau: no row can be longer than any row above it, and no column exists with a number of boxes $> n$.

b) The numbers in a row must not decrease from left to right: only different numbers are allowed in a column and they must increase from top to bottom.

c) Counting the numbers n_1, n_2, $\ldots n_k \ldots$ of the boxes of type 1, 2, $\ldots k \ldots$ row by row from the top, and from right to left in a row, the condition $n_1 \geq n_2 \ldots \geq n_k \ldots$ must be satisfied.

The above procedure is illustrated by a simple example.

Example 3
Suppose we have to decompose the direct product $8 \otimes 8$, where 8 is the eight-dimensional IR of $SU(3)$. Making use of the Young tableaux, we can write

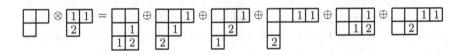

In the tableaux on the r.h.s., we can get rid of the complete (3 boxes) columns, and the numbers (p_1, p_2) of boxes in the remaining two rows identify completely the IR's into which the direct product is decomposed.

Labelling each IR by the dimension N (see Eq. (C.29)), and the corresponding adjoint IR by $\overline{N}$, one has:

$$8 \otimes 8 = 1 \oplus 8 \oplus 8 \oplus 10 \oplus \overline{10} \oplus 27$$

C.4 Decomposition of the IR's of $SU(n)$ with respect to given subgroups

It is a well known fact that a representation, which is irreducible with respect to a group $\mathcal{G}$, is in general reducible with respect to one of its subgroups $\mathcal{H}$. It is then useful to decompose this representation into the IR's of $\mathcal{H}$.

We are mainly interested in subgroups of the type $SU(\ell) \otimes SU(m)$ of $SU(n)$, where either $\ell + m = n$ or $\ell \cdot m = n$. The corresponding decomposition will give authomatically the content of an IR of $SU(n)$ in terms of IR's of the subgroup $SU(m)$ with $m < n$.

We shall consider the two cases separately.

a) $\ell + m = n$

Let us start from the controvariant vector of $SU(n)$ ξ^A ($A = 1, 2, \ldots n$). We can split it in the following way

$$\xi^A = \delta^A_\alpha (\xi^\alpha, 0) + \delta^A_a (0, \xi^a) \; ; \quad 1 \leq \alpha \leq \ell, \quad \ell + 1 \leq a \leq n, \qquad \text{(C.34)}$$

i.e.

$$\xi = x + y = \begin{pmatrix} \xi^1 \\ \xi^2 \\ \vdots \\ \xi^\ell \\ 0 \\ \vdots \\ 0 \end{pmatrix} + \begin{pmatrix} 0 \\ 0 \\ \vdots \\ 0 \\ \xi^{\ell+1} \\ \vdots \\ \xi^n \end{pmatrix}, \qquad \text{(C.35)}$$

The two vectors $x = (\xi^\alpha, 0)$ and $y = (0, \xi^a)$ are the bases of the self IR's of $SU(\ell)$ and $SU(m)$, respectively.

In terms of Young tableaux the decomposition (C.34) corresponds to

$$n = (\ell, 1) \oplus (1, m) \quad,$$

where the tableaux in parenthesis refer to $SU(\ell)$ and $SU(m)$, respectively.

For a general tensor, we can proceed in the following way. We split the corresponding Young tableau into two pieces in such a way that each piece is an allowed Young tableau relative to the subgroups $SU(\ell)$ or $SU(m)$. We perform all allowed splittings: all pairs of the sub-Young tableaux so obtained correspond to irreducible tensors and then to IR's of the subgroup $SU(\ell) \otimes SU(m)$.

The procedure is illustrated by the two following examples:

Example 4.
Let us consider the IR's of $SU(6)$ corresponding to the Young tableaux

whose dimensions, according to Eqs. (C.15) and (C.17), are 56 and 20, respectively. Their decomposition with respect to the subgroup $SU(4) \otimes SU(2)$ is obtained by writing

$$\boxed{}\boxed{}\boxed{} = \left(\boxed{}\boxed{}\boxed{}, \bullet\right) \oplus \left(\boxed{}\boxed{}, \boxed{}\right) \oplus \left(\boxed{}, \boxed{}\boxed{}\right) \oplus \left(\bullet, \boxed{}\boxed{}\boxed{}\right)$$

$$56 \quad = \quad (20\,,\,1) \quad \oplus \quad (10\,,\,2) \quad \oplus \quad (4\,,\,3) \quad \oplus \quad (1\,,\,4)$$

where the first tableau in parenthesis refers to $SU(4)$ and the second to $SU(2)$. In a similar way

$$\text{(column of 3)} = \left(\text{col 3}, \bullet\right) \oplus \left(\text{col 3}, \boxed{}\right) \oplus \left(\boxed{}, \text{col 3}\right)$$

$$20 \quad = \quad (\overline{4}\,,\,1) \quad \oplus \quad (6\,,\,2) \quad \oplus \quad (4\,,\,1)$$

Under the tableaux we have written the dimension of the corresponding IR. We note that, in the case of $SU(2)$, the Young tableau consisting of a column of two boxes corresponds to the identity, and a column of three boxes is not allowed.

Example 5
Let us consider now the same IR's of $SU(6)$ and their decomposition with respect to the group $SU(3) \otimes SU(3)$. One obtains in this case:

$$\square\square\square = \left(\square\square\square,\bullet\right)\oplus\left(\square\square,\square\right)\oplus\left(\square,\square\square\right)\oplus\left(\bullet,\square\square\square\right)$$

$$56 \quad = \quad (10\,,\,1) \quad \oplus \quad (6\,,\,3) \quad \oplus \quad (3\,,\,6) \quad \oplus \quad (1\,,\,10)$$

and

$$\square\!\!\square\!\!\square = \left(\square\!\!\square\!\!\square,\bullet\right)\oplus\left(\square\!\!\square\!\!\square,\square\right)\oplus\left(\square,\square\!\!\square\!\!\square\right)\oplus\left(\bullet,\square\!\!\square\!\!\square\right)$$

$$20 \quad = \quad (1\,,\,1) \quad \oplus \quad (\bar{3}\,,\,3) \quad \oplus \quad (3\,,\,\bar{3}) \quad \oplus \quad (1\,,\,1)$$

As a byproduct of the above recipe, taking $m = 1$ ($\ell = n - 1$), one can obtain immediately the content of an IR of $SU(n)$ in terms of the IR's of $SU(n-1)$.

b) $\ell \cdot m = n$

In this case the index A of the vector component ξ^A can be put in the one-to-one correspondence with a pair of indeces (α, a) ($\alpha = 1, \ldots, \ell; a = \ell+1, \ldots, n$), i.e.

$$\xi^A = x^\alpha y^a \quad \text{or} \quad \xi = x \otimes y\,, \tag{C.36}$$

where

$$x = \begin{pmatrix} x^1 \\ x^2 \\ \vdots \\ x^\ell \end{pmatrix}\,, \qquad y = \begin{pmatrix} y^1 \\ y^2 \\ \vdots \\ y^m \end{pmatrix}\,, \tag{C.37}$$

and in terms of the corresponding Young tableaux

$$\square = \left(\square,\square\right)$$

$$n \quad = \quad (\ell\,,\,m)$$

For a general tensor $\xi^{ABC\cdots}$ we consider the corresponding Young tableau, which specifies the symmetry properties of the indeces $A, B, C \ldots$. Each index is split into a pair of indeces according to the rule (C.36): $A = (\alpha, a)$, $B = (\beta, b)$, $C = (\gamma, c)$, ... Then we consider pairs of Young tableaux which refer independently to symmetry properties of the sets of indeces $(\alpha, \beta, \gamma, \ldots)$ and $(a, b, c, \ldots)$. However, one has to keep only those pairs of Young tableaux such that the global symmetry in $(A, B, C, \ldots)$ corresponding to the original tableau is preserved.

Example 6

Let us consider again the IR's 56 and 20 of $SU(6)$ and their decomposition with respect to the group $SU(3) \otimes SU(2)$. We can write

$$\boxed{A\,B\,C} = \left(\boxed{\alpha\,\beta\,\gamma}\,,\,\boxed{a\,b\,c}\right) \oplus \left(\boxed{\begin{smallmatrix}\alpha&\beta\\\gamma\end{smallmatrix}}\,,\,\boxed{\begin{smallmatrix}a&b\\c\end{smallmatrix}}\right)$$

$$56 \quad = \quad (10\,,\,4) \quad \oplus \quad (8\,,\,2)$$

and

$$\boxed{\begin{smallmatrix}A\\B\\C\end{smallmatrix}} = \left(\boxed{\begin{smallmatrix}\alpha\\\beta\\\gamma\end{smallmatrix}}\,,\,\boxed{a\,b\,c}\right) \oplus \left(\boxed{\begin{smallmatrix}\alpha&\beta\\\gamma\end{smallmatrix}}\,,\,\boxed{\begin{smallmatrix}a&c\\b\end{smallmatrix}}\right)$$

$$20 \quad = \quad (1\,,\,4) \quad \oplus \quad (8\,,\,2)$$

For the sake of convenience the boxes in the tableaux have been labelled with different letters. The order in which they are written is immaterial, provided the symmetry is preserved by the correspondence $A \leftrightarrow (\alpha, a)$, $B \leftrightarrow (\beta, b)$, $C \leftrightarrow (\gamma, c)$. Moreover, each pair of tableaux must appear only once.

Finally we remind that only allowed Young tableaux have to be included. For instance, in the general case $SU(\ell) \otimes SU(m)$, with $\ell \geq 3$, $m \geq 3$, the previous decompositions would be replaced by:

Obviously, the last term in each decomposition is not allowed in the case $m = 2$.

Solutions

Problems of Chapter 2

2.1 Making use of the definition (2.20), Eq. (2.21) can be written as

$$\sum_k \sigma_k x'_k = \sum_{kj} \sigma_k R_{kj} x_j = \sum_j u \sigma_j u^\dagger x_j.$$

Multiplying the above expression on the left by σ_i and taking the trace, one gets:

$$R_{ij} = \tfrac{1}{2}\mathrm{Tr}(\sigma_i u \sigma_j u^\dagger).$$

Finally, inserting the expressions of the Pauli matrices (2.22) and of the u matrix (2.15), one obtains:

$$R = \begin{pmatrix} \Re(a^2 - b^2) & -\Im(a^2 + b^2) & -2\Re(a^*b) \\ \Im(a^2 - b^2) & \Re(a^2 + b^2) & -2\Im(ab) \\ 2\Re(ab) & 0 & |a|^2 - |b|^2 \end{pmatrix}.$$

2.2 Let us denote by $U(R)$ the unitary operator in the Hilbert space corresponding to the rotation R. If H is invariant under rotations and $|\psi>$ is a solution of the Schrödinger equation, also $U(R)|\psi>$ is a solution and H commutes with $U(R)$. Then, according to Eq. (2.29), an infinitesimal rotation is given by $R \simeq 1 - i\phi \mathbf{J} \cdot \mathbf{n}$ and we get: $[H, J_k] = 0$, with $k = 1, 2, 3$.

2.3 The πN states corresponding to the resonant states with $J = \frac{3}{2}$ and $J_z = \pm\frac{1}{2}$, making use of the Eq. (2.71), can be written as:

$$\left|1, \tfrac{1}{2}; \tfrac{3}{2}, +\tfrac{1}{2}\right> = C\left(1, \tfrac{1}{2}, \tfrac{3}{2}; 0, \tfrac{1}{2}, \tfrac{1}{2}\right)\left|0, \tfrac{1}{2}\right> + C\left(1, \tfrac{1}{2}, \tfrac{3}{2}; 1, -\tfrac{1}{2}, \tfrac{1}{2}\right)\left|1, -\tfrac{1}{2}\right>$$

and

G. Costa and G. Fogli, *Symmetries and Group Theory in Particle Physics*,
Lecture Notes in Physics 823, DOI: 10.1007/978-3-642-15482-9,
© Springer-Verlag Berlin Heidelberg 2012

$$|1, \tfrac{1}{2}; \tfrac{3}{2}, -\tfrac{1}{2}> = C\left(1, \tfrac{1}{2}, \tfrac{3}{2}; 0, -\tfrac{1}{2}, -\tfrac{1}{2}\right)|0, -\tfrac{1}{2}> + C\left(1, \tfrac{1}{2}, \tfrac{3}{2}; 1, -\tfrac{1}{2}, \tfrac{1}{2}\right)| -1, \tfrac{1}{2}>$$

where all the states on the r.h.s. are understood to be referred to the case $j_1 = \ell = 1$, $j_2 = s = \tfrac{1}{2}$.

Inserting the explicit values of the CG-coefficients and the expressions for the spherical harmonics Y_1^0 and Y_1^{-1} (see Table A.3 and Eq. (A.11) in Appendix A), the final πN states can be represented by

$$|1, \tfrac{1}{2}; \tfrac{3}{2}, +\tfrac{1}{2}> \sim \sqrt{2} \cos\theta\, |\alpha> - \tfrac{1}{\sqrt{2}} e^{i\phi} \sin\theta\, |\beta> ,$$

and

$$|1, \tfrac{1}{2}; \tfrac{3}{2}, -\tfrac{1}{2}> \sim \sqrt{2} \cos\theta\, |\beta> + \tfrac{1}{\sqrt{2}} e^{-i\phi} \sin\theta\, |\alpha> .$$

where $|\alpha>$, $|\beta>$ stand for the spin $\tfrac{1}{2}$ states with $s_z = +\tfrac{1}{2}$ and $-\tfrac{1}{2}$, respectively. Projecting each state onto itself, one gets the angular distribution, which appears in the differential cross-section

$$\frac{d\sigma}{d\Omega} \sim |A_{\frac{3}{2}}|^2 \{1 + 3\cos^2\theta\},$$

where $A_{\frac{3}{2}}$ is the πN scattering amplitude in the $J = \tfrac{3}{2}$ state.

2.4 We start from Eq. (2.60), which we re-write here

$$R = R''_\gamma R'_\beta R_\alpha = e^{-i\gamma J_{z''}} e^{-i\beta J_{y'}} e^{-i\alpha J_z},$$

and make use of the appropriate unitary transformations $R'_\kappa = U R_\kappa U^{-1}$ that express each of the first two rotations in terms of the same rotation as seen in the previous coordinate system. Specifically,

$$e^{-i\gamma J_{z''}} = e^{-i\beta J_{y'}} e^{-i\gamma J_{z'}} e^{i\beta J_{y'}} ,$$

and

$$e^{-i\beta J_{y'}} = e^{-i\alpha J_z} e^{-i\beta J_y} e^{i\alpha J_z} .$$

By inserting the above expressions in the first relation and taking into account a similar expression for $e^{-i\gamma J_{z'}}$, one gets Eq. (2.61).

2.5 We denote by

$$\zeta_+ = \begin{pmatrix} 1 \\ 0 \end{pmatrix} \qquad \text{and} \qquad \zeta_- = \begin{pmatrix} 0 \\ 1 \end{pmatrix}$$

the two eigenstates $|\tfrac{1}{2}, +\tfrac{1}{2}>$ and $|\tfrac{1}{2}, -\tfrac{1}{2}>$ and apply to them the rotation

$$\exp\left\{i\frac{\pi}{2}\frac{\sigma_1}{2}\right\} = 1\cos(\tfrac{\pi}{4}) + i\sigma_1 \sin(\tfrac{\pi}{4}) = \sqrt{\tfrac{1}{2}} \begin{pmatrix} 1 & i \\ i & 1 \end{pmatrix}.$$

Then we obtain the requested (normalized) eigenstates:

$$\xi_+ = \sqrt{\tfrac{1}{2}} \begin{pmatrix} 1 \\ i \end{pmatrix} \qquad \text{and} \qquad \xi_- = \sqrt{\tfrac{1}{2}} \begin{pmatrix} i \\ 1 \end{pmatrix}.$$

Problems of Chapter 3

3.1 Starting from the general rotation (2.9), one can derive the rotation $R_\mathbf{n}$ which transforms the unit vector along the x^3-axis into the generic vector $\mathbf{n}$ of components (n_1, n_2, n_3). Written in terms of the components of $\mathbf{n}$, it reads

$$R_\mathbf{n} = \begin{pmatrix} 1 - \dfrac{(n_1)^2}{1+n_3} & -\dfrac{n_1 n_2}{1+n_3} & n_1 \\[2ex] -\dfrac{n_1 n_2}{1+n_3} & 1 - \dfrac{(n_2)^2}{1+n_3} & n_2 \\[2ex] -n_1 & -n_2 & n_3 \end{pmatrix}.$$

By assuming $\mathbf{n} = \boldsymbol{\beta}/|\boldsymbol{\beta}|$ and introducing the rotation $R_\mathbf{n}$ as a 4×4 matrix in Eq. (3.28), one can easily derive Eq. (3.27).

3.2 Making use of Eq. (3.42), one gets

$$\Lambda^\mu{}_\nu(AB) = \tfrac{1}{2}\mathrm{Tr}\left[\underline{\sigma}^\mu AB\sigma_\nu(AB)^\dagger\right] = \tfrac{1}{2}\mathrm{Tr}\left[A^\dagger\underline{\sigma}^\mu AB\sigma_\nu B^\dagger\right]$$

and

$$[\Lambda(A)\Lambda(B)]^\mu{}_\nu = \Lambda^\mu{}_\rho(A)\Lambda^\rho{}_\nu(B) = \tfrac{1}{4}\mathrm{Tr}\left[A^\dagger\underline{\sigma}^\mu A\sigma_\rho\right]\mathrm{Tr}\left[\underline{\sigma}^\rho B\sigma_\nu B^\dagger\right]$$

The identity $\Lambda(AB) = \Lambda(A)\Lambda(B)$ follows immediately from the property of the trace

$$\sum_\mu \mathrm{Tr}\left[G\sigma_\mu\right]\mathrm{Tr}\left[\underline{\sigma}^\mu H\right] = 2\,\mathrm{Tr}\left[GH\right]$$

valid for two arbitrary matrices G and H, and which can be easily checked. As an immediate consequence of the proved identity, one gets $[\Lambda(A)]^{-1} = \Lambda(A^{-1})$.

3.3 By inserting (3.39) into Eq. (3.42) one obtains the explicit expression

$$\Lambda(A) = \frac{1}{2} \cdot \begin{pmatrix} |\alpha|^2 + |\beta|^2 & \alpha\beta^* + \beta\alpha^* & i(-\alpha\beta^* + \beta\alpha^* & |\alpha|^2 - |\beta|^2 \\ +|\gamma|^2 + |\delta|^2 & +\gamma\delta^* + \delta\gamma^* & -\gamma\delta^* + \delta\gamma^*) & +|\gamma|^2 - |\delta|^2 \\[4pt] \alpha\gamma^* + \beta\delta^* & \alpha\delta^* + \beta\gamma^* & i(-\alpha\delta^* + \beta\gamma^* & \alpha\gamma^* - \beta\delta^* \\ +\gamma\alpha^* + \delta\beta^* & +\gamma\beta^* + \delta\alpha^* & -\gamma\beta^* + \delta\alpha^*) & +\gamma\alpha^* - \delta\beta^* \\[4pt] i(\alpha\gamma^* + \beta\delta^* & i(\alpha\delta^* + \beta\gamma^* & \alpha\delta^* - \beta\gamma^* & i(\alpha\gamma^* - \beta\delta^* \\ -\gamma\alpha^* - \delta\beta^*) & -\gamma\beta^* - \delta\alpha^*) & -\gamma\beta^* + \delta\alpha^* & -\gamma\alpha^* + \delta\beta^*) \\[4pt] |\alpha|^2 + |\beta|^2 & \alpha\beta^* + \beta\alpha^* & i(-\alpha\beta^* + \beta\alpha^* & |\alpha|^2 - |\beta|^2 \\ -|\gamma|^2 - |\delta|^2 & -\gamma\delta^* - \delta\gamma^* & +\gamma\delta^* - \delta\gamma^*) & -|\gamma|^2 + |\delta|^2 \end{pmatrix}$$

which can be written as the product of two matrices, as follows

$$\Lambda(A) = \frac{1}{2} \begin{pmatrix} \alpha & \beta & \gamma & \delta \\ \gamma & \delta & \alpha & \beta \\ -i\gamma & -i\delta & i\alpha & i\beta \\ \alpha & \beta & -\gamma & -\delta \end{pmatrix} \cdot \begin{pmatrix} \alpha^* & \beta^* & -i\beta^* & \alpha^* \\ \beta^* & \alpha^* & i\alpha^* & -\beta^* \\ \gamma^* & \delta^* & -i\delta^* & \gamma^* \\ \delta^* & \gamma^* & i\gamma^* & -\delta^* \end{pmatrix}$$

The calculation of the determinant (making use of the minors of the second order) gives, since $\det A = 1$,

$$\det \Lambda(A) = \frac{1}{16} \left[-4i(\det A)^2 \right] \left[4i(\det A^\dagger)^2 \right] = 1$$

3.4 From the previous problem one gets

$$\Lambda^0{}_0(A) = \frac{1}{2} \left(|\alpha|^2 + |\beta|^2 + |\gamma|^2 + |\delta|^2 \right)$$

and one can write

$$\Lambda^0{}_0(A) = \frac{1}{4} \left(|\alpha + \delta^*|^2 + |\alpha - \delta^*|^2 + |\beta + \gamma^*|^2 + |\beta - \gamma^*|^2 \right) \geq$$
$$\geq \frac{1}{4} \left(|\alpha + \delta^*|^2 - |\alpha - \delta^*|^2 + |\beta + \gamma^*|^2 - |\beta - \gamma^*|^2 \right) =$$
$$= \mathrm{Re}(\alpha\delta - \beta\gamma) \, .$$

Taking into account that

$$\det A = \alpha\delta - \beta\gamma = 1 \, ,$$

one has $\Lambda^0{}_0(A) \geq 1$.

3.5 If one takes the matrix A of the particular form

$$U = \cos \tfrac{1}{2}\phi - i\boldsymbol{\sigma} \cdot \mathbf{n} \sin \tfrac{1}{2}\phi \, ,$$

Eq. (3.38) becomes

$$X' = \sigma_\mu x'^\mu = (\cos \tfrac{1}{2}\phi - i\boldsymbol{\sigma} \cdot \mathbf{n} \sin \tfrac{1}{2}\phi)\, \sigma_\mu x^\mu \, (\cos \tfrac{1}{2}\phi + i\boldsymbol{\sigma} \cdot \mathbf{n} \sin \tfrac{1}{2}\phi) =$$
$$= x^0 + \boldsymbol{\sigma} \cdot \mathbf{x} \cos^2 \tfrac{1}{2}\phi + (\boldsymbol{\sigma} \cdot \mathbf{n})(\boldsymbol{\sigma} \cdot \mathbf{x})(\boldsymbol{\sigma} \cdot \mathbf{n}) \sin^2 \tfrac{1}{2}\phi -$$
$$- i \sin \tfrac{1}{2}\phi \cos \tfrac{1}{2}\phi \, [(\boldsymbol{\sigma} \cdot \mathbf{n})(\boldsymbol{\sigma} \cdot \mathbf{x}) - (\boldsymbol{\sigma} \cdot \mathbf{x})(\boldsymbol{\sigma} \cdot \mathbf{n})] \ .$$

Making use of the well-known identity

$$(\boldsymbol{\sigma} \cdot \mathbf{a})(\boldsymbol{\sigma} \cdot \mathbf{b}) = \mathbf{a} \cdot \mathbf{b} + i\boldsymbol{\sigma} \cdot (\mathbf{a} \times \mathbf{b}) \ ,$$

the above relation becomes

$$x^{0'} + (\boldsymbol{\sigma} \cdot \mathbf{x}') = x^0 + \boldsymbol{\sigma} \cdot \{(\mathbf{n} \cdot \mathbf{x})\mathbf{n} + \cos \phi \, [\mathbf{x} - (\mathbf{n} \cdot \mathbf{x})\mathbf{n}] + \sin \phi (\mathbf{n} \times \mathbf{x})\} \ ,$$

which clearly represents the application of the rotation matrix given by the Eqs. (3.23), (2.9) to the four-vector $x = (x^0, \mathbf{x})$.

3.6 Taking for A the matrix

$$H = \cosh \tfrac{1}{2}\psi - \boldsymbol{\sigma} \cdot \mathbf{n} \sinh \tfrac{1}{2}\psi \ ,$$

Eq. (3.38) becomes

$$X' = \sigma_\mu x'^\mu = (\cosh \tfrac{1}{2}\psi - \boldsymbol{\sigma} \cdot \mathbf{n} \sinh \tfrac{1}{2}\psi)\, \sigma_\mu x^\mu \, (\cosh \tfrac{1}{2}\psi - \boldsymbol{\sigma} \cdot \mathbf{n} \sinh \tfrac{1}{2}\psi) =$$
$$= x^0 \cosh \psi - \mathbf{n} \cdot \mathbf{x} \sinh \psi +$$
$$+ \boldsymbol{\sigma} \cdot \left\{\mathbf{x} - (\mathbf{n} \cdot \mathbf{x})\mathbf{n} + \mathbf{n} \left[(\mathbf{n} \cdot \mathbf{x}) \cosh \psi - x^0 \sinh \psi\right]\right\} \ .$$

The above relation corresponds to the transformations

$$x^{0'} = x^0 \cosh \psi - \mathbf{n} \cdot \mathbf{x} \sinh \psi \ ,$$
$$\mathbf{x}' = \mathbf{x} - (\mathbf{n} \cdot \mathbf{x})\mathbf{n} + \mathbf{n} \left[(\mathbf{n} \cdot \mathbf{x}) \cosh \psi - x^0 \sinh \psi\right] \ .$$

which are immediately identified with the application of the pure Lorentz matrix (3.27) to the four-vector x.

3.7 First let us check that an infinitesimal Lorentz transformation can be written in the form

$$\Lambda^\rho{}_\sigma = g^\rho{}_\sigma + \delta\omega^\rho{}_\sigma \quad \text{with} \quad \delta\omega^{\rho\sigma} = -\delta\omega^{\sigma\rho} \ .$$

The condition (3.7) gives

$$\left(g^\mu{}_\rho + \delta\omega^\mu{}_\rho\right) g_{\mu\nu} \left(g^\nu{}_\sigma + \delta\omega^\nu{}_\sigma\right) = g_{\rho\sigma} \ ,$$

i.e. $\delta\omega^{\rho\sigma} = -\delta\omega^{\sigma\rho}$. Next, let us consider an infinitesimal transformation obtained from (3.61):

$$\Lambda^\rho{}_\sigma = g^\rho{}_\sigma - \tfrac{1}{2} i \delta \omega^{\mu\nu} (M_{\mu\nu})^\rho{}_\sigma \;.$$

If we compare this expression of $\Lambda^\rho{}_\sigma$ with the one given above, we get immediately

$$(M_{\mu\nu})^\rho{}_\sigma = i \left(g_\mu{}^\rho g_{\nu\sigma} - g_\nu{}^\rho g_{\mu\sigma} \right) \;.$$

Writing explicitly $M_{\mu\nu}$ in matrix form, one gets the six independent matrices J_i, K_i given in (3.50) and (3.54).

3.8 According to the definition given in Section 1.2, $SO(4)$ is the group of four-dimensional real matrices α satisfying the condition:

$$\alpha \tilde{\alpha} = 1 \;, \qquad\qquad \det \alpha = 1 \;.$$

The group leaves invariant the lenght of a four-vector x^μ ($\mu = 1, 2, 3, 4$) in a four-dimensional euclidean space. Using the notation

$$x^2 = \delta_{\mu\nu} x^\mu x^\nu \;,$$

where $\delta_{\mu\nu}$ is the Kronecker symbol, one gets in fact:

$$(x')^2 = \delta_{\mu\nu} x'^\mu x'^\nu = \delta_{\mu\nu} \alpha^\mu{}_\rho \alpha^\nu{}_\sigma x^\rho x^\sigma = (x)^2 \;.$$

For this reason, $SO(4)$ can be regarded as the group of rotations in a four-dimensional space; they are *proper* rotations since the group contains only unimodular matrices ($\det \alpha = 1$).

In analogy with Problem 3.7, the infinitesimal transformation of $SO(4)$ can be written as

$$\alpha^\mu{}_\nu = \delta^\mu{}_\nu + \epsilon^\mu{}_\nu \;.$$

where $\epsilon^\mu{}_\nu = -\epsilon^\nu{}_\mu$ (the group is characterized by six parameters, since its order is 6), as can be easily checked from the condition $\alpha \tilde{\alpha} = 1$.

One can find the infinitesimal generators $J_{\mu\nu}$ by writing

$$\alpha^\rho{}_\sigma = \delta^\rho{}_\sigma - \tfrac{1}{2} i \epsilon^{\mu\nu} (J_{\mu\nu})^\rho{}_\sigma \;,$$

so that:

$$(J_{\mu\nu})^\rho{}_\sigma = i \left(\delta^\rho{}_\mu \delta_{\nu\sigma} - \delta^\rho{}_\nu \delta_{\mu\sigma} \right) \;.$$

From these, one can obtain the commutation relations

$$[J_{\mu\nu}, J_{\rho\sigma}] = -i \left(\delta_{\mu\rho} J_{\nu\sigma} - \delta_{\mu\sigma} J_{\nu\rho} - \delta_{\nu\rho} J_{\mu\sigma} + \delta_{\nu\sigma} J_{\mu\rho} \right) \;.$$

Let us now introduce the following linear combinations

$$
\begin{aligned}
M_i &= \tfrac{1}{2} \left(J_{i4} + J_{jk} \right) \;, \\
N_i &= -\tfrac{1}{2} \left(J_{i4} - J_{jk} \right) \;.
\end{aligned}
\qquad (i, j, k = 1, 2, 3 \text{ and cyclic permutations})
$$

Since these combinations are *real*, also the quantities M_i, N_i, as $J_{\mu\nu}$, can be taken as basic elements of the Lie algebra of $SO(4)$. It is easy to verify that the commutators of $J_{\mu\nu}$, expressed in terms of M_i, N_i, become:

$$[M_i, M_j] = i\epsilon_{ijk}M_k \ ,$$

$$[N_i, N_j] = i\epsilon_{ijk}N_k \ ,$$

$$[M_i, N_j] = 0 \ ,$$

The quantities M_i and N_i can be considered as the components of two independent angular momentum operators $\mathbf{M}$ and $\mathbf{N}$; each of them generates a group $SO(3)$. Then the group $SO(4)$ correponds to the direct product

$$SO(4) \sim SO(3) \otimes SO(3) \ .$$

This correspondence is not strictly an isomorphism, due to the arbitrariness in the choice of the signs for the two subgroups $SO(3)$. It is instructive to compare the above properties of $SO(4)$ with those given for $\mathcal{L}_+^\uparrow$. In particular, the IR's of $SO(4)$ can still be labelled by the eigenvaslues of the Casimir operators M^2 and N^2 (see Section 3.4); however, the IR $D^{(j,j')}(\alpha)$ of $SO(4)$, which is of order $(2j+1)(2j'+1)$, is now *unitary* since the group $SO(4)$ is *compact*, in contrast with the finite IR's of the non-compact group $\mathcal{L}_+^\uparrow$, which are not unitary.

3.9 From Eq. (3.70) one gets the following representation for a rotation about x^3 and a pure Lorentz transformation along x^3

$$A(R_3) = e^{-\frac{1}{2}i\phi\sigma_3} = \begin{pmatrix} e^{-\frac{1}{2}i\phi} & 0 \\ 0 & e^{\frac{1}{2}i\phi} \end{pmatrix} \ , \qquad A(L_3) = e^{-\frac{1}{2}\psi\sigma_3} = \begin{pmatrix} e^{-\frac{1}{2}\psi} & 0 \\ 0 & e^{\frac{1}{2}\psi} \end{pmatrix} \ .$$

The relative transformations of the spinor ξ are given from (3.68) and are

$$\begin{cases} \xi'^1 = e^{-\frac{1}{2}i\phi} \, \xi^1 \\ \xi'^2 = e^{\frac{1}{2}i\phi} \, \xi^2 \end{cases} , \qquad \begin{cases} \xi'^1 = e^{-\frac{1}{2}\psi} \, \xi^1 \\ \xi'^2 = e^{\frac{1}{2}\psi} \, \xi^2 \end{cases} ,$$

respectively. The transformations of the spinor ξ^*, according to (3.69), are given by the complex conjugate relations.

3.10 Any antisymmetric tensor can always be decomposed as

$$A_{\mu\nu} = \tfrac{1}{2}\left(A_{\mu\nu} + A_{\mu\nu}^D\right) + \tfrac{1}{2}\left(A_{\mu\nu} - A_{\mu\nu}^D\right) \ .$$

where the dual tensor $A_{\mu\nu}^D = \tfrac{1}{2}\epsilon_{\mu\nu\sigma\tau}A^{\sigma\tau}$ is clearly anti-symmetric. The two tensor $\left(A_{\mu\nu} + A_{\mu\nu}^D\right)$ and $\left(A_{\mu\nu} - A_{\mu\nu}^D\right)$ are selfdual and anti-selfdual, respectively, since $(A_{\mu\nu}^D)^D = A_{\mu\nu}$ (remember that $\tfrac{1}{2}\epsilon^{\mu\nu\sigma\tau}\epsilon_{\mu\nu\alpha\beta} = g^\sigma{}_\alpha g^\tau{}_\beta - g^\sigma{}_\beta g^\tau{}_\alpha$).

If $A'_{\mu\nu}$ is the transformed of $A_{\mu\nu}$ under an element of $\mathcal{L}_+^\uparrow$, the transformed of $A^D_{\mu\nu}$ is $A'^D_{\mu\nu} = \frac{1}{2}\epsilon_{\mu\nu\sigma\tau}A'^{\sigma\tau}$; so that the selfdual and anti-selfdual tensors are irreducible under $\mathcal{L}_+^\uparrow$.

The electromagnetic field tensor $F^{\mu\nu}$ and its dual $F^{D\mu\nu}$ are given in terms of the field components E^i, B^i by

$$
F^{\mu\nu} = \begin{pmatrix} 0 & E^1 & E^2 & E^3 \\ -E^1 & 0 & B^3 & -B^2 \\ -E^2 & -B^3 & 0 & B^1 \\ -E^3 & B^2 & -B^1 & 0 \end{pmatrix}, \quad
F^{D\mu\nu} = \begin{pmatrix} 0 & -B^1 & -B^2 & -B^3 \\ B^1 & 0 & -E^3 & E^2 \\ B^2 & E^3 & 0 & -E^1 \\ B^3 & -E^2 & E^1 & 0 \end{pmatrix},
$$

We see that the selfdual and anti-selfdual tensors correspond to the field combinations $E^i - B^i$ and $E^i + B^i$, respectively.

3.11 As seen in Section 3.4, the four-vector x^μ is the basis of the IR $D^{(\frac{1}{2},\frac{1}{2})}$ of $\mathcal{L}_+^\uparrow$. The basis of the direct product representation $D^{(\frac{1}{2},\frac{1}{2})} \otimes D^{(\frac{1}{2},\frac{1}{2})}$ can then be taken to be the tensor $x^\mu y_\nu$. This tensor is not irreducible and one can decompose it in the following way:

$$
x^\mu y_\nu = \left\{ \tfrac{1}{2}\left(x^\mu y_\nu + y^\mu x_\nu\right) - \tfrac{1}{4}(x \cdot y) \right\} + \tfrac{1}{2}\left(x^\mu y_\nu - y^\mu x_\nu\right) + \tfrac{1}{4}(x \cdot y) .
$$

It is easy to check that each of the terms on the r.h.s (the 9-component traceless symmetric tensor, the 6-component anti-symmetric tensor and the trace) transform into itself under $\mathcal{L}_+^\uparrow$. Moreover, we learnt from Problem 3.10 that the anti-symmetric tensor can be decomposed further into two irreducible 3-component tensors. One gets in this way four irreducible tensors. It is natural to take them as the bases of the IR's $D^{(1,1)}$, $D^{(0,0)}$, $D^{(1,0)}$ and $D^{(0,1)}$, which have, respectively, dimensions 9, 1, 3, and 3.

Problems of Chapter 4

4.1 The composition law of the Poincaré group is given by Eq. (4.2). One must show that the group properties given in Section 1.1 are satisfied. The multiplication is associative:

$$
\begin{aligned}
(a_3, \Lambda_3)\left[(a_2, \Lambda_2)(a_1, \Lambda_1)\right] &= [a_3 + \Lambda_3(a_2 + \Lambda_2 a_1), \Lambda_3\Lambda_2\Lambda_1] = \\
&= (a_3 + \Lambda_3 a_2, \Lambda_3\Lambda_2)(a_1, \Lambda_1) = \\
&= \left[(a_3, \Lambda_3)(a_2, \Lambda_2)\right](a_1, \Lambda_1) .
\end{aligned}
$$

The identity is $(0, I)$ and the inverse of (a, Λ) is $(-\Lambda^{-1}a, \Lambda^{-1})$. In fact:

$$(a, \Lambda)(-\Lambda^{-1}a, \Lambda^{-1}) = (a - \Lambda\Lambda^{-1}a, \Lambda\Lambda^{-1}) = (0, I) \ ,$$
$$(-\Lambda^{-1}a, \Lambda^{-1})(a, \Lambda) = (-\Lambda^{-1}a + \Lambda^{-1}a, \Lambda^{-1}\Lambda) = (0, I) \ .$$

In order to show that the translation group $\mathcal{S}$ is an invariant subgroup, it is required that, for any translation (b, I) and any element (a, Λ) of $\mathcal{P}_+^\uparrow$, the product

$$(a, \Lambda)(b, I)(a, \Lambda)^{-1} \ ,$$

is a translation. In fact, one gets immediately:

$$(a, \Lambda)(b, I)(-\Lambda^{-1}a, \Lambda^{-1}) = (a + \Lambda b, \Lambda)(-\Lambda^{-1}a, \Lambda^{-1}) = (\Lambda b, I) \ .$$

4.2 We have to show that P^2 and W^2 commute with all the generators of $\mathcal{P}_+^\uparrow$, i.e. with P^μ and $M^{\mu\nu}$. First, making use of the commutation relations (4.16) and (4.17), we get

$$[P^2, P^\nu] \ = [P_\mu P^\mu, P^\nu] = 0 \ ,$$

$$[P^2, M^{\mu\nu}] = g_{\sigma\tau}[P^\sigma P^\tau, M^{\mu\nu}] = g_{\sigma\tau}[P^\sigma, M^{\mu\nu}]P^\tau + g_{\sigma\tau}P^\sigma[P^\tau, M^{\mu\nu}] =$$
$$= ig_{\sigma\tau}(g^{\mu\sigma}P^\nu - g^{\nu\sigma}P^\mu)P^\tau + ig_{\sigma\tau}P^\sigma(g^{\mu\tau}P^\nu - g^{\nu\tau}P^\mu) =$$
$$= 2i[P^\mu, P^\nu] = 0 \ .$$

Then, making use of (4.21), we obtain

$$[W^2, P^\mu] \ = [W_\nu W^\nu, P^\mu] = 0 \ ,$$

$$[W^2, M^{\mu\nu}] = g_{\sigma\tau}[W^\sigma, M^{\mu\nu}]W^\tau + g_{\sigma\tau}W^\sigma[W^\tau, M^{\mu\nu}] = -2i[W^\mu, W^\nu] = 0 \ .$$

4.3 Writing explicitly the translation in (4.8), one gets

$$U^{-1}(\Lambda)e^{-ia^\mu P_\mu}U(\Lambda) = e^{-i(\Lambda^{-1})^\mu{}_\nu a^\nu P_\mu} = e^{-ia^\mu \Lambda_\mu{}^\nu P_\nu} \ ,$$

where $U(\Lambda)$ stands for $U(0, \Lambda)$. In the case of an infinitesimal translation, one obtains

$$U^{-1}(\Lambda)P_\mu U(\Lambda) = \Lambda_\mu{}^\nu P_\nu \ ,$$

which coincides with Eq. (4.35). Making use explicitly of the Lorentz transformation (3.61), one gets for $\omega^{\mu\nu}$ infinitesimal

$$[M_{\mu\nu}, P_\rho] = -(M_{\mu\nu})_\rho{}^\sigma P_\sigma \ ,$$

which are equivalent to the commutation relations (4.17) when one takes into account the expression of $(M_{\mu\nu})_\rho{}^\sigma$ derived in Problem 3.7.

4.4 Starting from

$$U(\Lambda') = e^{-\frac{i}{2}\omega^{\mu\nu}M_{\mu\nu}} ,$$

and assuming $\omega^{\mu\nu}$ infinitesimal, one gets

$$U(\Lambda)U(\Lambda')U(\Lambda^{-1}) = I - \frac{i}{2}\omega^{\mu\nu}U(\Lambda)M_{\mu\nu}U^{-1}(\Lambda) = I - \frac{i}{2}\overline{\omega}^{\alpha\beta}M_{\alpha\beta} ,$$

where $\overline{\omega}^{\alpha\beta}$ is given by

$$g^\alpha{}_\beta + \overline{\omega}^\alpha{}_\beta = (\Lambda\Lambda'\Lambda^{-1})^\alpha{}_\beta = \Lambda^\alpha{}_\mu(g^\mu{}_\nu + \omega^\mu{}_\nu)\Lambda^\nu{}_\beta = g^\alpha{}_\beta + \Lambda^\alpha{}_\mu\Lambda^\nu{}_\beta\omega^\mu{}_\nu .$$

One then obtains

$$U(\Lambda)M_{\mu\nu}U^{-1}(\Lambda) = \Lambda^\alpha{}_\mu\Lambda^\beta{}_\nu M_{\alpha\beta} ,$$

which are the transformation properties of a tensor of rank two.

Taking also for $U(\Lambda)$ a generic infinitesimal transformation, one gets

$$[M_{\mu\nu}, M_{\rho\sigma}] = -(M_{\rho\sigma})^\alpha{}_\mu M_{\alpha\nu} - (M_{\rho\sigma})^\beta{}_\nu M_{\mu\beta} ,$$

which corresponds to the commutation relation (3.60), as can be seen by use of the explicit expression of $(M_{\mu\nu})^\rho{}_\sigma$ derived in Problem 3.7.

4.5 The operators W_μ transform as P_μ, i.e. according to (4.35):

$$W'_\mu = U^{-1}(\Lambda)W_\mu U(\Lambda) = \Lambda_\mu{}^\nu W_\nu .$$

We use for Λ the pure Lorentz transformation L_p^{-1} which brings a state $|p, \zeta>$ at rest; its explicit expression is obtained immediately from (3.27) by changing the sign of the space-like components:

$$L_p^{-1} = \begin{pmatrix} \dfrac{p^0}{m} & \dfrac{p^i}{m} \\ -\dfrac{p_j}{m} & \delta^i{}_j - \dfrac{p^i p_j}{m(p^0 + m)} \end{pmatrix} .$$

One then obtains

$$W'_0 = \frac{1}{m}\left(p^0 W_0 + p^i W_i\right) = 0 \qquad \text{(see Eq. (4.20))} ,$$

$$W'_i = \frac{1}{m}\left(-p_i W_0 - \frac{p_i p^j W_j}{p^0 + m}\right) + W_i = W_i - \frac{p_i}{p^0 + m}W_0 .$$

Let us write the following commutator

$$[W'_i, W'_j] = [W_i, W_j] - \frac{p_i}{p^0 + m}[W_0, W_j] - \frac{p_j}{p^0 + m}[W_i, W_0] .$$

With the definition $J_k = \dfrac{1}{m} W'_k$, making use of (4.21) and having in mind that the operators are applied to momentum eigenstates $|p, \zeta>$, one gets

$$[J_i, J_j] = i\epsilon_{ijk} J_k \ .$$

The calculation is straightforward but lengthy: for simplicity we fix $i = 1$, $j = 2$:

$$[W'_1, W'_2] =$$

$$= -i(W_0 p_3 - p_0 W_3) + \frac{ip_1}{p^0 + m}(W_1 p_3 - p_1 W_3) + \frac{ip_2}{p^0 + m}(W_2 p_3 - p_2 W_3) =$$

$$= -iW_0 p_3 + iW_3 \left(p^0 - \frac{p_1^2 + p_2^2}{p^0 + m} \right) + i\frac{p_3}{p^0 + m} (W_1 p_1 + W_2 p_2) =$$

$$= -iW_0 p_3 + imW_3 - i\frac{p_3}{p^0 + m} \left(W^i p_i \right) =$$

$$= im \left(W_3 - \frac{p_3}{p^0 + m} W_0 \right) = imW'_3 \ . \qquad\qquad \text{q. d. e.}$$

4.6 From the identity (see Eqs. (4.19), (3.61)),

$$U(\Lambda) = e^{-in^\sigma W_\sigma} = e^{-\frac{1}{2} in^\sigma \epsilon_{\sigma\rho\mu\nu} P^\rho M^{\mu\nu}} = e^{-\frac{1}{2} i\omega_{\mu\nu} M^{\mu\nu}} \ ,$$

it follows

$$\omega_{\mu\nu} = \epsilon_{\sigma\rho\mu\nu} n^\sigma p^\rho \ ,$$

which shows that the given transformation belongs to $\mathcal{L}_+^\uparrow$ (the operators are considered acting on a one-particle state $|p, \zeta >$). The matrix elements of Λ are then

$$\Lambda^{\mu\nu} = g^{\mu\nu} + \omega^{\mu\nu} = g^{\mu\nu} + \epsilon^{\sigma\rho\mu\nu} n_\sigma p_\rho \ ,$$

so that

$$p'^\mu = \Lambda^\mu{}_\nu p^\nu = p^\mu + \epsilon^{\sigma\rho\mu\nu} n_\sigma p_\rho p_\nu = p^\mu \ ,$$

as follows from the antisymmetry of the Levi-Civita tensor.

Problems of Chapter 5

5.1 We note that the generic element of the little group of a given four-vector $\bar{p}$ depends on Λ and p ($p^2 = \bar{p}^2$), arbitrarily chosen in $\mathcal{L}_+^\uparrow$ and H_{p^2}, respectively. It is then easy to show that the set of elements $\mathcal{R}$ satisfies the group properties.

a) Product:

$$\mathcal{R}_a \;=\; \mathcal{R}(\Lambda_a, p_a) = L^{-1}_{p'_a\bar{p}}\Lambda_a L_{p_a\bar{p}} \qquad\qquad \text{with}\quad p'_a = \Lambda_a p_a \,,$$

$$\mathcal{R}_b \;=\; \mathcal{R}(\Lambda_b, p_b) = L^{-1}_{p'_b\bar{p}}\Lambda_b L_{p_b\bar{p}} \qquad\qquad \text{with}\quad p'_b = \Lambda_b p_b \,,$$

$$\mathcal{R}_b\mathcal{R}_a = L^{-1}_{p'_b\bar{p}}\Lambda_b L_{p_b\bar{p}} L^{-1}_{p'_a\bar{p}}\Lambda_a L_{p_a\bar{p}} =$$

$$= L^{-1}_{p'_b\bar{p}}\Lambda_{ba} L_{p_a\bar{p}} = \mathcal{R}(\Lambda_{ba}, p_a) \qquad\qquad \text{with}\quad p'_b = \Lambda_{ba} p_a \,.$$

It is easy to check that the product is associative.

b) Identity: it corresponds to the choice $\Lambda = I$; in fact

$$\mathcal{R}(I,p) = L^{-1}_{p\bar{p}} L_{p\bar{p}} = I \,.$$

c) Inverse: it is given by:

$$\mathcal{R}^{-1} = \left(L^{-1}_{p'\bar{p}}\Lambda L_{p\bar{p}} \right)^{-1} = L^{-1}_{p\bar{p}}\Lambda^{-1} L_{p'\bar{p}} =$$

$$= \mathcal{R}\left(\Lambda^{-1}, p'\right) \qquad\qquad \text{with}\quad p' = \Lambda p \,.$$

5.2 We suppose that a different standard vector $\bar{p}'$ has been fixed, instead of $\bar{p}$, belonging to the little Hilbert space H_{p^2} (i.e. $\bar{p}'^2 = \bar{p}^2 = p^2$). Then for any given element $\mathcal{R} = \mathcal{R}(\Lambda, p)$ of the little group of $\bar{p}$ we can build

$$\mathcal{R}' = L^{-1}_{\bar{p}\bar{p}'}\mathcal{R}L_{\bar{p}\bar{p}'} \,,$$

which is the element of the little group of $\bar{p}'$ corresponding to Λ and p. In fact, we can invert Eq. (5.6) as follows

$$\Lambda = L_{p'\bar{p}}\mathcal{R}(\Lambda, p)L^{-1}_{p\bar{p}} = L_{p'\bar{p}'}\mathcal{R}'L^{-1}_{p\bar{p}'} \,,$$

so that

$$\mathcal{R}' = L^{-1}_{p\bar{p}'}\Lambda L_{p\bar{p}'} = \mathcal{R}'(\Lambda, p) \,.$$

There is a one-to-one correspondence between $\mathcal{R}(\Lambda, p)$ and $\mathcal{R}'(\Lambda, p)$, which is preserved under multiplication:

$$\mathcal{R}'_2\mathcal{R}'_1 = L^{-1}\mathcal{R}_2 L L^{-1}\mathcal{R}_1 L \,.$$

Therefore there is an isomorphism between the two little groups relative to the standard vectors $\bar{p}$ and $\bar{p}'$. One can remove the restrictions that $\bar{p}$ and $\bar{p}'$ are in the same little Hilbert space, keeping in mind that the little group of a standard vector $\bar{p}$ is the little group also of any standard vector $c\bar{p}$ ($c \neq 0$), being

$$\mathcal{R}(c\bar{p}) = c\bar{p} \qquad\qquad \text{if} \qquad\qquad \mathcal{R}(\bar{p}) = \bar{p} \,.$$

Combining the results, we notice that the isomorphism is extended to all little groups of vectors in the same class, i.e. time-like, space-like and light-like.

5.3 From what proved in the previous problem, the little group is defined by any standard vector $\bar{p}$ in the class. Let us choose the vector $\bar{p} = (0,0,0,p_3)$: it is clear that any rotation about the third axis (i.e. in the $(1,2)$ plane), any pure Lorentz transformation in the $(1,2)$ plane, and any combination of such rotations and pure Lorentz transformations leave the chosen standard vector unchanged. In fact, the required transformation matrices must be a subset of $\mathcal{L}_+^\uparrow$ (i.e. $\tilde{\Lambda}g\Lambda = g$ and $\det \Lambda = +1$) satisfying the condition $\Lambda\bar{p} = \bar{p}$. Then such matrices have the form

$$\Lambda = \left(\begin{array}{c|c} \lambda & 0 \\ \hline 0 & 1 \end{array}\right),$$

where λ is a 3×3 matrix which satisfies the condition

$$\tilde{\lambda}g\lambda = g, \quad \det \lambda = 1 \quad \text{with} \quad g = \begin{pmatrix} 1 & & \\ & -1 & \\ & & -1 \end{pmatrix}.$$

It is easy to check that the set of the matrices λ can be identified with the group $SO(1,2)$ (for the definition, see Section 1.2). We note that the transformations of this group leave invariant the quantity $(x^0)^2 - (x^1)^2 - (x^2)^2$.

5.4 We start from the explicit expression for a generic pure Lorentz transformation (see Eq. (3.27))

$$L = \left(\begin{array}{c|c} \gamma & \gamma\beta_j \\ \hline -\gamma\beta^i & \delta^i{}_j - \frac{\beta^i\beta_j}{\beta^2}(\gamma-1) \end{array}\right),$$

and of the particular ones

$$L_p = \left(\begin{array}{c|c} \dfrac{p^0}{m} & -\dfrac{p_j}{m} \\ \hline \dfrac{p^i}{m} & \delta^i{}_j - \dfrac{p^i p_j}{m(p^0+m)} \end{array}\right)$$

and

$$L_{p'}^{-1} = \left(\begin{array}{c|c} \dfrac{p'^0}{m} & \dfrac{p'_j}{m} \\ \hline -\dfrac{p'^i}{m} & \delta^i{}_j - \dfrac{p'^i p'_j}{m(p'^0+m)} \end{array}\right).$$

After some algebra, making use of the relation $p' = Lp$, one obtains

$$\mathcal{R}_{p'p} = L_{p'}^{-1} L L_p =$$

$$= \left(\begin{array}{c|c} 1 & 0 \\ \hline & \\ 0 & \delta^i{}_j - \dfrac{\beta^i \beta_j}{\beta^2}(\gamma-1) + \dfrac{\gamma \beta^i p_j}{p^0+m} - \dfrac{\gamma p'^i \beta_j}{p'^0+m} + \dfrac{(\gamma-1)p'^i p_j}{(p^0+m)(p'^0+m)} \end{array} \right) \quad (a) .$$

When applied to p, $\mathcal{R}_{p'p}$ gives

$$(\mathcal{R}_{p'p})^i{}_j \, p^j = \frac{p^0 + \gamma m}{p'^0 + m} \, p'^i + \gamma m \beta^i .$$

In the ultrarelativistic limit ($p^0 \gg m$, $p'^0 \gg m$) this corresponds to

$$(\mathcal{R}_{p'p})^i{}_j \, \frac{p^j}{p^0} = \frac{p'^i}{p'^0} ,$$

so that $\mathcal{R}_{p'p}$ represents the rotation of the velocity vector (note that $|\mathbf{p}|/p^0 \simeq |\mathbf{p}'|/p'^0 \simeq 1$).

In the non-relativistic limit one gets simply $(\mathcal{R}_{p'p})^i{}_j \simeq \delta^i{}_j$ and there is no rotation at all.

It is instructive to express $\mathcal{R}_{p'p}$ in terms of the 2×2 representation of L, L_p, $L_{p'}^{-1}$. In this case one has

$$L = e^{-\frac{1}{2}\alpha \boldsymbol{\sigma} \cdot \mathbf{e}} \quad , \quad L_p = e^{-\frac{1}{2}\psi \boldsymbol{\sigma} \cdot \mathbf{n}} \quad , \quad L_{p'} = e^{-\frac{1}{2}\psi' \boldsymbol{\sigma} \cdot \mathbf{n}'} ,$$

where

$$\cosh \alpha = \gamma \quad , \quad \cosh \psi = \frac{p^0}{m} \quad , \quad \cosh \psi' = \frac{p'^0}{m} ,$$

and

$$\mathbf{e} = \frac{\boldsymbol{\beta}}{|\boldsymbol{\beta}|} \quad , \quad \mathbf{n} = \frac{\mathbf{p}}{|\mathbf{p}|} \quad , \quad \mathbf{n}' = \frac{\mathbf{p}'}{|\mathbf{p}'|} .$$

Then we can write

$$\mathcal{R}_{p'p} = e^{\frac{1}{2}\psi' \boldsymbol{\sigma} \cdot \mathbf{n}'} e^{-\frac{1}{2}\alpha \boldsymbol{\sigma} \cdot \mathbf{e}} e^{-\frac{1}{2}\psi \boldsymbol{\sigma} \cdot \mathbf{n}} ,$$

which, with some algebra, can be expressed in the form

$$\mathcal{R}_{p'p} = \frac{1}{\cosh \frac{\psi'}{2}} \left\{ \cosh \frac{\alpha}{2} \cosh \frac{\psi}{2} + \sinh \frac{\alpha}{2} \sinh \frac{\psi}{2} [(\mathbf{e} \cdot \mathbf{n}) + i \boldsymbol{\sigma} \cdot (\mathbf{e} \times \mathbf{n})] \right\} .$$

If one write

$$\mathcal{R}_{p'p} = e^{-\frac{1}{2}\theta \boldsymbol{\sigma} \cdot \boldsymbol{\nu}} ,$$

one can determine the angle θ and the direction $\boldsymbol{\nu}$ through the relations

$$\cos\frac{\theta}{2} = \frac{1}{\cosh\dfrac{\psi'}{2}}\left\{\cosh\frac{\alpha}{2}\cosh\frac{\psi}{2} + \sinh\frac{\alpha}{2}\sinh\frac{\psi}{2}(\mathbf{e}\cdot\mathbf{n})\right\},$$

(b)

$$\sin\frac{\theta}{2}\,\boldsymbol{\nu} = -\frac{1}{\cosh\dfrac{\psi'}{2}}\sinh\frac{\alpha}{2}\sinh\frac{\psi}{2}(\mathbf{e}\times\mathbf{n}).$$

5.5 In the present case Eq. (5.13) becomes

$$\mathcal{R}_{p'p} = L_{p'}^{-1}RL_p \quad , \quad \text{with } p' = Rp \quad \text{i.e.} \quad p'^0 = p^0, \ |\mathbf{p}'| = |\mathbf{p}| \ .$$

One can write (see Eq. (5.21)):

$$L_p = R_{\mathbf{p}}L_3(p)R_{\mathbf{p}}^{-1}\ ,$$

where $R_{\mathbf{p}}^{-1}$ is a rotation which transforms $\mathbf{p}$ along the direction of the x^3-axis, and $L_3(p)$ is the boost along the x^3-direction. Analogously

$$L_{p'} = R_{\mathbf{p}'}L_3(p')R_{\mathbf{p}'}^{-1}\ ,$$

with $L_3(p') = L_3(p)$ since $|\mathbf{p}| = |\mathbf{p}'|$. Making use of the above relations, one gets

$$\mathcal{R}_{p'p} = R_{\mathbf{p}'}L_3^{-1}(p')R_{\mathbf{p}'}^{-1}RR_{\mathbf{p}}L_3(p)R_{\mathbf{p}}^{-1}\ ,$$

and, since $R_{\mathbf{p}'}^{-1}RR_{\mathbf{p}}$ is a rotation about the x^3-axis (a vector along x^3 is not transformed) and therefore it commutes with the boost $L_3(p)$, one has the desired results

$$\mathcal{R} = R_{\mathbf{p}'}R_{\mathbf{p}'}^{-1}RR_{\mathbf{p}}R_{\mathbf{p}}^{-1} = R\ .$$

5.6 From Eqs. (5.22) and (5.14), one gets

$$U(a,\Lambda)|p,\lambda> = U(a,\Lambda)\sum_{\sigma}D_{\sigma\lambda}^{(s)}(R_{\mathbf{p}})|p,\sigma> =$$

$$= e^{-ip'a}\sum_{\sigma}D_{\sigma\lambda}^{(s)}(R_{\mathbf{p}})\sum_{\sigma'}D_{\sigma'\sigma}^{(s)}\left(L_{p'}^{-1}\Lambda L_p\right)|p',\sigma'> =$$

$$= e^{-ip'a}\sum_{\sigma'}D_{\sigma'\lambda}^{(s)}\left(L_{p'}^{-1}\Lambda L_p R_{\mathbf{p}}\right)|p',\sigma'> =$$

$$= e^{-ip'a}\sum_{\sigma'}D_{\sigma'\lambda}^{(s)}\left(R_{\mathbf{p}'}R_{\mathbf{p}'}^{-1}L_{p'}^{-1}\Lambda L_p R_{\mathbf{p}}\right)|p',\sigma'>\ .$$

Making use of (5.21) and again of (5.22), one finally gets

$$U(a, \Lambda)|p, \lambda> = e^{-ip'a} \sum_{\sigma' \lambda'} D^{(s)}_{\sigma' \lambda'}(R_{\mathbf{p}}) D^{(s)}_{\lambda' \lambda}\left(L_3^{-1}(p')R_{\mathbf{p}'}^{-1}\Lambda R_{\mathbf{p}}L_3(p)\right) |p', \sigma'> =$$

$$= e^{-ip'a} \sum_{\lambda'} D^{(s)}_{\lambda' \lambda}\left(L_3^{-1}(p')R_{\mathbf{p}'}^{-1}\Lambda R_{\mathbf{p}}L_3(p)\right) |p', \lambda'> ,$$

which concides with Eq. (5.20).

5.7 The transformation properties of $|p, \lambda>$ under a rotation are given by Eq. (5.20), where the little group matrices are of the type

$$\mathcal{R} = L_3^{-1}(p')R_{\mathbf{p}'}^{-1}RR_{\mathbf{p}}L_3(p) ,$$

with $p' = Rp$. We observe that $R_{\mathbf{p}'}^{-1}RR_{\mathbf{p}}$ is a rotation about the x^3-axis (in fact, it leaves unchanged the vectors along this axis), which commutes with a boost along the same axis (see (3.58)). Then, since $L_3(p') = L_3(p)$ (being $|\mathbf{p}| = |\mathbf{p}'|$), the little group matrix reduces just to that rotation

$$\mathcal{R} = R_{\mathbf{p}'}^{-1}RR_{\mathbf{p}} .$$

Now we use for R the explicit expression (2.9) of a generic rotation in terms of an angle ϕ and a unit vector $\mathbf{n}$. An explicit expression for $R_{\mathbf{p}}$ can be obtained in the form (it corresponds to $n_3 = 0$)

$$R_{\mathbf{p}} = \begin{pmatrix} (1 - \cos\theta)\sin^2\beta + \cos\theta & -(1 - \cos\theta)\sin\beta\cos\beta & \sin\theta\cos\beta \\ -(1 - \cos\theta)\sin\beta\cos\beta & (1 - \cos\theta)\cos^2\beta + \cos\theta & \sin\theta\sin\beta \\ -\sin\theta\cos\beta & -\sin\theta\sin\beta & \cos\theta \end{pmatrix} ,$$

and one has

$$R_{\mathbf{p}} \begin{pmatrix} 0 \\ 0 \\ |\mathbf{p}| \end{pmatrix} = \begin{pmatrix} |\mathbf{p}|\sin\theta\cos\beta \\ |\mathbf{p}|\sin\theta\sin\beta \\ |\mathbf{p}|\cos\theta \end{pmatrix} .$$

One can express $R_{\mathbf{p}'}$ in the same form, introducing two angles θ' and β'. The rotation $R_{\mathbf{p}'}$ can also be expressed in terms of θ, β and $\mathbf{n}$, ϕ making use of the condition

$$R_{\mathbf{p}'} \begin{pmatrix} 0 \\ 0 \\ |\mathbf{p}| \end{pmatrix} = RR_{\mathbf{p}} \begin{pmatrix} 0 \\ 0 \\ |\mathbf{p}| \end{pmatrix} .$$

In fact, with the notation $S = RR_{\mathbf{p}}$ the above relation gives $\sin\theta'\cos\beta' = S_{13}$, $\sin\theta'\sin\beta' = S_{23}$, $\cos\theta' = S_{33}$, so that one can write (compare with the expression given in Problem 3.1)

$$R_{\mathbf{p'}} = \begin{pmatrix} \dfrac{S_{23}^2}{1+S_{33}} + S_{33} & -\dfrac{S_{13}S_{23}}{1+S_{33}} & S_{13} \\ -\dfrac{S_{13}S_{23}}{1+S_{33}} & \dfrac{S_{13}^2}{1+S_{33}} + S_{33} & S_{23} \\ -S_{13} & -S_{23} & S_{33} \end{pmatrix} .$$

Performing the product $\mathcal{R} = R_{\mathbf{p'}}^{-1} S$ one then obtains

$$\mathcal{R} = \begin{pmatrix} \cos\alpha & -\sin\alpha & 0 \\ \sin\alpha & \cos\alpha & 0 \\ 0 & 0 & 1 \end{pmatrix} .$$

where

$$\cos\alpha = \frac{S_{11}+S_{22}}{1+S_{33}} \quad , \quad \sin\alpha = \frac{S_{21}-S_{12}}{1+S_{33}} .$$

Finally, expressing the elements of the matrix S in terms of the parameters θ, ϕ and $\mathbf{n}$, one finds[2], after a lenghty calculation, the following expression for the rotation $\mathcal{R}$:

$$e^{-i a} = \frac{(1+\cos\theta)\sin\phi + \hat{\mathbf{p}}^0 \cdot (\hat{\mathbf{p}} \times \mathbf{n}) - i(\hat{\mathbf{p}}_0 + \hat{\mathbf{p}}) \cdot \mathbf{n}(1-\cos\phi)}{(1+\cos\theta)\sin\phi + \hat{\mathbf{p}}^0 \cdot (\hat{\mathbf{p}} \times \mathbf{n}) + i(\hat{\mathbf{p}}_0 + \hat{\mathbf{p}}) \cdot \mathbf{n}(1-\cos\phi)} ,$$

where

$$\hat{\mathbf{p}}_0 = (0,0,1) \quad , \quad \hat{\mathbf{p}} = (\sin\theta\cos\beta, \sin\theta\sin\beta, \cos\theta) .$$

It is interesting to note that, in the specific case $\mathbf{n} = (0,0,1)$, the above relation, as expected, reduces to the identity $\alpha = \phi$.

5.8 We can consider the little group matrix given by Eq. (5.30) taking for Λ a rotation R:

$$\mathcal{R} = R_{\mathbf{p'}}^{-1} L_{p'\breve{p}}^{-1} R L_{p\breve{p}} R_{\mathbf{p}} = L_{\breve{p}'\breve{p}}^{-1} R_{\mathbf{p'}}^{-1} R R_{\mathbf{p}} L_{\breve{p}\breve{p}} \quad , \quad (p' = Rp) .$$

Since $R_{\mathbf{p'}}^{-1} R R_{\mathbf{p}}$ is a rotation about the x^3-axis, it commutes with any boost along the same axis; moreover, since $p' = Rp$, one has $\breve{p}' = \breve{p}$. The two boosts cancel and one gets

$$\mathcal{R} = R_{\mathbf{p'}}^{-1} R R_{\mathbf{p}} ,$$

as in the case of a massive particle (see Problem 5.5).

5.9 We can take $\bar{p} = (\rho, 0, 0, \rho)$ as standard vector. The element E of the little group has to satisfy the conditions

---
[2] See F.R. Halpern, *Special Relativity and Quantum Mechanics*, Prentice Hall, 1968, p. 102.

$$E\bar{p} = \bar{p} \quad , \quad \tilde{E}gE = g \quad , \quad \det E = +1 \ .$$

By applying the above conditions to a generic 4×4 real matrix, E can be written in terms of 3 independent real parameters, which are conveniently denoted by x, y and $\cos\alpha$. A sign ambiguity is eliminated by the condition $\det E = +1$. The final expression of the matrix E is the following:

$$E = \begin{pmatrix} 1 + \frac{1}{2}(x^2 + y^2) & -(x\cos\alpha + y\sin\alpha) & x\sin\alpha - y\cos\alpha & -\frac{1}{2}(x^2 + y^2) \\ -x & \cos\alpha & -\sin\alpha & x \\ -y & \sin\alpha & \cos\alpha & y \\ \frac{1}{2}(x^2 + y^2) & -(x\cos\alpha + y\sin\alpha) & x\sin\alpha - y\cos\alpha & 1 - \frac{1}{2}(x^2 + y^2) \end{pmatrix} .$$

If we introduce a two-dimensional vector r and a rotation R of the form

$$r = \begin{pmatrix} x \\ y \end{pmatrix} \quad , \quad R = \begin{pmatrix} \cos\alpha & -\sin\alpha \\ \sin\alpha & \cos\alpha \end{pmatrix} ,$$

the matrix E can be written in the form

$$E = \begin{pmatrix} 1 + \frac{1}{2}r^2 & -\tilde{r}R & -\frac{1}{2}r^2 \\ -r & R & r \\ \frac{1}{2}r^2 & -\tilde{r}R & 1 - \frac{1}{2}r^2 \end{pmatrix} .$$

The elements (r, R) form a group with the composition law

$$(r_2, R_2)(r_1, R_1) = (r_2 + R_2 r_1, R_2 R_1) ,$$

which is analogous to Eq. (4.3), except that now r is two-dimensional and R is a 2×2 matrix with the condition $R\tilde{R} = I$. The group is then the group of translations and rotations in a plane, which is called *Euclidean group in two dimensions*.

It is easy to find the generators from three independent infinitesimal transformations, i.e. translation along x, y and rotation in the (x, y) plane:

$$J_3 = \begin{pmatrix} 0 & 0 & 0 & 0 \\ 0 & 0 & -i & 0 \\ 0 & i & 0 & 0 \\ 0 & 0 & 0 & 0 \end{pmatrix} , \quad \Pi_1 = \begin{pmatrix} 0 & -i & 0 & 0 \\ -i & 0 & 0 & i \\ 0 & 0 & 0 & 0 \\ 0 & -i & 0 & 0 \end{pmatrix} , \quad \Pi_2 = \begin{pmatrix} 0 & 0 & -i & 0 \\ 0 & 0 & 0 & 0 \\ -i & 0 & 0 & i \\ 0 & 0 & -i & 0 \end{pmatrix} ,$$

Their commutation relations are given by

$$\begin{aligned} [\Pi_1, \Pi_2] &= 0 , \\ [J_3, \Pi_1] &= i\Pi_2 , \\ [J_3, \Pi_2] &= -i\Pi_1 , \end{aligned}$$

in agreement with Eq. (5.25).

5.10 The generic element of the little group, which is the two-dimensional Euclidean group, is given by Eq. (5.30), with Λ taken as a pure Lorentz transformation L. Accordingly, it can be written as

$$E = L_{\check{p}'\bar{p}}^{-1} R_{\mathbf{p}'}^{-1} L R_{\mathbf{p}} L_{\bar{p}\check{p}} , \tag{a}$$

where the general form for E has been obtained in Problem 5.9 in terms of a rotation and a two-dimensional translation. In principle, since we are interested only in the rotation (see Section 5.3), we could neglect the translation from the beginning. Moreover, one can realize that the two boosts are irrelevant for the calculation of the rotation angle. However, for the sake of completeness, we perform the full calculation. Assuming

$$\bar{p} = (\rho, 0, 0, \rho) , \quad \check{p} = (|\mathbf{p}|, 0, 0, |\mathbf{p}|) , \quad \check{p}' = (|\mathbf{p}'|, 0, 0, |\mathbf{p}'|) ,$$

from the condition

$$L_{\bar{p}\check{p}}\,\bar{p} = \check{p} ,$$

one gets

$$L_{\bar{p}\check{p}} = \begin{pmatrix} \dfrac{1+\delta^2}{2\delta} & 0 & 0 & \dfrac{1-\delta^2}{2\delta} \\[2mm] 0 & 1 & 0 & 0 \\[2mm] 0 & 0 & 1 & 0 \\[2mm] \dfrac{1-\delta^2}{2\delta} & 0 & 0 & \dfrac{1+\delta^2}{2\delta} \end{pmatrix} \qquad \text{with } \delta = \frac{\rho}{|\mathbf{p}|} ,$$

with $L_{\bar{p}\check{p}'}$ of the same form in terms of $\delta' = \dfrac{\rho}{|\mathbf{p}'|}$.

The rotation $R_{\mathbf{p}}$ can be written in the form (compare with Problem 3.1)

$$R_{\mathbf{p}} = \begin{pmatrix} 1 - \dfrac{(n_1)^2}{1+n_3} & -\dfrac{n_1 n_2}{1+n_3} & n_1 \\[3mm] -\dfrac{n_1 n_2}{1+n_3} & 1 - \dfrac{(n_2)^2}{1+n_3} & n_2 \\[3mm] -n_1 & -n_2 & n_3 \end{pmatrix}$$

in terms of the components of $\mathbf{n} = \mathbf{p}/|\mathbf{p}|$, and similarly for $R_{\mathbf{p}'}$ expressed in terms of $\mathbf{n}' = \mathbf{p}'/|\mathbf{p}'|$.

Performing the matrix product in (a), taking for L a generic Lorentz transformation of the type (3.27), one gets the matrix E, in the form reported in Problem 5.9. In particular, the parameters x and y of the translation are given by

$$x = E^1{}_3 = \frac{\gamma\delta}{1+n_3'}\,[\beta_1 - (\boldsymbol{\beta}\cdot\mathbf{n}')n_1' - k_2']$$

$$y = E^2{}_3 = \frac{\gamma\delta}{1+n_3'}\,[\beta_2 - (\boldsymbol{\beta}\cdot\mathbf{n}')n_2' - k_1']$$

and the rotation angle α by

$$\cos\alpha = E^1{}_1 = 1 - \frac{1}{(1+n_3)(1+n_3')}\frac{\gamma-1}{\beta^2}k_3 k_3' \,,$$

$$\sin\alpha = E^2{}_1 = \frac{1}{(1+n_3)(1+n_3')}\left[\frac{1}{\gamma}\frac{\gamma-1}{\beta^2}(k_3 + k_3' + \gamma\beta_3 k_3) + \gamma n_3' k_3\right]\,,$$

where

$$\mathbf{k} = \boldsymbol{\beta}\times\mathbf{n}\,,\qquad \mathbf{k'} = \boldsymbol{\beta}\times\mathbf{n'} = \frac{\delta'}{\delta}\mathbf{k}\,.$$

5.11 Instead of performing the products in Eq. (5.35), we make use of Eqs. (5.21) and (5.13) that allow to rewrite $R_1^{-1}(\epsilon)$ in the form

$$R_1^{-1}(\epsilon) = R_1^{-1}(\theta)L_{p'}^{-1}L_2(-v')L_3(-v) = R_1^{-1}(\theta)R_{p'p}\,,$$

where we have

$$p = (p^0, 0, 0, |\mathbf{p}|)\qquad\text{with}\qquad \left(p^0 = \gamma m\,,\ |\mathbf{p}| = p^0 v = \beta\gamma m\right)$$

and

$$p' = L_2(-v')p = (\gamma p^0, 0, -\gamma p^0 v', |\mathbf{p}|)\quad\text{i.e.}\quad p' = (\gamma\gamma' m, 0, \beta'\gamma\gamma' m, \beta\gamma m)\,.$$

In the above relations the usual definitions of the relativistic parameters γ, γ' and β, β' in terms of v and v' are adopted.

Then we evaluate $R_{p'p}$ according to Eq. (a) of Problem 5.4; in the present case $R_{p'p}$ reduces to a rotation about the x^1-axis through the angle δ. One finds

$$\cos\delta = \frac{\gamma+\gamma'}{1+\gamma\gamma'} = \frac{p^0+\gamma m}{p'^0+m}\,,$$

$$\sin\delta = \frac{\beta\gamma\beta'\gamma'}{1+\gamma\gamma'} = \frac{\gamma|\mathbf{p}|v'}{p'^0+m}\,.$$

It follows

$$\tan\delta = \frac{\gamma|\mathbf{p}|v'}{p^0+\gamma m} = \frac{\beta\gamma\beta'\gamma'}{\gamma+\gamma'} = \frac{\beta\beta'}{\sqrt{1-\beta^2}+\sqrt{1-\beta'^2}}\,,$$

and from $R_1^{-1}(\delta)R_1(\theta)$, making use of Eq. (5.34), one gets

$$\tan\epsilon = \tan(\theta-\delta) = \frac{v'}{v}\sqrt{1-v^2}\,.$$

We note that the same result for $\tan\delta$ can be obtained directly from the formulae (b) of Problem 5.4, that we re-write here as

$$\cos \frac{\delta}{2} = \frac{1}{\cosh \dfrac{\psi'}{2}} \left[\cosh \frac{\alpha}{2} \cosh \frac{\psi}{2} - \sinh \frac{\alpha}{2} \sinh \frac{\psi}{2} (\mathbf{e} \cdot \mathbf{n}) \right]$$

$$\sin \frac{\delta}{2} \boldsymbol{\nu} = \frac{1}{\cosh \dfrac{\psi'}{2}} \sinh \frac{\alpha}{2} \sinh \frac{\psi}{2} (\mathbf{e} \times \mathbf{n}) ,$$

by identifying the Wigner rotation as

$$\mathcal{R}_{p'p} = e^{\frac{1}{2}\psi' \boldsymbol{\sigma} \cdot \mathbf{n}'} e^{-\frac{1}{2}\alpha \boldsymbol{\sigma} \cdot \mathbf{e}} e^{-\frac{1}{2}\psi \boldsymbol{\sigma} \cdot \mathbf{n}} = L_{p'}^{-1} L_2(-v') L_3(-v) ,$$

where now $n = (1, 0, 0)$, $\mathbf{e} = (0, 1, 0)$ and

$$\cosh \psi = \gamma \quad , \quad \cosh \alpha = \gamma' \quad , \quad \cosh \psi' = \gamma \gamma' .$$

By applying the previous relations one easily obtains $\boldsymbol{\nu} = (1, 0, 0)$ and the expression of $\tan \delta$ reported above.

Problems of Chapter 6

6.1 Knowing that the spin of the π^0 is zero, the quantities which enter the matrix element of the decay $\pi^0 \to \gamma\gamma$ are the relative momentum $\mathbf{k}$ of the two photons in the c.m. system and their polarization (three-component) vectors ϵ_1 and ϵ_2. The matrix element is linear in the vectors ϵ_1 and ϵ_2, and it must be a *scalar* quantity under rotations.

One can build the two combinations, which are scalar and pseudoscalar under parity, respectively, and therefore correspond to even and odd parity for the π^0:

$$\epsilon_1 \cdot \epsilon_2 \quad , \quad \mathbf{k} \cdot \epsilon_1 \times \epsilon_2 .$$

In the first case, the polarization vectors of the two photon tend to be parallel, in the second case, perpendicular to each other. These correlations can be measured in terms of the planes of the electron-positron pairs in which the photons are converted, and then one can discriminate between the two cases. The parity of the π^0 was determined in this way to be odd (see R. Plano et al., Phys. Rev. Lett. **3**, 525 (1959)).

The positronium is an $e^+ e^-$ system in the state $^1 S_0$ (spin zero and S-wave): since $\ell = 0$, the parity of this state is determined by the intrinsic parity. The situation is similar to the case of the π^0; a measurement would detect that the polarization vectors of the two photons produced in the annihilation tend to be perpendicular, corresponding to the case of odd relative parity for electron and positron.

6.2 Knowing that all the particles in the reaction $K^- + He^4 \rightarrow {}_\Lambda H^4 + \pi^0$ have spin zero, angular momentum conservation implies that the orbital momentum is the same in the initial and final states. Then, the parity of the He^4 and ${}_\Lambda H^4$ being the same, the occurrence of the above reaction is in itself a proof that the parity of K^- is the same of that of π^0, i.e. odd.

6.3 The system of particles 1 and 2 can be considered a single particle of spin ℓ and intrinsic parity $\eta_{12} = \eta_1 \eta_2 (-1)^\ell$. The parity of the total system is simply given by the product:

$$\eta = \eta_3 \eta_{12} (-1)^L = \eta_1 \eta_2 \eta_3 (-1)^{\ell+L} .$$

6.4 In the decay $\rho^0 \rightarrow \pi^0 \pi^0$, the two spinless pions would be in a state of orbital momentum $\ell = 1$, and therefore antisymmetric under the exchange of the two π^0. This state is not allowed by Bose statistics, which require that the two identical pions can be only in symmetrical state.

In a similar way, in the decay $\rho^0 \rightarrow \gamma\gamma$ the final state must be symmetric under the interchange of the two photons. Then we have to build a matrix element in terms of the two polarization vectors ϵ_1, ϵ_2 and of the relative momentum $\mathbf{k}$ of the two γ's in the c.m. frame. The matrix element is linear in ϵ_1, ϵ_2, symmetric under the two γ exchange (i.e. under $\epsilon_1 \leftrightarrow \epsilon_2$ and $\mathbf{k} \leftrightarrow -\mathbf{k}$) and it transforms as a vector (since the spin of the ρ^0 is 1). Out of the three independent combinations of ϵ_1, ϵ_2, $\mathbf{k}$ which transform as vectors, i.e.

$$\epsilon_1 \times \epsilon_2 \quad , \quad (\epsilon_1 \cdot \epsilon_2)\,\mathbf{k} \quad \text{and} \quad (\mathbf{k} \cdot \epsilon_2)\,\epsilon_1 - (\mathbf{k} \cdot \epsilon_1)\,\epsilon_2 ,$$

only the last one is symmetric; however it is excluded by the transversality condition $(\mathbf{k} \cdot \epsilon_1 = \mathbf{k} \cdot \epsilon_2 = 0)$ required by electromagnetic gauge invariance. Therefore the decay is forbidden.

6.5 We notice that the electric dipole moment of a particle is proportional to its spin. For a spin $\frac{1}{2}$ particle (the electron e, for instance) one has $\boldsymbol{\mu}_e = \mu_e \boldsymbol{\sigma}$ which is an axial vector (it does not change under parity, see Eq. (6.2)). The Hamiltonian of the particle in an external electric field $\mathbf{E}$ contains the interaction term $H_I = -\mu_e \boldsymbol{\sigma} \cdot \mathbf{E}$, which is pseudoscalar, since $\mathbf{E}$ is a polar vector. The existence of an electric dipole moment would then indicate violation of parity.

6.6 From Problem 6.3 and the assignment $J^P = 0^-$ for the π-meson, we can determine the parity of the $J = 0$ $\pi^+ \pi^+ \pi^-$ state; in fact, if ℓ is the relative momentum of the two identical π^+, ℓ must be even and $\mathbf{J} = \boldsymbol{\ell} + \mathbf{L}$ implies that also L is even. The parity is then given by $-(-1)^{\ell+L} = -1$. On the other hand, the system $\pi^+\pi^0$ must have parity $+1$, being $J = \ell = 0$. Then the K^+ would decay into two systems of opposite parity. The occurrence of these decays, since the K^+ has spin zero and a *definite* parity $(J^P = 0^-)$, shows that parity is violated.

6.7 In general, in order to detect parity violation, one has to look for some quantity which is odd under parity (e.g. pseudoscalar); a non-vanishing expectation value of this quantity indicates that parity is violated. Suppose that the Λ^0 hyperon has been produced in the reaction $\pi^- + p = \Lambda^0 + K^0$; one can determine the normal versor to the production plane

$$n = \frac{\mathbf{p}_\pi \times \mathbf{p}_\Lambda}{|\mathbf{p}_\pi \times \mathbf{p}_\Lambda|} \ ,$$

where $\mathbf{p}_\pi$ and $\mathbf{p}_\Lambda$ are the momenta of the incident π^- and of the outgoing Λ^0, respectively. The quantity

$$\frac{\mathbf{n} \cdot \mathbf{k}}{|\mathbf{k}|} = \cos\theta \ ,$$

where $\mathbf{k}$ is the momentum of the π^- in the decay $\Lambda^0 \to \pi^- + p$, is clearly pseudoscalar; it was found that the average value of $\cos\theta$ is different from zero (which means asymmetry with respect to the production plane), and therefore that parity is violated in the decay.

The appearance of the $\cos\theta$ can be understood by noting that, if parity is not conserved, the Λ^0 $(J^P = \frac{1}{2}^+)$ can decay both in a $J^P = \frac{1}{2}^-$ (S-wave) and in a $J^P = \frac{1}{2}^+$ (P-wave) $\pi^- p$ state; the interference between the S- and P-waves gives rise to the $\cos\theta$ term.

6.8 The Maxwell equations, in natural units $(\hbar = c = 1)$, are:

$$\nabla \cdot \mathbf{E} = \rho \quad , \quad \nabla \cdot \mathbf{B} = 0 \ ,$$

$$\nabla \times \mathbf{B} = \mathbf{j} + \frac{\partial \mathbf{E}}{\partial t} \quad , \quad \nabla \times \mathbf{E} = -\frac{\partial \mathbf{B}}{\partial t} \ .$$

The last equation shows that the vectors $\mathbf{E}$ and $\mathbf{B}$ behave in opposite ways under time reversal. From the usual definitions of the charge density ρ and the current density $\mathbf{j}$, one realizes that ρ is unchanged under time reversal, while the current density changes direction: $\mathbf{j} \to -\mathbf{j}$. Therefore, assuming that the Maxwell equations are invariant under time reversal, one has $\mathbf{E} \to \mathbf{E}$ and $\mathbf{B} \to -\mathbf{B}$.

6.9 As discussed in Problem 6.5, the interaction Hamiltonian is given by $H_I = -\mu_e \boldsymbol{\sigma} \cdot \mathbf{E}$. Since under time reversal $\boldsymbol{\sigma} \to -\boldsymbol{\sigma}$, $\mathbf{E} \to \mathbf{E}$, this interaction term changes its sign under time reversal. The presence of an electric dipole moment would then indicate, besides parity violation, non-invariance under time reversal.

Problems of Chapter 7

7.1 The IR's $D^{(1,0)}$ and $D^{(0,1)}$ of $\mathcal{L}_+^\uparrow$ are irreducible also with respect to the rotation group $SO(3)$: within this subgroup they are both equivalent to $D^{(1)}$, so that they describe a vector particle. In a covariant description, a three-vector is then replaced by a selfdual (or an anti-selfdual) antisymmetric tensor (see also Problems 3.10, 3.11).

7.2 By differentiation of the antisymmetric tensor $f^{\mu\nu}$ one gets

$$\partial_\mu f^{\mu\nu} = \partial_\mu \partial^\mu \Phi^\nu - \partial_\mu \partial^\nu \Phi^\mu = \Box \Phi^\nu - \partial^\nu \partial_\mu \Phi^\mu = -m^2 \Phi^\nu \, ,$$

which are the Proca equations. Conversely, by differentiation of the Proca equations

$$\partial_\nu (\partial_\mu f^{\mu\nu} + m^2 \Phi^\nu) = \partial_\nu \partial_\mu f^{\mu\nu} + m^2 \partial_\nu \Phi^\nu = 0$$

the subsidiary condition $\partial_\nu \Phi^\nu = 0$ follows, since $f^{\mu\nu}$ is antisymmetric. If the subsidiary condition is now used in Proca equations, one gets

$$\partial_\mu (\partial^\mu \Phi^\nu - \partial^\nu \Phi^\mu) + m^2 \Phi^\nu = \Box \Phi^\nu - \partial^\nu \partial_\mu \Phi^\mu + m^2 \Phi^\nu = 0 \, ,$$

which is the Klein-Gordon equation (7.17).

We observe that the above equivalence no longer holds if $m = 0$, since the condition $m \neq 0$ is used in the derivation.

7.3 Under a Lorentz transformation $x'^\mu = \Lambda^\mu{}_\nu x^\nu$, the Dirac equation (7.79)

$$(i\gamma^\mu \partial_\mu - m)\psi(x) = 0 \tag{a}$$

will be transformed into

$$(i\gamma'^\mu \partial'_\mu - m)\psi'(x') = 0 \, , \tag{b}$$

where $\psi'(x') = S(\Lambda)\psi(x)$ (see Eq. (7.52)), and

$$\partial_\mu = \Lambda^\nu{}_\mu \partial'_\nu \qquad \left(\partial_\mu \equiv \frac{\partial}{\partial x^\mu} \, , \quad \partial'_\nu \equiv \frac{\partial}{\partial x'^\nu} \right) .$$

The matrices γ'^μ in (b) are equivalent to the matrices γ^μ in (a) up to a unitary transformation, so that we can simply replace γ' by γ (see J.D. Bjorken and S.D. Drell, *Relativistic Quantum Mechanics*, McGraw-Hill, New York, 1964, p.18). Making use of the above relations, Eq. (a) becomes

$$i\Lambda^\nu{}_\mu S\gamma^\mu S^{-1}\partial'_\nu\psi'(x') - m\psi'(x') = 0\,,$$

which coincides with Eq. (b), if one makes use of

$$\Lambda^\nu{}_\mu S\gamma^\mu S^{-1} = \gamma^\nu \qquad \text{i.e.} \qquad S^{-1}\gamma^\nu S = \Lambda^\nu{}_\mu\gamma^\mu\,.$$

7.4 Let us start from the infinitesimal Lorentz transformations written as (compare with Problem 3.7)

$$\Lambda^\rho{}_\sigma = g^\rho{}_\sigma + \delta\omega^\rho{}_\sigma \qquad \text{with} \qquad \delta\omega_{\rho\sigma} = -\delta\omega_{\sigma\rho}\,,$$

and assume for $S(\Lambda)$ the infinitesimal form

$$S = I - \frac{i}{4}\sigma_{\mu\nu}\delta\omega^{\mu\nu}\,,$$

where $\sigma_{\mu\nu}$ are six 4×4 antisymmetric matrices. Inserting both the above infinitesimal transformations into Eq. (7.76), at the first order one finds

$$\frac{i}{4}\delta\omega^{\mu\nu}(\sigma_{\mu\nu}\gamma^\alpha - \gamma^\alpha\sigma_{\mu\nu}) = \delta\omega^\alpha{}_\beta\gamma^\beta\,,$$

that, taking into account the antisymmetry of the $\delta\omega^{\mu\nu}$, can be rewritten in the form

$$[\gamma^\alpha, \sigma_{\mu\nu}] = 2i(g^\alpha{}_\mu\gamma_\nu - g^\alpha{}_\nu\gamma_\mu)\,,$$

satisfied by

$$\sigma_{\mu\nu} = \frac{i}{2}[\gamma_\mu, \gamma_\nu]\,.$$

7.5 Under a Lorentz transformation (neglecting for simplicity the argument x) the spinor ψ transforms as

$$\psi \to \psi' = S\psi\,,$$

so that for $\overline{\psi}'$ one gets

$$\overline{\psi}' = \psi^\dagger\gamma^0 S^{-1} = \overline{\psi}S^{-1}\,.$$

Then, for the scalar quantity, it follows immediately:

$$\overline{\psi}'\psi' = \overline{\psi}S^{-1}S\psi = \overline{\psi}\psi\,. \tag{a}$$

For the pseudoscalar quantity, one obtains:

$$\overline{\psi}'\gamma_5\psi' = \overline{\psi}S^{-1}\gamma_5 S\psi \ ,$$

and, since

$$S^{-1}\gamma_5 S = iS^{-1}\gamma^0\gamma^1\gamma^2\gamma^3 S = i\Lambda^0{}_\mu\Lambda^1{}_\nu\Lambda^2{}_\sigma\Lambda^3{}_\tau\gamma^\mu\gamma^\nu\gamma^\sigma\gamma^\tau =$$
$$= i\epsilon^{\mu\nu\sigma\tau}\Lambda^0{}_\mu\Lambda^1{}_\nu\Lambda^2{}_\sigma\Lambda^3{}_\tau\gamma^0\gamma^1\gamma^2\gamma^3 = (\det\Lambda)\gamma_5 \ ,$$

it follows

$$\overline{\psi}'\gamma_5\psi' = (\det\Lambda)\overline{\psi}\gamma_5\psi \ . \tag{b}$$

Space inversion corresponds to $\det\Lambda = -1$, so that $\overline{\psi}\gamma_5\psi$ behaves as a pseudoscalar under $\mathcal{L}^\uparrow$.

A quantity that transforms as a four-vector satisfies the relation

$$\overline{\psi}'\gamma^\mu\psi' = \overline{\psi}S^{-1}\gamma^\mu S\psi = \Lambda^\mu{}_\nu\overline{\psi}\gamma^\nu\psi \ . \tag{c}$$

Similarly, one finds that $\overline{\psi}\gamma_5\gamma^\mu\psi$ transforms as a pseudo-vector and

$$\overline{\psi}[\gamma_\mu,\gamma_\nu]\psi \tag{d}$$

as an antisymmetric tensor.

7.6 From Eq. (7.85) one gets

$$\overline{\psi} \xrightarrow{I_t} \overline{\psi}' = \overline{\psi}\gamma_5\gamma^0 \ .$$

Then the transformation properties under time reversal of the quantities (7.87) can be easily derived as follows:

$$\overline{\psi}\psi \qquad \longrightarrow -\overline{\psi}\gamma_5\gamma^0\gamma^0\gamma_5\psi = -\overline{\psi}\psi \ , \tag{a}$$

$$\overline{\psi}\gamma_5\psi \qquad \longrightarrow -\overline{\psi}\gamma_5\gamma^0\gamma_5\gamma^0\gamma_5\psi = \overline{\psi}\gamma_5\psi \ , \tag{b}$$

$$\overline{\psi}\gamma^\mu\psi \qquad \longrightarrow -\overline{\psi}\gamma_5\gamma^0\gamma^\mu\gamma^0\gamma_5\psi = \begin{cases} \overline{\psi}\gamma^0\psi & (\mu = 0) \ , \\ -\overline{\psi}\gamma^k\psi & (\mu = k = 1,2,3) \ , \end{cases} \tag{c}$$

$$\overline{\psi}\gamma_5\gamma^\mu\psi \longrightarrow -\overline{\psi}\gamma_5\gamma^0\gamma_5\gamma^\mu\gamma^0\gamma_5\psi = \begin{cases} -\overline{\psi}\gamma_5\gamma^0\psi & (\mu = 0) \ , \\ \overline{\psi}\gamma_5\gamma^k\psi & (\mu = k = 1,2,3) \ , \end{cases} \tag{d}$$

$$\overline{\psi}\gamma^\mu\gamma^\nu\psi \longrightarrow -\overline{\psi}\gamma^\mu\gamma^\nu\psi \ . \tag{e}$$

Problems of Chapter 8

8.1 The state $|pd>$ is a pure $|I = \frac{1}{2}, I_3 = \frac{1}{2}>$ state. Since the pion is an isotriplet and ^3He and ^3H form an isodoublet with $I_3 = \frac{1}{2}$ and $I_3 = -\frac{1}{2}$, respectively, the final states can have either $I = \frac{1}{2}$ or $I = \frac{3}{2}$. Making use of the relevant Clebsh-Gordan coefficients (see Table A.3), one gets

$$|\pi^+ \, {}^3H > = |1,1> \otimes |\tfrac{1}{2}, -\tfrac{1}{2}> = \sqrt{\tfrac{1}{3}} \, |\tfrac{3}{2}, \tfrac{1}{2}> + \sqrt{\tfrac{2}{3}} \, |\tfrac{1}{2}, \tfrac{1}{2}>,$$

$$|\pi^0 \, {}^3He> = |1,0> \otimes |\tfrac{1}{2}, \ \tfrac{1}{2}> = \sqrt{\tfrac{2}{3}} \, |\tfrac{3}{2}, \tfrac{1}{2}> - \sqrt{\tfrac{1}{3}} \, |\tfrac{1}{2}, \tfrac{1}{2}> \ .$$

From isospin invariance, one obtains for the S-matrix elements:

$$<\pi^+ \, {}^3H| \, S \, |pd> = \sqrt{\tfrac{2}{3}} A_{\frac{1}{2}},$$

$$<\pi^0 \, {}^3He| \, S \, |pd> = -\sqrt{\tfrac{1}{3}} A_{\frac{1}{2}},$$

where $A_{\frac{1}{2}}$ is the amplitude for pure $I = \frac{1}{2}$ state. Then the ratio of the relative cross-sections is:

$$R = \frac{\sigma(pd \rightarrow \pi^+ \, {}^3H)}{\sigma(pd \rightarrow \pi^0 \, {}^3He)} = 2 \ .$$

8.2 Since the pion is an isotriplet, we have three independent isospin amplitudes: A_0, A_1, A_2, which refer to $I = 0, 1, 2$, respectively. The isospin analysis of the various states, making use of the Clebsch-Gordan coefficients (see Section A.2 of Appendix A), gives

$$|\pi^+ \pi^+ > = |2, 2>,$$

$$|\pi^+ \pi^- > = \sqrt{\tfrac{1}{6}} |2,0> + \sqrt{\tfrac{1}{2}} |1,0> + \sqrt{\tfrac{1}{3}} |0,0>,$$

$$|\pi^0 \pi^0 > = \sqrt{\tfrac{2}{3}} |2,0> - \sqrt{\tfrac{1}{3}} |0,0> \ .$$

Then one obtains

$$<\pi^+ \pi^+| \, S \, |\pi^+ \pi^+ > = A_2 ,$$

$$<\pi^+ \pi^-| \, S \, |\pi^+ \pi^- > = \tfrac{1}{6} A_2 + \tfrac{1}{2} A_1 + \tfrac{1}{3} A_0 ,$$

$$<\pi^0 \pi^0| \, S \, |\pi^+ \pi^- > = \tfrac{1}{3} A_2 - \tfrac{1}{3} A_0 ,$$

$$<\pi^0 \pi^0| \, S \, |\pi^0 \pi^0 > = \tfrac{2}{3} A_2 + \tfrac{1}{3} A_0 .$$

8.3 By identifying the basis of the IR $D^{(1)}$ with

$$x^1 = \xi^1 \xi^1 ,$$
$$x^2 = \tfrac{1}{\sqrt{2}}(\xi^1 \xi^2 + \xi^2 \xi^1) ,$$
$$x^3 = \xi^2 \xi^2 ,$$

and taking into account the trasformation properties of the controcovariant vector ξ under $I_i = \tfrac{1}{2}\sigma_i$, i.e.

$$I_1 \begin{pmatrix} \xi^1 \\ \xi^2 \end{pmatrix} = \tfrac{1}{2} \begin{pmatrix} \xi^2 \\ \xi^1 \end{pmatrix}, \quad I_2 \begin{pmatrix} \xi^1 \\ \xi^2 \end{pmatrix} = i\tfrac{1}{2} \begin{pmatrix} -\xi^2 \\ \xi^1 \end{pmatrix}, \quad I_3 \begin{pmatrix} \xi^1 \\ \xi^2 \end{pmatrix} = \tfrac{1}{2} \begin{pmatrix} \xi^1 \\ -\xi^2 \end{pmatrix},$$

one can apply them independently to the factors of the products which appear in x^1, x^2, x^3. For example, one obtains

$$I_1 x^1 = I_1(\xi^1 \xi^1) = (I_1\xi^1)\xi^1 + \xi^1(I_1\xi^1) = \tfrac{1}{2}(\xi^2\xi^1 + \xi^1\xi^2) = \tfrac{1}{\sqrt{2}} x^2 ;$$

similarly

$$I_1 x^2 = \tfrac{1}{\sqrt{2}}(x^1 + x^3) ,$$
$$I_1 x^3 = \tfrac{1}{\sqrt{2}} x^2 ,$$

and analogously for I_2 and I_3. From these relations the following matrix structure of the generators in the three-dimensional IR can be easily derived:

$$I_1 = \tfrac{1}{\sqrt{2}} \begin{pmatrix} 0 & 1 & 0 \\ 1 & 0 & 1 \\ 0 & 1 & 0 \end{pmatrix} , \quad I_2 = i\tfrac{1}{\sqrt{2}} \begin{pmatrix} 0 & -1 & 0 \\ 1 & 0 & -1 \\ 0 & 1 & 0 \end{pmatrix} , \quad I_3 = \begin{pmatrix} 1 & 0 & 0 \\ 0 & 0 & 0 \\ 0 & 0 & -1 \end{pmatrix} .$$

8.4 From Eqs. (8.54) it is easy to find the matrix S which transforms the vector $\boldsymbol{\pi} = (\pi_1, \pi_2, \pi_3)$ into the vector of components π^+, π^0, π^-:

$$S = \sqrt{\tfrac{1}{2}} \begin{pmatrix} -1 & i & 0 \\ 0 & 0 & \sqrt{2} \\ 1 & i & 0 \end{pmatrix} .$$

By transforming, by a similarity transformation, the usual form of the $SU(2)$ generators

$$(I_i)_{jk} = -i\epsilon_{ijk} ,$$

one finds the expressions for I_1, I_2, I_3 given in the solution of the previous problem, so that π^+, π^0, π^- are eigenstates of I_3 with eigenvalues $+1, 0, -1$, respectively. By introducing the raising and lowering operators $I_\pm = I_1 \pm iI_2$, given explicitly by

$$I_+ = \sqrt{2} \begin{pmatrix} 0 & 1 & 0 \\ 0 & 0 & 1 \\ 0 & 0 & 0 \end{pmatrix} , \quad I_- = \sqrt{2} \begin{pmatrix} 0 & 0 & 0 \\ 1 & 0 & 0 \\ 0 & 1 & 0 \end{pmatrix} .$$

it is easy to verify:

$$I_\pm \pi^0 = \sqrt{2}\,\pi^\pm \ .$$

8.5 Following the recipe given in Appendix C, we obtain the following decompositions, besides the $8 \otimes 8$, already reported in Section C.3:

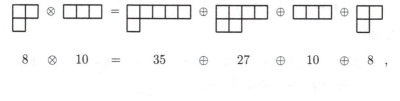

$$8 \quad \otimes \quad 10 \quad = \quad 35 \quad \oplus \quad 27 \quad \oplus \quad 10 \quad \oplus \quad 8 \ ,$$

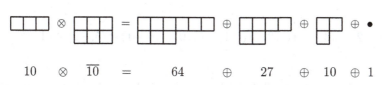

$$10 \quad \otimes \quad \overline{10} \quad = \quad 64 \quad \oplus \quad 27 \quad \oplus \quad 10 \oplus 1 \ .$$

Let us now examine the $SU(2)_I \otimes U(1)_Y$ content of some of the above IR's. We start with the IR's 8 and 10, obtained from the decomposition of the product $3 \otimes 3 \otimes 3$. It is convenient to make use of the strangeness S since the $U(1)_Y$ counts the non-strange $S = 0$ and the strange ($S = -1$) quarks. We obtain:

$$\square\!\square \ = \ \left(\square,\bullet\right) \ \oplus \left(\square\!\square,\square\right) \oplus \ \left(\bullet,\square\right) \ \oplus \left(\square,\square\!\square\right)$$

$$8 \ = \ (2, S=0) \quad \oplus \quad (3,-1) \quad \oplus \quad (1,-1) \quad \oplus \quad (1,-2) \ ,$$

$$\square\!\square\!\square \ = \ \left(\square\!\square\!\square,\bullet\right) \ \oplus \left(\square\!\square,\square\right) \oplus \left(\square,\square\!\square\right) \oplus \left(\bullet,\square\!\square\!\square\right)$$

$$10 \ = \ (4, S=0) \quad \oplus \quad (3,-1) \quad \oplus \quad (2,-2) \quad \oplus \quad (1,-3) \ .$$

Since each quark has $B = \frac{1}{3}$, in the present case we have $Y = S + B = S + 1$, and the above relations, in terms of Y, read

$$8 = (2, Y = 1) \oplus (3,0) \oplus (1,0) \oplus (2,-1) \,,$$
$$10 = (4, Y = 1) \oplus (3,0) \oplus (2,-1) \oplus (1,-2) \,.$$

Next we consider the representation 27:

$27 = (3, S = 0) \oplus (2, -1) \oplus (4, -1) \oplus (1, -2) \oplus (3, -2) \oplus (5, -2) \oplus (2, -3) \oplus$
$\qquad \oplus (4, -3) \oplus (3, -4)$.

Taking into account that the initial tableau contains 6 boxes, we can go from S to Y by means of $Y = S + B = S + 2$. Then we get:

$27 = (3, Y = 2) \oplus (2, 1) \oplus (4, 1) \oplus (1, 0) \oplus (3, 0) \oplus (5, 0) \oplus (2, -1) \oplus$
$\qquad \oplus (4, -1) \oplus (3, -2)$.

Similarly:

$35 = (5, Y = 2) \oplus (6, 1) \oplus (4, 1) \oplus (5, 0) \oplus (3, 0) \oplus (4, -1) \oplus (2, -1) \oplus$
$\qquad \oplus (3, -2) \oplus (1, -2) \oplus (2, -3)$

and

$64 = 27 \oplus (4, 3) \oplus (5, 2) \oplus (6, 1) \oplus (7, 0) \oplus (6, -1) \oplus (5, -2) \oplus (4, -3)$.

8.6 According to their definition, it is easy to find the following commutation relations among the shift operators:

$$[I_3, I_\pm] = \pm I_\pm , \qquad\qquad [Y, I_\pm] = 0 ,$$
$$[I_3, U_\pm] = \mp \tfrac{1}{2} U_\pm , \qquad\qquad [Y, U_\pm] = \pm U_\pm ,$$
$$[I_3, V_\pm] = \pm \tfrac{1}{2} V_\pm , \qquad\qquad [Y, V_\pm] = \pm V_\pm ,$$

and

$$[I_+, I_-] = 2I_3 , \quad [U_+, U_-] = 2U_3 = \tfrac{3}{2} Y - I_3 , \quad [V_+, V_-] = 2V_3 = \tfrac{3}{2} Y + I_3 ,$$
$$[I_\pm, U_\pm] = \pm V_\pm , \quad [I_\mp, V_\pm] = \pm U_\pm , \qquad\qquad [U_\pm, V_\mp] = \pm I_\mp ,$$
$$[I_+, U_-] = [I_+, V_+] = [U_+, V_+] = 0 .$$

It follows that $I_\pm$, $U_\pm$ and $V_\pm$ act as raising and lowering operators, specifically:

$I_\pm$ connects states with $\Delta I_3 = \pm 1$, $\Delta Y = 0$;

$U_\pm$ connects states with $\Delta I_3 = \mp \tfrac{1}{2}$, $\Delta Y = \pm 1$ $(\Delta Q = 0)$;

$V_\pm$ connects states with $\Delta I_3 = \pm \tfrac{1}{2}$, $\Delta Y = \pm 1$ $(\Delta S = \Delta Q)$.

This action on the states in the (I_3, Y) plane has been shown in Fig. 8.8. For each IR, all states can be generated, starting from whatever of them, by a repeated application of the shift operators.

In order to obtain the matrix elements between two given states, it is usual to fix the relative phase according to

$$I_{\pm}|I,I_3,Y> = [I \mp I_3)(I \pm I_3 + 1)]^{\frac{1}{2}}|I,I_3 \pm 1,Y>, \qquad (a)$$

$$V_{\pm}|V,V_3,Q> = \sum_{I'} a^{\pm}(I,I',I_3,Y)|I',I_3 \pm \tfrac{1}{2},Y \pm 1>, \qquad (b)$$

by requiring the coefficients $a^{\pm}$ always real non-negative numbers[3]. Let us note that at this point the action of $U^{\pm}$ is uniquely fixed by the commutation relations. The following convention is also adopted, which connects the eigenstates of two conjugate IR's $D(p_1,p_2)$, $D(p_2,p_1)$:

$$|(p_1,p_2);I,I_3,Y>^* = (-1)^{I_3+\frac{Y}{2}}|(p_1,p_2);I,-I_3,-Y> .$$

Let us now consider the 8 IR, whose isospin eigenstates are represented in Fig. 8.11 in terms of the $\tfrac{1}{2}^{+}$ baryon octet. Making use of the previous convention, it is easy to find the matrix elements of the shift operators reported in the figure.

The transition operated by $I_{\pm}$ are expressed by:

$$I_- p = n, \qquad I_- \Lambda^0 = 0, \qquad I_- \Sigma^0 = \sqrt{2}\,\Sigma^- .$$

which can be easily derived from Eq. (a). The action of the $V_{\pm}$ operators can be obtained making use of the commutation relations and of Eq. (b), e.g.

$$V_- p = a_0^- \Lambda^0 + a_1^- \Sigma^0$$

with a_0^-, a_1^- both positive and normalized according to

$$(a_0^-)^2 + (a_1^-)^2 = 1 ;$$

then from

$$[I_-,V_-] = 0 ,$$

being

$$I_- V_- p = a_1^- \sqrt{2}\,\Sigma^- , \qquad V_- I_- p = \Sigma^- ,$$

one finds

$$V_- p = \sqrt{\tfrac{3}{2}} \Lambda^0 + \tfrac{1}{\sqrt{2}} \Sigma^0 .$$

Let us now consider $U_{\pm}$ and the Σ_u^0, Λ_u^0 combinations of the Σ^0, Λ^0 states. Assuming, in analogy with the isospin eigenstates,

$$\sqrt{2}\,\Sigma_u^0 = U_- n ,$$

from

$$U_- = [V_-,I_+]$$

one finds

$$U_- p = V_- p - I_+ \Sigma^- = \left[\sqrt{\tfrac{3}{2}} \Lambda^0 + \tfrac{1}{\sqrt{2}} \Sigma^0\right] - \sqrt{2}\,\Sigma^0 = \sqrt{\tfrac{3}{2}} \Lambda^0 - \tfrac{1}{\sqrt{2}} \Sigma^0 ,$$

---

[3] J.J. de Swart, Rev. of Mod. Phys. 35, 916 (1963).

so that, accordingly with Eq. (8.124),

$$\Sigma_u^0 = \sqrt{\tfrac{3}{2}}\,\Lambda^0 - \tfrac{1}{\sqrt{2}}\,\Sigma^0\ ,$$

Λ_u^0 being the orthogonal combination.

In a similar way one can derive the matrix elements of the shift operators for the decuplet.

8.7 We recall that the photon is U-spin singlet ($U = 0$). On the other hand, π^0 and η^0 are superpositions of $U = 0$ (η_U^0) and $U = 1, U_3 = 0$ (π_0^U) states; in analogy with Eq. (8.124) one has

$$\pi_U^0 = -\tfrac{1}{2}\pi^0 + \tfrac{\sqrt{3}}{2}\eta^0\ ,$$

$$\eta_U^0 = \tfrac{\sqrt{3}}{2}\pi^0 + \tfrac{1}{2}\eta^0\ ,$$

i.e.

$$\pi^0 = -\tfrac{1}{2}\pi_U^0 + \tfrac{\sqrt{3}}{2}\eta_U^0\ ,$$

$$\eta^0 = \tfrac{\sqrt{3}}{2}\pi_U^0 + \tfrac{1}{2}\eta_U^0\ .$$

Only the η_U^0 term can contribute to the decays $\pi^0 \to 2\gamma$ and $\eta^0 \to 2\gamma$; therefore, the corresponding amplitudes satisfy the relation

$$A(\pi^0 \to 2\gamma) = \sqrt{3}\,A(\eta^0 \to 2\gamma)$$

Taking into account the phase space corrections, the ratio of the decay widths is given by

$$\frac{\Gamma(\pi^0 \to 2\gamma)}{\Gamma(\eta^0 \to 2\gamma)} = \frac{1}{3}\left(\frac{m_\eta}{m_\pi}\right)^3 .$$

8.8 We recall Eq. (8.154):

$$\omega = \cos\theta\,\omega_1 + \sin\theta\,\omega_8\ ,$$

$$\phi = -\sin\theta\,\omega_1 + \cos\theta\,\omega_8\ ,$$

i.e.

$$\begin{pmatrix}\omega \\ \phi\end{pmatrix} = R(\theta)\begin{pmatrix}\omega_1 \\ \omega_8\end{pmatrix}$$

with

$$R(\theta) = \begin{pmatrix} \cos\theta & \sin\theta \\ -\sin\theta & \cos\theta \end{pmatrix} .$$

The two mass matrices

$$M^2_{\omega_1,\omega_8} = \begin{pmatrix} m_1{}^2 & m_{18}{}^2 \\ m_{18}{}^2 & m_8{}^2 \end{pmatrix}, \qquad M^2_{\omega,\phi} = \begin{pmatrix} m_\omega{}^2 & 0 \\ 0 & m_\phi{}^2 \end{pmatrix}$$

are connected by

$$R(\theta) M^2_{\omega_1,\omega_8} R^{-1}(\theta) = M^2_{\omega,\phi}$$

from which Eqs. (8.156) and (8.157) follow.

The two reactions $\omega \to e^+ e^-$, $\phi \to e^+ e^-$ occur through an intermediate photon, and

$$\omega_1 \not\to \gamma, \qquad \qquad \omega_8 \to \gamma,$$

since γ transforms as the $U = Q = 0$ component of an octet. Then, neglecting the phase space correction, one obtains for the ratio of the decay amplitudes:

$$\frac{A(\omega \to e^+ e^-)}{A(\phi \to e^+ e^-)} = \tan \theta.$$

8.9 Taking into account Eq. (8.124) one gets, from U-spin invariance:

$$\mu_{\Sigma_u^0} = <\Sigma_u^0|\,\mu\,|\Sigma_u^0> = <-\tfrac{1}{2}\Sigma^0 + \tfrac{\sqrt{3}}{2}\Lambda^0|\,\mu\,|-\tfrac{1}{2}\Sigma^0 + \tfrac{\sqrt{3}}{2}\Lambda^0> =$$

$$= \tfrac{1}{4}\mu_{\Sigma^0} + \tfrac{3}{4}\mu_{\Lambda_0} - \tfrac{\sqrt{3}}{2}\mu_{\Lambda^0\Sigma^0},$$

$$<\Sigma_u^0|\,\mu\,|\Lambda_u^0> = <-\tfrac{1}{2}\Sigma^0 + \tfrac{\sqrt{3}}{2}\Lambda^0|\,\mu\,|\tfrac{\sqrt{3}}{2}\Sigma^0 + \tfrac{1}{2}\Lambda^0> =$$

$$= \tfrac{\sqrt{3}}{4}\mu_{\Sigma_0} + \tfrac{\sqrt{3}}{4}\mu_{\Lambda_0} + \tfrac{3}{2}\mu_{\Lambda^0\Sigma^0}.$$

The required relations follow immediately from the above equations.

8.10 We can write the Casimir operator in terms of the shift operators (see Problem 8.6) and of I_3, Y, in the form

$$F^2 = F_i F_i = \tfrac{1}{2}\{I_+, I_-\} + \tfrac{1}{2}\{U_+, U_-\} + \tfrac{1}{2}\{V_+, V_-\} + I_3^2 + \tfrac{3}{4}Y^2.$$

Since F^2 is an invariant operator, its eigenvalue in a given IR can be obtained by applying it to a generic state. It is convenient to take into account the so-called *maximum state*, ψ_{max}, defined as the state with the maximum eigenvalue of I_3. It is easy to verify that for each IR there is only one such a state, with a specific eigenvalue of Y. Moreover, because of its properties

$$I_+\psi = V_+\psi = U_-\psi = 0.$$

Looking at the eight dimensional IR, ψ_{max} corresponds to the eigenvalues 1 and 0 for I_3 and Y, respectively. Since

$$I_+ I_- \psi_{\max} = 2\psi_{\max} , \qquad U_+ U_- \psi_{\max} = \psi_{\max} , \qquad V_+ V_- \psi_{\max} = \psi_{\max} ,$$

$$I_3^2 \psi_{\max} = \psi_{\max} , \qquad Y^2 \psi_{\max} = 0 ,$$

the eigenvalue of F^2 in the adjoint IR is 3. Making use of the definition of F^2 and of its eigenvalue, Eq. (8.145) is then checked, once Eqs. (8.140) and (8.94) are taken into account.

More generally, it is possible to obtain the expression of the eigenvalues of F^2 in terms of the two integers p_1 and p_2 which characterize a given IR. Since p_1 and p_2 represent the number of times that the representations 3 and $\bar{3}$ are present in the direct product from which $D(p_1, p_2)$ is obtained, then the maximun eigenvalue of I_3 is given by

$$I_3 \psi_{\max} = \tfrac{1}{2}(p_1 + p_2)\psi_{\max} ,$$

while the eigenvalue of Y is

$$Y \psi_{\max} = \tfrac{1}{3}(p_1 - p_2)\psi_{\max} .$$

From the commutation relations of the shifting operators, one obtains:

$$\{ I_+ , I_- \}\psi_{\max} = I_+ I_- \psi_{\max} = [I_+ , I_-]\psi_{\max} = 2I_3\psi_{\max} = (p_1 + p_2)\psi_{\max} ,$$

$$\{ U_+ , U_- \}\psi_{\max} = U_- U_+ \psi_{\max} = [U_+ , U_-]\psi_{\max} = -2U_3\psi_{\max} = p_2\psi_{\max} ,$$

$$\{ V_+ , V_- \}\psi_{\max} = V_+ V_- \psi_{\max} = [V_+ , V_-]\psi_{\max} = 2V_3\psi_{\max} = p_1\psi_{\max} .$$

By inserting these relations in the expression of F^2, one finds the general expression for its eigenvalues:

$$\tfrac{1}{3}(p_1^2 + p_2^2 + p_1 p_2) + p_1 + p_2 .$$

8.11 By inserting the λ matrices in the Jacobi identity

$$[A, [B, C]] + [B, [C, A]] + [C, [A, B]] = 0$$

and making use of the commutation relations (8.86), one finds:

$$f_{j\ell k}[\lambda_i, \lambda_k] + f_{\ell i k}[\lambda_j, \lambda_k] + f_{ijk}[\lambda_\ell, \lambda_k] = 0 .$$

By multiplying by λ_m and taking the traces according to the relation (8.89), one obtains the identity

$$f_{j\ell k} f_{ikm} + f_{\ell i k} f_{jkm} + f_{ijk} f_{\ell km} = 0 , \qquad (a)$$

and, taking into account the definition (8.137), one gets the commutation relations (8.92)

$$[F_i, F_j] = i f_{ijk} F_k .$$

In analogous way, making use of the identity

$$[A, \{B, C\}] = \{[A, B], C\} + \{[A, C], B\} ,$$

and of the relations (8.89) and (8.90), one finds

$$d_{jk\ell} f_{i\ell m} = f_{ij\ell} d_{\ell km} + f_{ik\ell} d_{\ell jm} , \qquad (b)$$

which, taking into account the definitions (8.137) and (8.138), corresponds to Eq. (8.139):

$$[F_i, D_j] = i f_{ijk} D_k .$$

By multiplying the previous relation by $-if_{ij\ell}$ and summing over (i, j) one finds

$$-if_{ij\ell}[F_i, D_j] = F^2 D_\ell , \qquad (c)$$

where $F^2 = F_i F_i$ is the quadratic Casimir operator in the adjoint $D(1,1)$ representation; it is equal to 3 times the unit matrix (make use of the expression obtained for F^2 in Problem 8.10 with $p_1 = p_2 = 1$). On the other hand, the l.h.s., making use of the relation (b) and of the symmetry properties of the f and d coefficients, can be written in the form:

$$-if_{ij\ell}[F_i, D_j] = 2F^2 D_k - d_{\ell mn}\{F_m, F_n\} = 2F^2 D_\ell - 2d_{\ell mn} F_m F_n . \qquad (d)$$

By comparison of (c) and (d) one gets

$$F^2 D_\ell = 2d_{\ell mn} F_m F_n ,$$

which coincides with Eq. (8.140).

8.12 In $SU(6)$ the diquark states d correspond to

$$6 \ \otimes \ 6 \ = \ 21 \ \oplus \ 15 ,$$

so that the S-wave symmetric states belong to the IR 21. Its content in term of the subgroup $SU(3) \otimes SU(2)_S$ is given by

$$21 = (6, 3) \oplus (\bar{3}, 1) .$$

The baryon states will be described as bound dq systems:

$$21 \quad \otimes \quad 6 \quad = \quad 56 \quad \oplus \quad 70 \, .$$

Including a relative $d-q$ orbital momentum L, one can easily verify that the symmetric states belong to the multiplets $(56, L_{\text{even}}^{+})$ and $(70, L_{\text{odd}}^{-})$, which are the only ones definitively established.

On the other hand, if one build the $d\bar{d}$ mesons, one obtains lots of exotic states. In fact, according to

$$21 \otimes \overline{21} = 1 \oplus 35 \oplus 405 \, ,$$

one gets, besides the IR 1 and 35, the $SU(6)$ multiplet 405 whose $SU(3) \otimes SU(2)_S$ content is:

$$405 = (1 + 8 + 27, 1) \oplus (8 + 8 + 10 + \overline{10} + 27, 3) \oplus (1 + 8 + 27, 5) \, .$$

8.13 Taking into account the spin, the four quarks (u, d, s, c) belong to the IR 8 of $SU(8)$, which corresponds, in term of the subgroup $SU(4) \otimes SU(2)_S$, to

$$8 = (4, 2) \, .$$

The meson states are assigned to the representation

$$8 \otimes \bar{8} = 1 \oplus 83 \, ,$$

and the 83-multiplet has the following content in terms of the above subgroups:

$$83 = (15, 1) \oplus (15 + 1, 3) \, .$$

We see that one can fit into the same multiplet both the 15 pseudoscalar mesons $(K, \overline{K}, \pi, \eta, F, D, \overline{F}, \overline{D})$ and the 16 vector mesons $(K^{*}, \overline{K}^{*}, \rho, \omega, \phi, F^{*}, D^{*}, \overline{F}^{*}, \overline{D}^{*})$, where F^{*} and D^{*} are the vector couterparts of the scalar mesons F and D.

The baryon states are classified according to the IR's:

$$8 \quad \otimes \quad 8 \quad \otimes \quad 8 \quad = \quad 120 \quad \oplus \quad 168 \quad \oplus \quad 168 \quad \oplus \quad 56 \, .$$

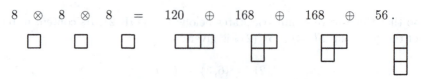

The lowest S-wave states can be fitted into the completely symmetric IR 120:

$$120 \quad = \quad (20,4) \quad \oplus \quad (20',2) \ .$$

$$\square\square\square \qquad (\square\square\square\ ,\ \square\square\square) \qquad \left(\begin{array}{c}\square\square\\\square\end{array}\ ,\ \square\right)$$

The content of the other two IR's is given by;

$$168 = (20,2) \oplus (20',4) \oplus (20',2) \oplus (\bar{4},2)$$

$$56 = (20',2) \oplus (\bar{4},4) \ .$$

8.14 In analogy with the strangeness S ($S = -1$ for the s-quark), a quantum number b is introduced for *beauty* ($b = -1$ for the b-quark). Then the Gell-Mann Nishijima formula (8.45) is replaced by (compare with Eqs. (8.180), (8.181))

$$Q = I_3 + \tfrac{1}{2}(B + S + b) \qquad \text{and} \qquad Y = B + S + b \ .$$

The 0^- b-mesons can be assigned to the 15-multiplet of $SU(4)$ and the situation is analogous to that represented in Fig. 8.15 (replacing C by $-b$): there are an iso-doublet $B^+(u\bar{b})$, $B^0(d\bar{b})$ and an iso-singlet $B_s^0(s\bar{b})$, all with $b = 1$, and the corresponding antiparticles with $b = -1$: $B^-(\bar{u}b)$, $\overline{B}^0(\bar{d}b)$ and $\overline{B}_s^0(\bar{s}b)$.

The situation for the 1^- b-mesons is similar to that of the charmed ones: they should be assigned to a $15 + 1$ multiplet with a mixing giving rise to a pure $Y(b\bar{b})$ state, which is the analogue of the $J/\psi(c\bar{c})$ state.

The $\tfrac{1}{2}^+$ b-baryons can be assigned to the $20'$ multiplet (we limit ourselves to this case, since many of these states have been observed experimentally) and the situation is the following (where q stands for u or d):

- Baryons with $b = -1$: a triplet $\Sigma_b(qqb)$, a doublet $\Xi_b(qsb)$ and two singlets $\Lambda_b(udb)$ and $\Omega_b(ssb)$.
- Baryons with $b = -2$: a doublet $\Xi_{bb}(qbb)$ and a singlet $\Omega_{bb}(sbb)$.

Problems of Chapter 9

9.1 We rewrite here the Lagrangian (9.46)

$$\mathcal{L}(x) = \sum_j \bar{q}^j(x)(i\gamma^\mu D_\mu - m_j)q^j(x) - \frac{1}{2}\text{Tr}(G_{\mu\nu}G^{\mu\nu}),$$

where D_μ is the *covariant derivative*

$$\partial_\mu \to D_\mu = \partial_\mu + ig_s G_\mu(x) \,,$$

and $G_{\mu\nu}$ the *field strength*

$$G_{\mu\nu} = \partial_\mu G_\nu - \partial_\nu G_\mu + ig_s[G_\mu, G_\nu] \,,$$

and require its invariance under the transformation

$$q(x) \to U(x)q(x).$$

In order to obtain the invariance of the first term of $\mathcal{L}$, we write the transformation properties to the quantity $D_\mu q(x)$:

$$(D_\mu q(x))' = U(D_\mu q(x)) = U\partial_\mu q + ig_s U G_\mu q.$$

and require that they correspond with what we expect from its invariance

$$(D_\mu q(x))' = (\partial_\mu + ig_s G'_\mu)Uq = (\partial_\mu U)q + U\partial_\mu q + ig_s G'_\mu Uq.$$

A comparison of these two relations gives the required transformation properties of G_μ

$$G'_\mu = U G_\mu U^\dagger + \frac{i}{g_s}(\partial_\mu U)U^\dagger \,,$$

which coincide with Eq. (9.50).

We can prove the invariance of the second term in $\mathcal{L}$ by considering the following commutator:

$$[D_\mu, D_\nu]q(x) = [\partial_\mu + ig_s G_\mu \,, \partial_\nu + ig_s G_\nu]q(x) =$$
$$= ig_s(\partial_\mu G_\nu - \partial_\nu G_\mu)q(x) - g_s^2[G_\mu, G_\nu]q(x).$$

Comparing with the expression of the field strength $G_{\mu\nu}$, we obtain the relation

$$G_{\mu\nu} = -\frac{i}{g_s}[D_\mu, D_\nu].$$

From this relation we can derive the transformation properties of the field strength $G_{\mu\nu}$

$$ig_s G'_{\mu\nu} = [D'_\mu, D'_\nu] = [U D_\mu U^\dagger, U D_\nu U^\dagger] = U[D_\mu, D_\nu]U^\dagger = ig_s U G_{\mu\nu} U^\dagger \,.$$

It follows

$$Tr(G_{\mu\nu}G^{\mu\nu}) = Tr(U G_{\mu\nu} U^\dagger U G^{\mu\nu} U^\dagger) = Tr(G_{\mu\nu}G^{\mu\nu}),$$

which concludes the proof of the gauge invariance of the Lagrangian.

9.2 If we adopt the compact notation

$$\Phi = \begin{pmatrix} \phi_1 \\ \phi_2 \\ \vdots \\ \phi_N \end{pmatrix}$$

$\mathcal{L}$ can be written in the form

$$\mathcal{L} = \tfrac{1}{2}\partial_\mu\tilde{\Phi}\,\partial^\mu\Phi - V(\Phi^2)\,,$$

where

$$V(\Phi^2) = \tfrac{1}{2}\mu^2\Phi^2 + \tfrac{1}{4}\lambda\Phi^4$$

is the "potential" and

$$\Phi^2 = \tilde{\Phi}\Phi = \sum_{i=1}^{N}\phi_i^2\,.$$

The minimum of the potential corresponds to the v.e.v.

$$(\Phi^2)_0 = -\frac{\mu^2}{\lambda} = v^2\,,$$

and one can choose a coordinate system such that

$$(\Phi)_0 = \begin{pmatrix} 0 \\ \vdots \\ 0 \\ v \end{pmatrix}\,.$$

With the definition

$$\Phi(x) = \begin{pmatrix} \xi_1(x) \\ \vdots \\ \xi_{N-1}(x) \\ \eta(x) + v \end{pmatrix}\,.$$

$\mathcal{L}$ can be rewritten in the form

$$\mathcal{L} = \tfrac{1}{2}\partial^\mu\eta\,\partial_\mu\eta - \tfrac{1}{2}\sum_{i=1}^{N-1}\partial^\mu\xi_i\,\partial_\mu\xi_i + \mu^2\eta^2 - \tfrac{1}{4}\lambda\eta^4 - \tfrac{1}{4}\lambda\sum_{i=1}^{N-1}\xi_i^4 - \tfrac{1}{2}\lambda\eta^2\sum_{i=1}^{N-1}\xi_i^2\,.$$

Only the field $\eta(x)$ has a mass different from zero, with the value $m_\eta^2 = -2\mu^2$, while the $N-1$ fields $\xi_i(x)$ are massless: they are the Goldstone bosons. According to the Goldstone theorem, their number is equal to $n - n'$, where $n = N(N-1)/2$ and $n' = (N-1)(N-2)/2$ are the numbers of generators of the group $O(N)$ and of the subgroup $O(N-1)$, respectively; in fact, $n-n' = N-1$.

9.3 The leading $W^\pm$-exchange contribution to the $\nu_e e^-$ invariant scattering amplitude is given by:

$$\mathcal{M}(W^\pm) = \left(\frac{g}{2\sqrt{2}}\right)^2 \left[\bar{\nu}_e \gamma^\mu (1-\gamma_5) e \frac{-ig_{\mu\nu}}{q^2 - M_W^2} \bar{e}\gamma_\nu (1-\gamma_5)\nu_e\right]$$

and, since the momentum transfer satisfies the condition $q^2 \ll M_W^2$, one can write:

$$\mathcal{M}(W^\pm) \approx -i \frac{g^2}{8M_W^2}\left[\bar{\nu}_e \gamma^\mu (1-\gamma_5)e\right]\left[\bar{e}\gamma_\mu (1-\gamma_5)\nu_e\right] .$$

Starting from the Fermi Lagrangian (9.105), one obtains a similar expression: the only difference is that the factor $\frac{g^2}{8M_W^2}$ is replaced by $\frac{G_F}{\sqrt{2}}$, so that the requested relation is the following:

$$G_F = \frac{g^2}{4\sqrt{2}M_W^2} = \frac{1}{\sqrt{2}v^2} .$$

The analogous contribution due to the Z^0-exchange is given by

$$\mathcal{M}(Z^0) \approx -i \frac{g^2}{8M_Z^2 \cos^2\theta_w}\left[\bar{\nu}_e \gamma^\mu (1-\gamma_5)\nu_e\right]\left[\bar{e}\gamma_\mu(1-\gamma_5)e + 4\sin^2\theta_w \bar{e}\gamma_\mu e\right] .$$

The ratio between the couplings of the neutral and the charged current is then given by (comparing with Eq. (9.135))

$$\rho = \frac{M_W^2}{M_Z^2 \cos^2\theta_w} = 1 .$$

9.4 The real scalar triplet and its v.e.v. can be written as follows

$$\Psi = \begin{pmatrix} \Psi_1 \\ \Psi_2 \\ \Psi_3 \end{pmatrix} , \qquad <\Psi>_0 = \begin{pmatrix} 0 \\ V \\ 0 \end{pmatrix}$$

and we have to include in the Lagrangian (9.118) the additional terms

$$\tfrac{1}{2}(D_\mu\tilde{\Psi})(D^\mu\Psi) - V(\Psi^2) ,$$

where in particular, being $Y_\psi = 0$,

$$D_\mu\Psi = \partial_\mu\Psi - ig I_i A_\mu^i \Psi .$$

The contribution to the vector boson masses comes from the covariant derivative when the fields are shifted and are substituted by their v.e.v.'s. Concerning the field Ψ, its contribution is then given by

$$\tfrac{1}{2}[(D_\mu\tilde{\Psi})(D^\mu\Psi)]_0 .$$

However, this term contributes only to the mass of the charged bosons: in fact, since the Ψ component which develops the v.e.v. different from zero must be a neutral component, from the assumption $Y_\Psi = 0$ we get also $I_{3,\Psi} = 0$. Accordingly

$$\frac{1}{2}[(D_\mu\tilde{\Psi})(D^\mu\Psi)]_0 = \frac{1}{2}g^2V^2\left(I_{1,\Psi}A^{(1)}_\mu + I_{2,\Psi}A^{(2)}_\mu\right)^\dagger\left(I_{1,\Psi}A^{(1)\mu} + I_{2,\Psi}A^{(2)\mu}\right) =$$
$$= \frac{1}{4}g^2V^2\left[I^-W^{(-)}_\mu + I^+W^{(+)}_\mu\right]\left[I^+W^{(+)\mu} + I^-W^{(-)\mu}\right] =$$
$$= \frac{1}{4}g^2V^2\left[I^+I^- + I^-I^+\right]W^{(+)}_\mu W^{(-)\mu} =$$
$$= \frac{1}{4}g^2V^2 2\{I^2_\Psi - I^2_{3,\Psi}\} = \frac{1}{2}g^2V^2\, I_\Psi(I_\Psi + 1) = g^2V^2 \ .$$

since $I_{3,\Psi} = 0$ and Ψ is a triplet ($I_\Psi = 1$) of weak isospin. This contribution must be added to the term $\frac{1}{4}g^2v^2$ coming from the isospin doublet Φ of Eq. (9.125): we see that the value of M_W of the Standard Model is modified into

$$M'_W = \frac{1}{2}g\sqrt{v^2 + 4V^2} \ .$$

The mass M_Z is not changed, so that the ratio M_W/M_Z is replaced by

$$\frac{M'_W}{M_Z} = \frac{g}{\sqrt{g^2 + g'^2}}\sqrt{1 + 4\frac{V^2}{v^2}} \ .$$

If only the triplet Ψ were present in the scalar sector, only $SU(2)_L$ would be broken and the vacuum symmetry would be $U(1)_{I_3} \otimes U(1)_Y$. As a consequence, only the charged vector fields $W^\pm_\mu$ acquire mass different from zero, while the two fields $A^{(3)}_\mu$ and B_μ remain massless.

9.5 Let us generalize the Lagrangian density $\mathcal{L}_{\text{gauge}}$ of Eq. (9.118) by introducing several Higgs fields ϕ_ℓ. The terms contributing to the masses of the vector bosons and to their couplings with the Higgs fields are given by

$$\sum_\ell \left\{\left[gI_{i,\ell}A^i_\mu + g'\frac{1}{2}Y_\ell B_\mu\right]\phi_\ell\right\}^\dagger \left\{\left[gI_{i,\ell}A^i_\mu + g'\frac{1}{2}Y_\ell B^\mu\right]\phi_\ell\right\} \ .$$

This expression can be rewritten in terms of the physical fields, introducing the charged vector fields $W^{(\pm)}_\mu$ of Eq. (9.128), the neutral fields A^μ and Z^μ through the relations (9.132), and the operators $I_\pm = I_1 \pm iI_2$. For the terms in the second bracket, we obtain

$$\left\{\left[gI_{i,\ell}A^i_\mu + g'\frac{1}{2}Y_\ell B^\mu\right]\phi_\ell\right\} = \left\{g\left[I^+_\ell W^{(+)}_\mu + I^-_\ell W^{(-)}_\mu\right]\phi_\ell + \right.$$
$$\left. + \frac{1}{\sqrt{g^2 + g'^2}}\left[gg'(I_{3\ell} + \frac{1}{2}Y_\ell)A_\mu + (g^2 I_{3,\ell} - g'^2\frac{1}{2}Y_\ell)Z_\mu\right]\phi_\ell\right\} \ .$$

We suppose that each Higgs field ϕ_ℓ develops a vacuum expectation value different from zero, defined by $\frac{1}{\sqrt{2}}v_\ell$, where v_ℓ must correspond to a neutral component of ϕ_ℓ, otherwise also the electromagnetic gauge invariance would be broken. This means, on the basis of the relation (9.112), that for each v.e.v. v_ℓ one gets $I_{3,\ell} + \frac{1}{2}Y_\ell = 0$, independently of the specific representation of ϕ_ℓ. It follows that the coefficient of the e.m. field A_μ goes to zero (A_μ is massless, as required) and the previous relation becomes

$$\left\{ \frac{1}{\sqrt{2}}g\left[I_\ell^+ W_\mu^{(+)} + I_\ell^- W_\mu^{(-)}\right] - \sqrt{g^2 + g'^2}\,\frac{1}{2}Y_\ell Z_\mu \right\}\phi_\ell .$$

Finally, introducing the v.e.v. v_ℓ, we get

$$\mathcal{L}_{\text{mass}} = \frac{1}{2}\sum_\ell v_\ell^2 \left\{ \frac{1}{2}g^2\left[I_\ell^+ I_\ell^- + I_\ell^- I_\ell^+\right]W_\mu^{(+)}W^{(-)\mu} + (g^2 + g'^2)\frac{1}{4}Y_\ell^2 Z_\mu Z^\mu \right\} =$$

$$= \frac{1}{2}\sum_\ell v_\ell^2 \left\{ g^2\left[I_\ell(I_\ell + 1) - \frac{1}{4}Y_\ell^2\right]W_\mu^{(+)}W^{(-)\mu} + (g^2 + g'^2)\frac{1}{4}Y_\ell^2 Z_\mu Z^\mu \right\},$$

and, by taking the ratio of the two squared masses, we obtain the required expression

$$\rho = \frac{\sum_\ell v_\ell^2[I_\ell(I_\ell + 1) - \frac{1}{4}Y_\ell^2]}{\frac{1}{2}\sum_\ell v_\ell^2 Y_\ell^2} .$$

9.6 The Lagrangian contains two terms

$$(D_{L\,\mu}\Phi_L)^\dagger(D_L^{\,\mu}\Phi_L) + (D_{R\,\mu}\Phi_R)^\dagger(D_R^{\,\mu}\Phi_R) ,$$

where

$$(D_{L,R})_\mu = \partial_\mu - ig\frac{1}{2}\tau_i(A_{L,R}^i)_\mu - \frac{1}{2}ig'B_\mu .$$

Introducing the charged vector fields

$$(W_{L,R}^\pm)_\mu = \frac{1}{\sqrt{2}}\left[(A_{L,R}^1)_\mu \mp i(A_{L,R}^2)_\mu\right] ,$$

one obtains

$$(D_{L,R})_\mu\Phi_{L,R} = -\frac{i}{2}\left\{ g(W_{L,R}^+)_\mu \begin{pmatrix} v_{L,R} \\ 0 \end{pmatrix} - \frac{1}{\sqrt{2}}\left[g(A_{L,R}^3)_\mu + g'B_\mu\right]\begin{pmatrix} 0 \\ v_{L,R} \end{pmatrix} \right\} .$$

It follows that all the charged vector bosons become massive with masses given by

$$M_{W_L} = \frac{1}{2}gv_L , \qquad M_{W_R} = \frac{1}{2}gv_R .$$

Since there is no experimental evidence of other vector bosons besides those of the Standard Model, one must assume : $v_R \gg v_L$.

For the squared masses of the neutral gauge bosons we obtain the matrix

$$M^2 = \frac{1}{4}g^2 v_R^2 \begin{pmatrix} y & 0 & -\kappa y \\ 0 & 1 & -\kappa \\ -\kappa y & -\kappa & (1+y)\kappa^2 \end{pmatrix} ,$$

where

$$\kappa = \frac{g'}{g} \qquad \text{and} \qquad y = \frac{v_L^2}{v_R^2} \,.$$

Since $\det M^2 = 0$, one eigenvalue is equal to zero. For the other two eigenvalues, in the case $v_R \gg v_L$, one gets the approximate values:

$$M_1^2 \approx \tfrac{1}{4}g^2 v_L^2 \frac{1+2\kappa}{\kappa^2+1} \,, \qquad M_2^2 \approx \tfrac{1}{4}g^2 v_R^2 (\kappa^2+1) \,.$$

9.7 Let us consider first the IR $\bar{5}$ of $SU(5)$; its content in terms of the subgroup $SU(3) \otimes SU(2)$, according to Eq. (9.169), is given by

$$\bar{5} = (1,2)_{-1} + (\bar{3},1)_{-\frac{2}{3}} \,,$$

where the subscripts stand for the values of Y which can be read from Tables 9.2 and 9.3.

The IR 24 can be obtained by the direct product

$$5 \otimes \bar{5} = 1 \oplus 24$$

and its content in terms of the subgroup G_{SM} follows from the above relation:

$$24 = (8,1)_0 \oplus (1,1)_0 \oplus (1,3)_0 \oplus (3,2)_{-\frac{5}{3}} \oplus (\bar{3},2)_{+\frac{5}{3}} \,.$$

Let us define

$$(\bar{3},2)_{+\frac{5}{3}} \rightarrow \begin{pmatrix} X_1 & X_2 & X_3 \\ Y_1 & Y_2 & Y_3 \end{pmatrix} \,.$$

From the usual relation $Q = I_3 + \tfrac{1}{2}Y$ we obtain the electric charges of the X and Y particles:

$$Q(X_i) = \tfrac{4}{3} \,, \qquad Q(Y_i) = \tfrac{1}{3} \,.$$

The multiplet $(3,2)_{-\frac{5}{3}}$ contains the antiparticles $\overline{X}$ and $\overline{Y}$ which have opposite values of Q. These particles are called lepto-quarks because they have the same quantum numbers of the lepton-quark pairs. They mediate new interactions which violate baryon and lepton numbers.

9.8 We recall the decomposition of the 24-multiplet Φ in terms of the subgroup $G_{SM} = SU(3)_c \otimes SU(2)_L \otimes U(1)_Y$:

$$24 = (1,1)_0 \oplus (1,3)_0 \oplus (8,1)_0 \oplus (3,2)_{-5/3} \oplus (\bar{3},2)_{+5/3} \,,$$

and write Φ as a 5×5 traceless tensor. It is than clear that, in order to preserve the G_{SM} symmetry, the v.e.v. $< \Phi >_0$ must behave as $(1,1)_0$ and then it must have the following form

$$<\Phi>_0 = \begin{pmatrix} a & & & & \\ & a & & & \\ & & a & & \\ & & & b & \\ & & & & b \end{pmatrix}$$

and, since the matrix is traceless, it can be written as

$$<\Phi>_0 = V \begin{pmatrix} 1 & & & & \\ & 1 & & & \\ & & 1 & & \\ & & & -\frac{3}{2} & \\ & & & & -\frac{3}{2} \end{pmatrix}$$

The vacuum symmetry is G_{SM}; therefore the 12 gauge vectors fields of G_{SM} remain massless, and the other twelve acquire mass: they are the multiplets $(3,2)$ and $(\overline{3},2)$ considered in Problem 9.7.

In order to break also

$$G_{SM} \rightarrow SU(3)_c \otimes U(1)_Q$$

one needs a scalar multiplet containing the doublet ϕ of the Standard Model. The minimal choice is then a field

$$\phi_5 \sim (3,1) \oplus (1,2)$$

with v.e.v. which transforms as the neutral component of (1,2).

Bibliography

We list here some of the books which were used while preparing the manuscript, and other books which can be consulted to broaden and deepen the different topics. The titles are collected according to the main subjects, that often concern more than one chapter. We are sure that there are many other good books on the same topics, and the list is limited to those we know better.

Group and Representation Theory

Baker, A., *Matrix Groups: An Introduction to Lie Group Theory*, Springer-Verlag, London, 2002.

Barut, A.O. and Raczka, R., *Theory of Group Representations and Applications*, Polish Scientific Publishers, Warsaw, 1980.

Bröcker, T. and Dieck, T.t., *Representations of Compact Lie Groups*, Springer-Verlag, Berlin, 1995.

Cahn, R.N., *Semi-Simple Lie Algebras and Their Representations*, Benjamin-Cummings, Menlo Park, California, 1984.

Chevalley, C., *Theory of Lie Groups*, Princeton University Press, Princeton, 1946.

Hamermesh, M., *Group Theory and its Application to Physical Problems*, Addison-Wesley, Reading, Mass, 1962.

Herman, R., *Lie Groups for Physicists*, Benjamin, New York, 1966

Humphreys, J.E., *Introduction to Lie Algebras and Representation Theory*, Springer-Verlag, New York, 1972.

Iachello, F., *Lie Algebras and Applications*, Springer, Berlin Heidelberg, 2006.

Mackey, G.W., *The Theory of Unitary Group Representations*, University of Chicago Press, Chicago, 1976.

Nash, C. and Sen, S., *Topology and Geometry for Physicists*, Academic Press, New York and London, 1983.

G. Costa and G. Fogli, *Symmetries and Group Theory in Particle Physics*, 281
Lecture Notes in Physics 823, DOI: 10.1007/978-3-642-15482-9,
© Springer-Verlag Berlin Heidelberg 2012

Sagle, A.A. and Walde, R.E., *Introduction to Lie Groups and Lie Algebras*, Academic Press, New York and London, 1973.

Varadarajan, V.S., *Lie Groups, Lie Algebras, and their Representations*, Springer, New York, 1984.

Wigner, E.P., *Group Theory and its Application to Quantum Mechanics of Atomic Spectra*, Academic Press, New York, 1959.

Rotation, Lorentz and Poincaré Groups

Carmeli, M. and Malin, S., *Representations of the Rotation and Lorentz Groups. An Introduction*, Marcel Dekker, New York, 1976.

Edmonds, A.R., *Angular Momentum in Quantum Mechanics*, Princeton University Press, Princeton, 1957.

Gelfand, I.M., Minlos, R.A. and Shapiro, Z.Ya., *Representations of the Rotation and Lorentz Groups and their Applications*, Pergamon Press, London, 1963.

Naimark, M.A., *Linear Representations of the Lorentz Group*, Pergamon Press, London, 1964.

Rose, M.E., *Elementary Theory of Angular Momentum*, J. Wiley and Sons, New York, 1957.

Quantum Mechanics and Relativity

Bjorken, J.D. and Drell, S.D., *Relativistic Quantum Mechanics*, McGraw-Hill, New York, 1964.

Greiner, W., *Quantum Mechanics, An Introduction*, Springer-Verlag, Berlin, 1989.

Halpern, F.R., *Special Relativity and Quantum Mechanics*, Prentice-Hall, Englewood Cliffs, N.J., 1968.

Landau, L.D. and Lifshitz, E.M., *Quantum Mechanics*, Pergamon Press, London, 1985.

Merzbacher, E., *Quantum Mechanics*, J. Wiley and Sons, New York, 2003.

Messiah, A., *Quantum Mechanics*, North Holland, Amsterdam, 1962.

Sakurai, J.J., *Advanced Quantum Mechanics*, Addison-Wesley, Reading, Mass, 1967.

Scadron, M.D., *Advanced Quantum Theory*, Springer-Verlag, Berlin, 1979.

Schiff, L.I., *Quantum Mechanics*, McGraw-Hill, New York, 1968.

Schwabl, F., *Quantum Mechanics*, Springer–Verlag, Berlin, 1992.

Schwartz, H.M., *Introduction to Special Relativity*, McGraw–Hill, New York, 1968.

Strocchi, F., *Elements of Quantum Mechanics of Infinite Systems*, World Scientific, Singapore, 1985.

Quantum Field Theory

Aitchison, I.J.R. and Hey, A.J.G., *Gauge Theories in Particle Physics*, Adam Hilger, Bristol, 1989.

Bailin, D. and Love, A., *Introduction to Gauge Field Theory*, Adam Hilger, Bristol, 1986.

Bailin, D. and Love, A., *Supersymmetric Gauge Field Theory and String Theory*, Institute of Physics Publishing, Bristol, 1994.

Bjorken, J.D. and Drell, S.D., *Relativistic Quantum Fields*, McGraw–Hill, New York, 1965.

Bogoliubov, N.N. and Shirkov, D.V., *Introduction to the Theory of Quantized Fields*, Interscience, New York, 1959.

Faddeev, L.D. and Slavnov, A.A., *Gauge Fields: Introduction to Quantum Theory*, Benjamin-Cummins, Reading, Mass, 1980.

Itzykson, C. and Zuber, J-B., *Quantum Filed Theory*, McGraw-Hill, New York, 1980.

Mandl, F. and Shaw, G., *Quantum Field Theory*, Wiley, New York, 1984.

Peskin, M.E. and Schroeder, D.V., *An Introduction to Quantum Field Theory*, Addison-Wesley, Menlo Park, California, 1995.

Quigg, C., *Gauge Theories of Strong, Weak and Electromagnetic Interactions*, Cambridge University Press, Cambridge, England, 1983.

Schweber, S.S., *An Introduction to Relativistic Quantum Field Theory*, Harper and Row, New York, 1961.

Taylor, J.C., *Gauge Theories of Weak Interactions*, Cambridge University Press, Cambridge, England, 1976.

Weinberg, S., *The Quantum Theory of Fields, Vol. I, Foundations*, Cambridge University Press, Cambridge, Mass, 1995.

Weinberg, S., *The Quantum Theory of Fields, Vol. II, Modern Applications*, Cambridge University Press, Cambridge, Mass, 1996.

Particle Physics

Bettini, A., *Introduction to Elementary Particle Physics*, Cambridge University Press, Cambridge, England, 2008.

Bigi, I.I. and Sanda, A.I., *CP Violation*, Cambridge University Press, Cambridge, England, 2000.

Bilenky, S,M., *Introduction to the Physics of Electroweak Interactions*, Pergamon Press, London, 1982.

Cahn, R.N. and Goldhaber, G., *The Experimental Foundations of Particle Physics*, Cambridge University Press, Cambridge, England, 1989.

Close, F.E., *An Introduction to Quarks and Partons*, Academic Press, London, 1979.

Commins, E.D. and Bucksbaum, P.H., *Weak Interactions of Leptons and Quarks*, Cambridge University Press, Cambridge, England, 1983.

Feld, B.T., *Models of Elementary Particles*, Blaisdell, Waltham, Mass, 1969.

Gasiorowicz, S., *Elementary Particle Physics*, Wiley, New York 1966.

Georgi, H., *Weak Interactions and Modern Particle Physics*, Benjamin-Cummings, Menlo Park, California, 1984.

Giunti, C. and Kim, C.W., *Fundamentals of Neutrino Physics and Astrophysics*, Oxford University Press, Oxford, 2007.

Halzen, F. and Martin, A.D., *Quarks and Leptons: An Introductory Course in Particle Physics*, Wiley, New York, 1984.

Huang, K., *Quarks, Leptons and Gauge Fields*, World Scientific, Singapore, 1982.

Kokkedee, J.J.J., *The Quark Model*, Benjamin, New York, 1969.

Lee, T.D., *Particle Physics and an Introduction to Field Theory*, Harwood Academic Publishers, New York, 1982.

Martin, B.R. and Shaw, G., *Particle Physics*, Wiley, New York, 1992.

Okun, L.B., *Leptons and Quarks*, North Holland, Amsterdam, 1981.

Perkins, D.H., *Introduction to High Energy Physics*, Addison-Wesley, Menlo Park, California, 1984.

Ross, G.G., *Grand Unified Theories*, Benjamin-Cummings, Menlo Park, California, 1984.

Sakurai, J.J., *Invariance Principles and Elementary Particles*, Princeton University Press, Princeton, N.J., 1964.

Streater, R.F. and Wightman, A.S., *PCT, Spin and Statistics, and all that*, Benjamin, New York, 1963.

Internal Symmetries

Dyson, F.J., *Symmetry Groups in Nuclear and Particle Physics*, Benjamin, New York, 1966.

Fonda, L. and Ghirardi, G., *Symmetry Principles in Quantum Physics*, Marcel Dekker, New York, 1970.

Gell-Mann, M. and Neeman, Y., *The Eightfold Way*, Benjamin, New York, 1964.

Georgi, H., *Lie Algebras in Particle Physics*, Benjamin-Cummings, Reading, Mass, 1982.

Gourdin, M., *Unitary Symmetries*, North Holland, Amsterdam, 1967.

Greiner, W. and Müller, B., *Quantum Mechanics: Symmetries*, Springer-Verlag, Berlin, 1994.

Lichtenberg, D.B., *Unitary Symmetries and Elementary Particles*, Academic Press, London, 1970.

Low, F., *Symmetries and Elementary Particles*, Gordon and Breach, 1967.

Ne'eman, Y., *Algebraic Theory of Particle Physics*, Benjamin, New York, 1967.

Sozzi, M.S., *Discrete Symmetries and CP Violation*, Oxford University Press, Oxford, 2008.

Strocchi, F., *Symmetry Breaking*, Springer, Berlin-Heidelberg, 2008.

Index

G. Costa and G. Fogli, *Symmetries and Group Theory in Particle Physics*,
Lecture Notes in Physics 823, DOI: 10.1007/978-3-642-15482-9,
© Springer-Verlag Berlin Heidelberg 2012